江苏大学英文教材基金资助出版

ADVANCED MATHEMATICS I

高等数学 上

主编

杜瑞瑾 （Du Ruijin） 江苏大学

董高高 （Dong Gaogao） 江苏大学

杨 洁 （Yang Jie） 江苏大学

参编

王明刚 南京师范大学

陈 琳 西北工业大学

江苏大学出版社
JIANGSU UNIVERSITY PRESS

镇 江

图书在版编目(CIP)数据

高等数学.上 = Advanced Mathematics Ⅰ：英文 / 杜瑞瑾,董高高,杨洁主编. —镇江：江苏大学出版社, 2018.12(2024.9重印)
ISBN 978-7-5684-0958-2

Ⅰ.①高… Ⅱ.①杜… ②董… ③杨… Ⅲ.①高等数学－高等学校－教材－英文 Ⅳ.①O13

中国版本图书馆 CIP 数据核字(2018)第 298307 号

高等数学(上)
Advanced Mathematics（Ⅰ）

主　　编	/杜瑞瑾　董高高　杨　洁
责任编辑	/吴昌兴
出版发行	/江苏大学出版社
地　　址	/江苏省镇江市京口区学府路 301 号(邮编：212013)
电　　话	/0511-84446464(传真)
网　　址	/http：//press.ujs.edu.cn
排　　版	/镇江文苑制版印刷有限责任公司
印　　刷	/广东虎彩云印刷有限公司
开　　本	/787 mm×1 092 mm　1/16
印　　张	/18.75
字　　数	/624 千字
版　　次	/2018 年 12 月第 1 版
印　　次	/2024 年 9 月第 2 次印刷
书　　号	/ISBN 978-7-5684-0958-2
定　　价	/58.00 元

如有印装质量问题请与本社营销部联系(电话：0511-84440882)

PREFACE

Advanced mathematics (Calculus) is a part of modern mathematics education. It has two major branches, differential calculus (concerning instantaneous rates of change and slopes of curves) and integral calculus (concerning accumulation of quantities and the areas under and between curves). These two branches are related to each other by fundamental theorem of calculus.

Due to the extensive application of Advanced Mathematics in Science, Engineering and Economics, this course has become one of the most important foundation courses for college students. Not only is it critical that the text be accurate, but readability and even page design also play an important role. The authors have the following purposes when writing the book:

1. To write a book that students can easily comprehend.

2. To enable students to apply what they have learned to solve practical problems.

3. To develop students' meticulous logical thinking skills.

And the features of the book are as follows:

1. The book is not only well defined, comprehensive, but also rigorously structured to improve readability.

2. Almost all knowledge points are accompanied by corresponding examples, which can help students gain solid knowledge of the basic topics.

3. The exercises after each section can help students gain better insight into the mathematical concepts and assess their skills.

4. At the end of each chapter, there exist a summary of the main contents and basic requirements.

This book aims to enable students to construct the links between what they are learning and how they may apply the knowledge. To cultivate students' logical thinking ability is helpful for them to think about mathematical problems from a multi-dimensional perspective, so as to develop a good habit of thinking. It is advisable for students to flexibly apply what they have learned in class and what they have expanded after class to the modeling and solving of practical problems. The book is suitable for students majoring in engineering, economics, life sciences, as well as mathematics and applied mathematics (Chinese-foreign cooperation). For any errors in the book, the authors would be grateful if they were sent to: dudo999@126.com.

CONTENTS

Chapter 1 Function and limit / 1

1.1 Function / 1
   1.1.1 Set / 1
   1.1.2 Function / 4
   1.1.3 Elementary functions / 10
Exercises 1-1 / 11
1.2 Limit / 13
   1.2.1 Limit of sequence / 13
   1.2.2 Limit of function / 18
Exercises 1-2 / 24
1.3 Rules of limit operations / 25
   1.3.1 Rational operation rules / 25
   1.3.2 Operation rule for composite function / 28
Exercises 1-3 / 29
1.4 Principle of limit existence and two important limits / 30
   1.4.1 Principle of limit existence / 30
   1.4.2 Two important limits / 32
Exercises 1-4 / 34
1.5 Infinitesimal and infinity / 35
   1.5.1 Infinitesimal / 35
   1.5.2 Infinity / 37
   1.5.3 Comparison of infinitesimal order / 39
Exercises 1-5 / 42
1.6 Continuity of function and the properties / 42
   1.6.1 Continuity of function and discontinuity / 42
   1.6.2 Operations on continuous functions and the continuity of elementary functions / 47
   1.6.3 Properties of continuous functions on a closed interval / 50
Exercises 1-6 / 52
Summary / 53
Quiz / 55
Exercises / 56

Chapter 2 Derivative and differential / 58

2.1 Definition of derivatives / 58
- 2.1.1 Cited examples / 59
- 2.1.2 Definition of derivatives / 62
- 2.1.3 Geometric interpretation of derivative / 65
- 2.1.4 Use the unit to explain derivative / 67
- 2.1.5 Relationship between derivability and continuity / 68
- 2.1.6 The application of derivative in natural science / 69

Exercises 2-1 / 70

2.2 Rules of finding derivatives / 72
- 2.2.1 Derivative formulas for several fundamental elementary functions / 72
- 2.2.2 Derivation rules of rational operations / 73
- 2.2.3 Derivative of inverse functions / 75
- 2.2.4 Derivation rules of composite functions / 76

Exercises 2-2 / 81

2.3 Higher-order derivatives / 82
- 2.3.1 The concept of higher order derivative / 82
- 2.3.2 The linear property and Leibniz formula / 84

Exercises 2-3 / 85

2.4 Derivation of implicit functions and parametric equations / 85
- 2.4.1 Derivation of implicit functions / 85
- 2.4.2 Logarithmic derivation method / 87
- 2.4.3 Derivation of parametric equations / 88
- 2.4.4 Derivative of the function expressed by the polar equation / 91

Exercises 2-4 / 93

2.5 Differential of the function / 94
- 2.5.1 Concept of the differential / 94
- 2.5.2 The necessary and sufficient condition for differential / 95
- 2.5.3 Geometric meaning of differential / 96
- 2.5.4 Differential rules of elementary functions / 97
- 2.5.5 Application of the differential / 99

Exercises 2-5 / 102

2.6 The application of derivative in economic analysis / 103
- 2.6.1 Marginal analysis / 103
- 2.6.2 Elasticity analysis / 108

Exercises 2-6 / 113

Summary / 114

Quiz / 116

Exercises / 117

Chapter 3 The mean value theorem / 119

3.1 Mean value theorems the differential calculus / 119
3.1.1 Fermat Lemma / 119
3.1.2 Rolle's Theorem / 120
3.1.3 Lagrange's Theorem / 121
3.1.4 Cauchy's Theorem / 124
Exercises 3-1 / 125
3.2 Taylor's Theorem / 126
Exercises 3-2 / 130
Summary / 130
Quiz / 131
Exercises / 131

Chapter 4 Applications of derivatives / 133

4.1 Indeterminate form limit / 133
4.1.1 The indeterminate form of type $\dfrac{0}{0}$ / 133
4.1.2 The indeterminate form of type $\dfrac{\infty}{\infty}$ / 136
4.1.3 Other indeterminate forms / 137
Exercises 4-1 / 139
4.2 Monotonicity and local extreme value / 140
4.2.1 Monotonicity of functions / 140
4.2.2 The extreme value of function / 142
4.2.3 Maximum value and minimum value of function / 145
Exercises 4-2 / 148
4.3 The convexity, concavity and inflection point of function / 150
Exercises 4-3 / 155
4.4 Description of the function graph / 155
4.4.1 Asymptote of curve / 155
4.4.2 Description of the function graph / 156
Exercises 4-4 / 158
4.5 Curvature / 159
4.5.1 Differentiation of an arc / 159
4.5.2 Calculation formula of curvature / 160
4.5.3 Curvature circle / 162
Exercises 4-5 / 163
*4.6 Approximation of equation / 164
4.6.1 Bisection method / 164
4.6.2 Tangent method / 165
Exercises 4-6 / 166

4.7　The application of extreme value of function in economic management　/　167
　　4.7.1　Demand analysis　/　167
　　4.7.2　Problems of minimum average cost　/　169
　　4.7.3　Problems of minimizing inventory cost　/　171
　　4.7.4　Problems of maximal profit　/　173
　　4.7.5　Problems of maximal revenue　/　175
　　4.7.6　Problems of maximal tax　/　176
　　4.7.7　Problems of compound interest　/　177
Exercises 4-7　/　179
Summary　/　180
Quiz　/　182
Exercises　/　183

Chapter 5　Indefinite integrals　/　186

5.1　Indefinite integrals　/　186
　　5.1.1　Antiderivative　/　186
　　5.1.2　The conception of indefinite integral　/　187
　　5.1.3　Table of basic indefinite integrals　/　188
　　5.1.4　Properties of the indefinite integral　/　190
Exercises 5-1　/　192
5.2　Integration by substitution　/　192
　　5.2.1　Integration by the first substitution (Gather differential together)　/　193
　　5.2.2　Integration by the second substitution　/　197
Exercises 5-2　/　200
5.3　Integration by parts　/　201
Exercises 5-3　/　205
5.4　Integration of rational functions　/　205
　　5.4.1　Integration of rational functions　/　205
　　5.4.2　Integration of rational trigonometric functions　/　209
　　5.4.3　Integration of simply irrational functions　/　210
Exercises 5-4　/　211
5.5　Application of indefinite integral in economy　/　212
Exercises 5-5　/　214
Summary　/　214
Quiz　/　216
Exercises　/　217

Chapter 6　Definite integrals　/　220

6.1　The concept of definite integral　/　220
　　6.1.1　Examples of definite integral problems　/　220
　　6.1.2　The definition of definite integral　/　222

Exercises 6-1 / 225
6.2　Properties of definite integrals / 225
Exercises 6-2 / 230
6.3　Fundamental formula of calculus / 230
　　6.3.1　The relationship between the displacement and the velocity / 230
　　6.3.2　A function of upper limit of integral / 231
　　6.3.3　Newton–Leibnitz formula / 232
Exercises 6-3 / 236
6.4　Integration by substitution and parts for definite integrals / 236
　　6.4.1　Integration by substitution for definite integrals / 237
　　6.4.2　Integration by parts for definite integrals / 241
Exercises 6-4 / 243
6.5　Improper integrals / 244
　　6.5.1　Improper integrals on an infinite interval / 245
　　6.5.2　Improper integral of unbounded functions / 247
Exercises 6-5 / 249
Summary / 249
Quiz / 252
Exercises / 253

Chapter 7　Applications of definite integral / 255

7.1　The element method of definite integral / 255
7.2　Application of definite integral in geometry / 257
　　7.2.1　The area of a region / 257
　　7.2.2　Volume / 260
　　7.2.3　Arc length of plane curve / 262
Exercises 7-2 / 264
7.3　Application of definite integral in economy / 265
　　7.3.1　Calculate total output if its rate of change is given / 265
　　7.3.2　Calculate amount function when marginal function is given / 266
　　7.3.3　Discount problems / 268
Exercises 7-3 / 269
Summary / 270
Quiz / 271
Exercises / 272

Answers / 273

Chapter 1 Function and limit

> Basically, the object of research in elementary mathematics is a constant quantity, but in advanced mathematics is a variable. In this chapter, we will introduce three fundamental concepts, which are functions, limits and continuity, and some of their properties.

1.1 Function

1.1.1 Set

1) Definition of set

A set is defined as the collection of all objects having some specified property. For example, all students of a class constitute a set, lines passing through a fixed point on a plane constitute a set, natural numbers can be divided by three constitute a set and so on. Each object belonging to a set is called an element of the set.

Sets are usually denoted by capital letters A, B, X, Y, $\cdots$. Elements are usually denoted by small letters $a, b, x, y, \cdots$. If a is a member (or element) of A, then we call a belongs to A, the notation $a \in A$ is used; if a is not a member (or element) of A, we call a doesn't belong to A, denoted as $a \notin A$. An element either belongs to A or not.

There are two methods to represent a set: one is the method of citation and the other is description.

The method of citation, that is, to list all the elements of the set. For example, a set consisting of finite elements a_1, a_2, $\cdots$, a_n is written as $A = \{a_1, a_2, \cdots, a_n\}$. A set consisted of infinite elements b_1, b_2, $\cdots$, b_n, $\cdots$ is denoted by $B = \{b_1, b_2, \cdots, b_n, \cdots\}$.

The method of description is to point out the determining property held in common by all the elements of a set, that is $X = \{x \mid x \text{ has property } p(x)\}$. For instance, solution set of inequality $x^2 - 5x + 6 > 0$ can be presented by
$$X = \{x \mid x^2 - 5x + 6 > 0\}.$$

A set consisting of finite number of elements is called a finite set. A set consisted of infinite elements is called an infinite set. A set containing no element is called the empty set, denoted by $\varnothing$.

Usually **N** stands for the set of natural numbers, **Z** stands for the set of all integers, and **Q** stands for the set of rational numbers.

$$N = \{0, 1, 2, \cdots, n, \cdots\};$$
$$Z = \{0, \pm 1, \pm 2, \cdots, \pm n, \cdots\};$$
$$Q = \left\{\frac{p}{q} \,\middle|\, p \in Z, q \in N, q \neq 0, \text{ and } p, q \text{ are co-prime numbers}\right\}.$$

In addition, the set of real numbers consists of all real numbers and is denoted by **R**, and **C** stands for the set of complex numbers.

" $*$ " marked in the upper right corner of the letter represents 0 is excluded out of the set. And " $+$ " indicates that 0 and negative numbers are removed from the set. For example,

$$Z^* = \{\pm 1, \pm 2, \cdots, \pm n, \cdots\};$$
$$Z^+ = \{1, 2, \cdots, n, \cdots\}.$$

2) Operations on sets

Suppose that A and B are two sets. If each element in A is also an element in B, then A is called a subset of the set B, denoted by $A \subseteq B$ or $B \supseteq A$, read as "A is contained by B (or B contains A)". For any set A, because $\varnothing \subseteq A$, $A \subseteq A$, then $\varnothing$ and A are both subsets of A. If $A \subseteq B$ but $A \neq B$, then A is called a proper subset of B, denoted by $A \subsetneq B$.

Some basic operations on sets are as follows.

Suppose that A and B are two sets. Union of the sets A and B, denoted by $A \cup B$, is the set of all subjects that are a member of A, or B, or both, namely

$$A \cup B = \{x \mid x \in A \text{ or } x \in B\}.$$

Intersection of the sets A and B, denoted $A \cap B$, is the set of all objects that are members of both A and B. A set consisted of all elements belonging to both A and B is called intersection of A and B, namely

$$A \cap B = \{x \mid x \in A \text{ and } x \in B\}.$$

Set difference of A and B, denoted $A \backslash B$, is the set of all members of A that are not members of B, namely

$$A \backslash B = \{x \mid x \in A \text{ but } x \notin B\}.$$

If all the sets are subsets of I, then I is called a universal set, and the set difference $I \backslash A$ is also called the complement of A in I, the notation A^c is sometimes used instead of $I \backslash A$, namely

$$A^c = I \backslash A = \{x \mid x \in I \text{ but } x \notin A\}.$$

The operations on sets are shown in Figure 1-1 (the shaded region in the figure are the results).

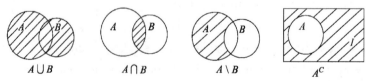

Figure 1-1

It is easily seen from the geometric meaning that operations on sets satisfy the following rules. Let A, B, C be any three sets, then

(1) Commutative law $A \cup B = B \cup A$, $A \cap B = B \cap A$;

(2) Associative law $A \cup (B \cup C) = (A \cup B) \cup C$, $A \cap (B \cap C) = (A \cap B) \cap C$;

(3) Distribution law $(A \cup B) \cap C = (A \cap C) \cup (B \cap C)$,
$(A \cap B) \cup C = (A \cup C) \cap (B \cup C)$;

(4) Idempotent law $A \cup A = A$, $A \cap A = A$;

(5) Absorption law $A \cup (A \cap B) = A$, $A \cap (A \cup B) = A$;

(6) Dualization law (De Morgan law)
$$(A \cup B)^c = A^c \cap B^c, \quad (A \cap B)^c = A^c \cup B^c.$$

3) Interval

Intervals are a class of important number sets. Let a and b are real numbers, $a < b$. The number set $\{x \mid a < x < b\}$ is called an open interval, denoted by (a, b), that is
$$(a, b) = \{x \mid a < x < b\},$$
a and b are called the end points of the open interval (a, b). The number set $\{x \mid a \leqslant x \leqslant b\}$ is said to be a closed interval, denoted by $[a, b]$, that is
$$[a, b] = \{x \mid a \leqslant x \leqslant b\}.$$
a and b are called the end points of the closed interval $[a, b]$. Similarly left-open right-closed interval is $(a, b]$, and left-closed right-open interval is $[a, b)$,
$$(a, b] = \{x \mid a < x \leqslant b\}, \quad [a, b) = \{x \mid a \leqslant x < b\}.$$

The above intervals are finite intervals. Notations $+\infty$ (read as positive infinity) and $-\infty$ (read as negative infinity) can be used to define infinite intervals:
$$(a, +\infty) = \{x \mid a < x < +\infty\},$$
$$(-\infty, b) = \{x \mid -\infty < x < b\},$$
$$[a, +\infty) = \{x \mid a \leqslant x < +\infty\},$$
$$(-\infty, b] = \{x \mid -\infty < x \leqslant b\},$$
$$(-\infty, +\infty) = \{x \mid -\infty < x < +\infty\} = \mathbf{R}.$$

4) Neighborhood

The definition of neighborhood is useful and widely used. An open interval with center a is said to be a neighborhood of point a, denoted by $U(a)$. Let δ be any positive number, then the open interval $(a - \delta, a + \delta)$ is a neighborhood of point a, which is called δ-neighborhood of a, denoted by $U(a, \delta)$. That is
$$U(a, \delta) = \{x \mid a - \delta < x < a + \delta\} = \{x \mid |x - a| < \delta\},$$
where a is the center of neighborhood, δ is called the radius of the neighborhood. For example
$$U\left(-2, \frac{1}{2}\right) = \left\{x \mid |x - (-2)| < \frac{1}{2}\right\} = \left(-\frac{5}{2}, -\frac{3}{2}\right).$$

The neighborhood is called a deleted neighborhood, if the center is removed from $U(a, \delta)$, denoted by

$$\overset{\circ}{U}(a, \delta) = \{x \mid 0 < |x-a| < \delta\}.$$

The open interval $(a-\delta, a)$ is called the left δ-neighborhood of a, and $(a, a+\delta)$ is called the right δ-neighborhood of a.

1.1.2 Function

1) Concept of function

In observing natural and social phenomena, there are many different quantities, some of which are constant quantity; and others are variables. Usually a, b, c are used to represent constants, x, y, z are used to represent variables.

In the same process, there are often several variables that are interrelated and interact with one another and follow certain objective laws. If the causal relationship of these changes can be accurately described, this will help us really understand the trend and the law of development of things. The function is the most basic mathematical tool to reflect the relationship of variables.

> **Definition 1** Let x, y are two variables, D is a nonempty set. If for every $x \in D$, there exists one and only one y corresponding to x according to some determined rule f, then y is called a function of x on set D, denoted by $y = f(x)$, where x is called the independent variable, and y is called the dependent variable. The set D is called domain of f, denoted by $D(f)$. The set consists of all functional value y is called the range of f, denoted by $W(f)$, namely
> $$W(f) = \{y \mid y = f(x), x \in D\}.$$

In the definition of function, there are two essential factors, namely the domain D and the corresponding rule f. If the domain and the corresponding rule are determined, then the function and the range of function are determined. If two functions have the same domain D and the same corresponding rule, then the two functions are the same. Therefore the symbols, which are used to denote the independent variable and dependent variable, are insignificant.

For example, the two functions $f(x) = x^2$ $(D(f) = [0,1])$ and $g(t) = t^2$ $(D(g) = [0,1])$ are the same, because they have the same domain and corresponding rule. The two functions $y = \ln x^2$ $(x \neq 0)$ and $y = 2\ln x$ $(x > 0)$ are different, because their domains are different.

Suppose the domain of definition of $y = f(x)$ is D, in the xOy plane, for every $x \in D$, there corresponds to a unique point $(x, y) = (x, f(x))$ corresponding to it. When we study a function given by an analytic representation, sketching its graph will often help us to obtain some intuitive information about the function. In the coordinate system, the point set $G = \{(x, y) \mid y = f(x), x \in D\}$ often a curve, is called the graph of the function $y = f(x)$ (Figure 1-2).

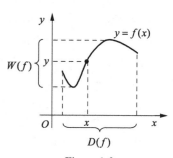

Figure 1-2

It should be noted that the analytic representation of a function sometimes consists of several components on different subsets of the domain of the function. A function expressed by this kind of representations is called a piecewise defined function.

Example 1 Absolute function (Figure 1-3)

$$y=|x|=\begin{cases} x, & x\geqslant 0, \\ -x, & x<0. \end{cases}$$

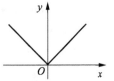

Figure 1-3

Example 2 Sign function (Figure 1-4)

$$y=\operatorname{sgn} x=\begin{cases} 1, & x>0, \\ 0, & x=0, \\ -1, & x<0, \end{cases}$$

The domain of the function is $D=(-\infty,+\infty)$, the range of function is $W=\{-1, 0, 1\}$.

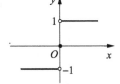

Figure 1-4

Example 3 The greatest integer function $y=[x]$, where $[x]$ represents the value at any number x is the largest integer smaller than or equal to x. The domain of the function is $D=(-\infty,+\infty)=\mathbf{R}$, the range of function is $W=\{0, \pm 1, \pm 2, \cdots\}=\mathbf{Z}$. Its graph is shown in Figure 1-5. For instance, $[2.5]=2, [\sqrt{2}]=1, [-\pi]=-4$.

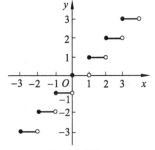

Figure 1-5

In Definition 1, for every $x \in D$, there is a unique y corresponding to x according to some determined rule f, then function $y=f(x)$ is called single-valued function. If relax restriction on the uniqueness of dependent variable: there are two or more y corresponding to x, then function $y=f(x)$ is called multi-valued function.

For multi-valued functions, we can limit the range of the dependent variable to make it become a single-valued function. For example, the function $y=\pm\sqrt{1-x^2}$ defined by $x^2+y^2=1$ is a multi-valued function. If $y\geqslant 0$, then $y=\sqrt{1-x^2}$. If $y\leqslant 0$, then $y=-\sqrt{1-x^2}$. These two functions are single-valued functions, each of which is called one branch of multiple-valued function. For another example, anti-trigonometric function $y=\operatorname{Arcsin} x$ is a multi-valued function, for any $x\in[-1,1]$ there are infinitely y corresponding to x. We define as $-\frac{\pi}{2}\leqslant y\leqslant\frac{\pi}{2}$, then $y=\arcsin x$, it is a single-valued function. And it is called the principal branch of $y=\operatorname{Arcsin} x$. Other anti-trigonometric functions have a similar situation.

The functions to be discussed in the future, if not specified, are single valued functions.

2) The properties of functions

(1) Monotone functions

Let the domain of function $y=f(x)$ is $D(f)$ and interval $I \subset D(f)$.

If $f(x_1)<f(x_2)$ for any x_1, $x_2 \in I$ with $x_1<x_2$, then $f(x)$ is said to be monotone increasing on the interval I (Figure 1-6a).

If $f(x_1)>f(x_2)$ for any x_1, $x_2 \in I$ with $x_1<x_2$, then $f(x)$ is said to be monotone decreasing on the interval I (Figure 1-6b).

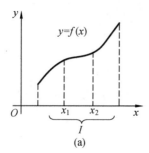
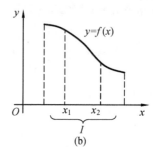

Figure 1-6

Monotone increasing functions and monotone decreasing functions are called by a joint name monotone function, I is called monotone interval. It should be noted that the monotonicity of the function is related to the range of independent variable. For example, $y=x^2$ is monotone decreasing on the interval $(-\infty,0]$ and monotone increasing on the interval $(0,+\infty]$. However, $y=x^2$ is not monotonous in the interval $(-\infty,+\infty)$.

(2) Even (odd) functions

Suppose the domain of function $y=f(x)$ is an interval D that is symmetric about the origin. If $f(-x)=f(x)$ for any $x \in D$, $y=f(x)$ is called an even function; If $f(-x)=-f(x)$ for any $x \in D$, $y=f(x)$ is called an odd function.

The graph of an even function is symmetric about the y-axis, and the graph of an odd function is symmetric about the origin (Figure 1-7).

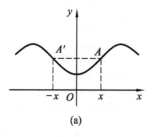
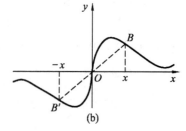

Figure 1-7

For example, both $y=x^2$ and $y=\cos x$ are even functions in the interval $(-\infty,+\infty)$. $y=x^3$ and $y=\sin x$ are odd functions in the interval $(-\infty,+\infty)$.

(3) Bounded functions

Let the domain of function $y=f(x)$ is $D(f)$ and interval $I \subset D(f)$. If there exists a constant M such that

$$f(x) \leqslant M$$

for any $x \in I$, then $f(x)$ is said to be bounded above on I. M is called the upper bound of $f(x)$.

If there exists a constant m such that

$$f(x) \geqslant m$$

for any $x \in I$, then $f(x)$ is said to be bounded below on I. m is called the lower bound of $f(x)$.

If there exists a positive number K such that

$$|f(x)| \leqslant K$$

for any $x \in I$, then $f(x)$ is said to be bounded on I (Figure 1-8). If for any such positive number K, there exists $x_1 \in I$ such that $|f(x_1)| > M$, then $f(x)$ is said to be unbounded on I.

For example, $y = \sin x$ and $y = \cos x$ are bounded in the interval $(-\infty, +\infty)$, $y = \tan x$ and $y = \cot x$ are unbounded in their domains.

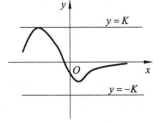

Figure 1-8

(4) Periodic functions

Let the domain of function $f(x)$ is D. If there exists a positive number l such that

$$f(x+l) = f(x)$$

holds for all $x \in D$ and $x + l \in D$, then $f(x)$ is said to be a periodic function and l is the period of $f(x)$. Usually, we mean the minimal positive period.

For example, $y = \sin x$ and $y = \cos x$ are both periodic functions with the period 2π, $y = \tan x$ and $y = \cot x$ are both periodic functions with the period π, $y = x - [x]$ is a periodic function with the period 1, the constant function $y = C$ can be regarded as a periodic function, but doesn't have a minimal positive period.

The graphs of a periodic function have the same shape in every interval whose length is l (Figure 1-9).

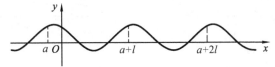

Figure 1-9

3) Inverse functions

In the function, the relationship between the independent variable and the dependent variable is relative. Sometimes it is necessary to change the position of them. For example, two issues can be considered when studying the free-fall motion.

On the one hand, if you want to understand how the distance s of the object falling changes with the time t, then

$$s = f(t) = \frac{1}{2}gt^2.$$

On the other hand, if you want to find how long will it take when the distance s of the object falling, then

$$t = \varphi(s) = \sqrt{\frac{2s}{g}}.$$

Depending on different study goals, sometimes the correspondence from x to y may be considered, sometimes the relationship from y to x should also be taken into account.

> **Definition 2** Let the function $f: D \to f(D)$, where D and $f(D)$ are both real number sets. If $\forall y \in f(D)$, there exists a unique $x \in D$ such that $f(x) = y$. Then we can define a function $f^{-1}: f(D) \to D$ which maps each $y \in f(D)$ to $x \in D$, that is $f^{-1}(y) = x$. The function f^{-1} is called the inverse function of f.

From the definition, we may see that if f^{-1} is the inverse function of f, then f must be the inverse function of f^{-1}, and that the domain of definition of f is the range of f^{-1} and the range of f is domain of f^{-1}. Corresponding to the inverse function $x = f^{-1}(y)$, the original function $y = f(x)$ is called the direct function.

For a function f, suppose that its inverse function exists. Because the function $y = f(x)$ and its inverse function $x = \varphi(y)$ are represented by the same relation between x and y, their graph is the same curve in the xOy plane. In this case, if $P(a, b)$ is a point on the graph of direct function $y = f(x)$, that is $b = f(a)$, then $a = \varphi(b)$, the point $Q(b, a)$ should be on the graph of inverse function $x = \varphi(y)$. It is easy to see that the points P and Q are symmetric about the straight line $y = x$. Hence, the graphs of the direct function $y = f(x)$ and the inverse function $x = \varphi(y)$ are symmetric with respect to the line $y = x$ (Figure 1-10).

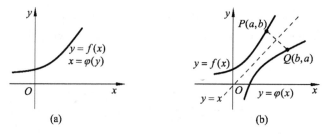

Figure 1-10

In general, x denotes independent variable and y denotes dependent variable. Therefore, the inverse function can also be written as $y = f^{-1}(x)$. And we always use this notation to represent inverse function.

Let f is a monotonous function on D, then there must exist an inverse function f^{-1}, and the function f^{-1} is also a monotonous function on the corresponding range $f(D)$.

4) Composite function

Definition 3 Let the domain of function $y=f(u)$ be $D(f)$, the range of function $u=\varphi(x)$ be $W(\varphi)$. If $D(f) \cap W(\varphi) \neq \varnothing$, the new function $y=f[\varphi(x)]$ is called the composite function of the functions $y=f(u)$ and $u=\varphi(x)$, where u is called an intermediate variable of the composite function.

The composite function of $y=f(u)$ and $u=\varphi(x)$ is denoted by $f \circ \varphi$, namely
$$(f \circ \varphi)(x) = f[\varphi(x)].$$
According to Definition 3, the domain of composite function $f[\varphi(x)]$ is
$$D(f \circ \varphi) = \{x \mid x \in D(\varphi), \varphi(x) \in D(f)\}.$$
Therefore, the domain $D(f \circ \varphi)$ of composite function $f[\varphi(x)]$ is either the proper subset of the domain of function $u=\varphi(x)$ (when $W(\varphi) \not\subset D(f)$), or the same as of the domain of function $u=\varphi(x)$ (when $W(\varphi) \subset D(f)$).

Example 4 There are two functions $y=\sin u$ and $u=1+x^2$ Then the domain of the composite function $y=\sin(1+x^2)$ is $(-\infty, +\infty)$, which is the same as the domain of the inside function $u=\sqrt{1+x^2}$.

Example 5 There are two functions $y=\sqrt{u}$ and $u=1-x^3$, then the domain of the composite function $y=\sqrt{1-x^3}$ is $(-\infty, 1]$, which is a proper subset of domain $(-\infty, +\infty)$ of the inside function $u=1-x^3$.

Example 6 Function $y=\sqrt{1-u}$ can't be composed with function $u=x^2+2$. Because the domain of the function $y=\sqrt{1-u}$ is $D(f)=(-\infty, 1]$, while the range of the function $u=x^2+2$ is $W(\varphi)=[2, +\infty)$. Because $D(f) \cap W(\varphi) = \varnothing$, so the two functions can't be composited.

Composition of functions can be extended to any finite number of functions. For example, $y=u^2$, $u=1+t^3$ and $t=e^x$ can be composed into $y=(1+e^{3x})^2$.

It is also important to note that, for a given composite function, which functions are composed into it. For example, $y=\ln(2+\sin x)$ is the composite function of functions $y=\ln u$, $u=2+v$, $v=\sin x$, and $y=\cos\sqrt{1+e^{x^2}}$ is the composite function of five simple functions: $y=\cos u$, $u=\sqrt{v}$, $v=1+w$, $w=e^t$, $t=x^2$.

5) Operations on functions

Let the domains of functions $f(x)$ and $g(x)$ be $D(f)$ and $D(g)$ respectively. If $D=D(f) \cap D(g) \neq \varnothing$, then we can define arithmetic operations as following:

(1) Addition and Subtraction $(f \pm g)(x) = f(x) \pm g(x)$, $x \in D$;

(2) Multiplication $(f \cdot g)(x) = f(x) \cdot g(x)$, $x \in D$;

(3) Division $\left(\dfrac{f}{g}\right)(x) = \dfrac{f(x)}{g(x)}$, $x \in D \setminus \{x \mid g(x)=0, x \in D\}$.

Example 7 Suppose the domain of function $f(x)$ is $[-a,a]$. Prove there exists an even function $g(x)$ and an odd function $h(x)$ on $[-a,a]$ such that $f(x)=g(x)+h(x)$.

Proof Let $g(x)=\dfrac{1}{2}[f(x)+f(-x)]$, $h(x)=\dfrac{1}{2}[f(x)-f(-x)]$, then we have

$$f(x)=g(x)+h(x).$$

Since

$$g(-x)=\dfrac{1}{2}[f(x)+f(-x)]=g(x),$$

$$h(-x)=\dfrac{1}{2}[f(-x)-f(x)]=-h(x),$$

$g(x)$ is an even function, and $h(x)$ is an odd function.

1.1.3 Elementary functions

We have learned the following functions in high school:

(1) Power function: $y=x^\mu$ (μ is a constant, $\mu\in\mathbf{R}$).

(2) Exponential function: $y=a^x$ ($a>0$, $a\neq 1$). If $a=e$, $y=e^x$, where $e=2.718\,281\,828\,459\,045\cdots$.

(3) Logarithmic function: $y=\log_a x$ ($a>0, a\neq 1$). If $a=e$, $y=\ln x$, where $\ln x$ is a logarithm to the base of the mathematical constant e.

(4) Trigonometric function: $y=\sin x$, $y=\cos x$, $y=\tan x$, $y=\cot x$. Besides there are the other two trigonometric functions $y=\sec x$ and $y=\csc x$, where $\sec x=\dfrac{1}{\cos x}$, $\csc x=\dfrac{1}{\sin x}$.

(5) Inverse trigonometric function: $y=\arcsin x$, $y=\arccos x$, $y=\arctan x$, $y=\text{arccot}\, x$.

The above functions are called the basic elementary function. An elementary function can be expressed by an analytic expression, which is made up of basic elementary functions and constants by a finite number of arithmetic operations and compositions of functions.

Most functions discussed in this book are all elementary functions. For example,

$$y=\sin^2(3x+1),\quad y=\sqrt{1-(\ln\cos x)^2},\quad y=\dfrac{\lg x+\sqrt[3]{x}+2\tan x}{10^x-x+1}.$$

But generally the piecewise is not an elementary function, because it can't be described by one analytic expression.

A particular class of elementary functions is commonly used in engineering application, called hyperbolic function. They are

Hyperbolic sine $\quad\text{sh}\,x=\dfrac{e^x-e^{-x}}{2}$, $x\in(-\infty,+\infty)$;

Hyperbolic cosine $\quad\text{ch}\,x=\dfrac{e^x+e^{-x}}{2}$, $x\in(-\infty,+\infty)$;

Hyperbolic tangent $\quad\text{th}\,x=\dfrac{\text{sh}\,x}{\text{ch}\,x}=\dfrac{e^x-e^{-x}}{e^x+e^{-x}}$, $x\in(-\infty,+\infty)$.

Their graphs are showed in Figure 1-11.

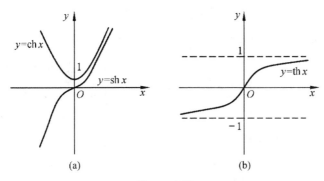

Figure 1-11

Obviously, sh x and th x are odd functions, and ch x is an even function. Since $|\text{th } x|<1$, so th x is a bounded function. sh x, ch x and th x are monotonic increasing in $[0, +\infty)$.

There are some identities for hyperbolic functions which are similar to those for trigonometric functions. We list them as follows:

$$\text{sh}(x \pm y) = \text{sh } x\text{ch } y \pm \text{ch } x\text{sh } y;$$
$$\text{ch}(x \pm y) = \text{ch } x\text{ch } y \pm \text{sh } x\text{sh } y;$$
$$\text{sh } 2x = 2\text{sh } x\text{ch } x;$$
$$\text{ch } 2x = \text{ch}^2 x + \text{sh}^2 x;$$
$$\text{ch}^2 x - \text{sh}^2 x = 1.$$

It is not difficult to prove them.

Exercises 1-1

1. Suppose A and B are two sets, prove dualization law: $(A \cap B)^C = A^C \cup B^C$.
2. Find the domains of the following functions:

(1) $f(x) = \dfrac{x+1}{x^2-x-2}$;

(2) $f(x) = \dfrac{\log_3(4-x)}{x^2}$;

(3) $f(x) = \dfrac{1}{x} - \sqrt{1-x^2}$;

(4) $f(x) = \sqrt{1-e^{\frac{x}{1-x}}}$;

(5) $f(x) = \dfrac{\sqrt{x+2}}{\sin \pi x}$;

(6) $f(x) = \arccos(2\sin x)$;

(7) $f(x) = \sqrt{3-x} + \arctan \dfrac{1}{x}$;

(8) $f(x) = \dfrac{1}{\sqrt{x-x^2}}$.

3. Are the following functions the same, why?

(1) $f(x) = \ln x^4, g(x) = 4\ln x$;

(2) $f(x) = \sin(3x^2+1), g(t) = \sin(3t^2+1)$;

(3) $f(x) = \dfrac{x^2-4}{x-2}, g(x) = x+2$.

4. Let $f(x-1) = x^2, \varphi(x) = x^2$, find $f[\varphi(x)]$ and $\varphi[f(x)]$.

5. Let the domain of function $f(x)$ be $[0, 1]$, find the domains of the following functions:

(1) $f(x+1)$;

(2) $f(\sin x)$;

(3) $f(\ln x)$; (4) $f(x+a)+f(x-a)\ (a>0)$.

6. Sketch the following functions:

(1) $y=\sqrt{x+1}$;

(2) $y=\dfrac{1}{2}(\sin x+|\sin x|)$;

(3) $y=x-[x]$;

(4) $y=1-|x-3|$.

7. Which of the following functions are even functions? Which are odd functions? Which are neither even nor odd?

(1) $y=|x|\sin\dfrac{1}{x^2}$;

(2) $y=x(x-1)(x+1)$;

(3) $y=\ln(x+\sqrt{x^2+1})$;

(4) $y=\sin x-\cos x+1$;

(5) $y=[x]$;

(6) $y=\operatorname{sgn} x=\begin{cases}1, & x>0,\\ 0, & x=0,\\ -1, & x<0.\end{cases}$

8. Judge the monotony of the following functions in the specified intervals:

(1) $y=\dfrac{2-x}{1-x},\ x\in(-\infty,1)$;

(2) $y=x+\ln x,\ x\in(0,+\infty)$.

9. Determine if each of the following functions is bounded or not in the domain:

(1) $y=\dfrac{1+2x}{1+x^2}$;

(2) $y=x+\ln x$;

(3) $y=e^{-x}$;

(4) $y=x+[x]$.

10. Find the inverse function of each of the following functions. Specify the domains of the inverse functions.

(1) $y=\dfrac{1-x}{1+x}$;

(2) $y=\sqrt[3]{x+1}$;

(3) $y=1+\ln(x-2)$;

(4) $y=\begin{cases}x^2, & x\geq 1,\\ 2x-1, & x<1.\end{cases}$

11. Let the function $f(x)$ is defined in $[-l,\ l]$, $f(x)$ is both even and odd. Prove $f(x)\equiv 0$.

12. Which basic elementary functions are composed into the following elementary functions?

(1) $y=(\sin x^2)^5$;

(2) $y=\ln(\tan^2 x^3)$;

(3) $y=\sqrt[3]{\arccos\dfrac{1}{x^3}}$;

(4) $y=\cos(\sin e^{\sqrt{x}})$.

13. When $x\neq 0$, let $f(x)=xg\left(\dfrac{1}{x}\right)$, prove $g(x)=xf\left(\dfrac{1}{x}\right)$.

14. Let $f(x)=ax^2+bx+c, f(0)=1$, and $f(x+1)=f(x)+2x$, find $f(x)$.

15. Cut off a sector with the central angle α from the center of a circular iron sheet, whose radius is R. The remaining part is rolled into a conical funnel. Find the function of the volume of the cone V with respect to angle α.

16. It is known that if exchange the US dollars into another currency, the currency value will increase 12%. if another currency is converted into US dollars, the value will

reduce 12%. Find the two functions respectively, and prove that they are not inverse functions. If a person will exchange $10 000 twice, how much dollars will be lost?

17. A transport company provides transportation charges for certain goods: no more than 200 kilometers, 6 Yuan per ton-kilometer; more than 200 kilometers, but not more than 500 kilometers, 4 Yuan per ton-kilometer; more than 500 kilometers, 3 Yuan per ton-kilometer. Try to find the function relation between the freight and distance.

18. There is a coverless cuboid wooden box. The volume is 1 cubic meter and the height is 2 meters. Suppose the length of one side of the bottom is x meters. Try to express the surface area of the box with respect to x.

1.2 Limit

Limit is the basic tool to study the changing trend of variables. The limit method is a basic method to study functions. Many basic concepts in advanced mathematics, such as continuous, derivative, definite integral, infinite series, are all based on the limit.

1.2.1 Limit of sequence

1) Concept of sequence

According to a given rule, we have a real number x_n, for each positive integer n, then the ordered list of numbers

$$x_1, x_2, x_3, \cdots, x_n, \cdots$$

is called a sequence, and denoted by $\{x_n\}$. x_n is called the general term of the sequence. For examples,

$$1, \frac{1}{2}, \frac{1}{3}, \cdots, \frac{1}{n}, \cdots;$$

$$1, 3, 9, \cdots, 3n, \cdots;$$

$$\frac{1}{2}, \frac{2}{3}, \frac{3}{4}, \cdots, \frac{n}{n+1}, \cdots;$$

$$\frac{2}{1}, \frac{1}{2}, \frac{4}{3}, \cdots, \frac{n+(-1)^{n-1}}{n}, \cdots;$$

$$-1, 1, -1, \cdots, (-1)^n, \cdots.$$

They are all sequences, and their general terms are

$$\frac{1}{n}, 3n, \frac{n}{n+1}, \frac{n+(-1)^{n-1}}{n}, (-1)^n, \text{ respectively.}$$

Geometrically, a sequence $\{x_n\}$ can be regarded as a motive particle on the number axis as shown in Figure 1-12.

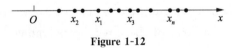

Figure 1-12

Because a sequence is just a special function, which is defined on the set $\mathbf{N}^+$, the concepts of monotonicity and boundedness of functions can be applied equally to sequences.

(1) Monotonicity

If the sequence $\{x_n\}$ satisfies $x_1 \leqslant x_2 \leqslant x_3 \leqslant \cdots \leqslant x_n \leqslant \cdots$, the sequence is called a monotonic increasing sequence.

If the sequence $\{x_n\}$ satisfies $x_1 \geqslant x_2 \geqslant x_3 \geqslant \cdots \geqslant x_n \geqslant \cdots$, the sequence is called a monotonic decreasing sequence.

Monotonic increasing sequence and monotonic decreasing sequence are called by a joint name monotonic sequence.

(2) Boundedness

If there exists a positive number M, for any positive integer n, $|x_n| \leqslant M$, holds, then the sequence $\{x_n\}$ is bounded. Otherwise, $\{x_n\}$ is unbounded.

Since $|x_n| \leqslant M$, we have $x_n \in [-M, M]$. From the number axis, all the points corresponding to bounded sequence are in the finite interval $[-M, M]$. On the contrary, if all the points of a sequence are in a finite interval, it is easy to know that this sequence is bounded.

Take infinite numbers from the sequence $\{x_n\}$ and keep them in the original order, the new sequence obtained is called the subsequence of $\{x_n\}$.

For example,
$$x_1, x_3, x_5, \cdots, x_{2n-1}, \cdots;$$
$$x_2, x_4, x_6, \cdots, x_{2n-1}, \cdots.$$
They are all subsequences of $\{x_n\}$. Generally, a subsequence of $\{x_n\}$ is denoted as
$$\{x_{n_k}\}: x_{n_1}, x_{n_2}, \cdots, x_{n_k}, \cdots.$$

2) Definition of limit of a sequence

The concept and method of limit is generated from the precise solution to some practical problems. For these problems, we can improve the accuracy of the approximate solution to approach the precise solution. In the process of gradually improving the accuracy, the variables will be infinitely close to a constant. The process of infinitely approaching the constant is the process of limit.

Liu Hui, an ancient mathematician in China, found the area of a circle by calculating the area of the inscribed equilateral polygons cyclotomic technique, which is the geometric application of the limit thought.

Make inscribed regular hexagon, regular dodecagon, $\cdots$, regular polygon with edges, $\cdots$ respectively inside the circle of radius of 3×2^n. The areas of these regular polygons are in turn,

Within the radius r of the circle, draw a regular hexagon, a regular dodecagon, $\cdots$, a 3×2^n regular polygon, $\cdots$. The area of these regular polygons are
$$A_1, A_2, \cdots, A_n, \cdots,$$
they form a sequence.

From a geometric point of view, when the number of edges increases, the difference between the inscribed regular polygon in the circle and the circle itself becomes smaller.

That is to say, as the number of edges of the regular polygon increases infinitely, its area is infinitely close to the area of the circle. Therefore, as n increases infinitely, the value of the sequence $\{A_n\}$ infinitely approaches to the area of circle. Namely, the sequence $\{A_n\}$ is infinitely close to the constant πr^2.

If a sequence $\{x_n\}$ approaches a constant a with n increasing to infinity, the constant a is called the limit of the sequence $\{x_n\}$. As an intuitive understanding of limit of a sequence, this expression can be regarded as a descriptive definition. But depending only on descriptive definition is not enough. We need to describe the trend precisely. How may we define the statement that n increases and tends to infinity and the distance x_n approaches a?

Example 1 Find the trend of change of the sequence $\{x_n\} = \left\{\dfrac{n+(-1)^{n-1}}{n}\right\}$, with n increasing to infinity.

By intuition, when n tends to infinity, the sequence x_n approaches 1. Actually, since $x_n = \dfrac{n+(-1)^{n-1}}{n}$, that is $x_n - 1 = \dfrac{(-1)^{n-1}}{n}$, then we have

$$|x_n - 1| = \frac{1}{n}.$$

As n increases, $\dfrac{1}{n}$ becomes smaller. From the perspective of geometry, the sequence x_n comes closer and closer to 1. We may say that as n increases, the distance $|x_n - 1| = \dfrac{1}{n}$ can be made arbitrarily small, that is $\dfrac{1}{n}$ becomes as small as may be required by making n large enough. To express it in mathematical language, we can use a very small number to describe it, for instance,

$|x_n - 1| < 0.1$, $\dfrac{1}{n} < 0.1$, that is $n > 10$, so when $n > 10$, we have $|x_n - 1| < 0.1$;

$|x_n - 1| < 0.01$, $\dfrac{1}{n} < 0.01$, that is $n > 100$, so when $n > 100$, we have $|x_n - 1| < 0.01$;

$|x_n - 1| < 10^{-6}$, $\dfrac{1}{n} < 10^{-6}$, that is $n > 10^6$, so when $n > 10^6$, we have $|x_n - 1| < 10^{-6}$;

....

Generally, for any given positive value $\varepsilon > 0$, no matter how small it is, $|x_n - 1| = \dfrac{1}{n} < \varepsilon$ holds only if $n > \dfrac{1}{\varepsilon}$. Hence we may take $N = \left[\dfrac{1}{\varepsilon}\right]$, when $n > N$, we have $|x_n - 1| < \varepsilon$.

Definition 1 Let $\{x_n\}$ be a sequence and let $a \in \mathbf{R}$ be a constant. If for any given $\varepsilon > 0$, there exists a positive integer N, such that $|x_n - a| < \varepsilon$ holds for all $n > N$, then we say that the sequence $\{x_n\}$ tends to a as n tends to infinity, or the sequence has a limit, and a is called the limit of $\{x_n\}$, or the sequence $\{x_n\}$ is convergent to a, denoted as

$$\lim_{n \to \infty} x_n = a \quad \text{or} \quad x_n \to a \ (n \to \infty).$$

If the constant a doesn't exist, we say the sequence $\{x_n\}$ does not have a limit, it is not convergent then it is said to be divergent.

The definition of limit for a sequence is often expressed as following:
$\lim\limits_{n\to\infty} x_n = a \Leftrightarrow \forall \varepsilon > 0$, $\exists N \in \mathbf{N}^+$ such that $|x_n - a| < \varepsilon$ holds for all $n > N$.

In the above definition, ε is positive. It can be chosen arbitrarily small, but it is a constant whenever it is given. It is used to describe the fact that $\{x_n\}$ may be arbitrarily close to a, or the distance $|x_n - a| < \varepsilon$ may be arbitrarily small. We also note that N is dependent on the given ε. Generally, when ε is taken smaller, N will have to be taken larger.

If we draw the sequence $\{x_n\}$ on the number axis, we get a range of points. If the sequence $\{x_n\}$ is convergent to a, this means for any given neighborhood of a, $(a-\varepsilon, a+\varepsilon)$ no matter how small, there is a positive integer N such that all the terms of $\{x_n\}$ beginning from x_{N+1} are located in the neighborhood $(a-\varepsilon, a+\varepsilon)$ (Figure 1-13).

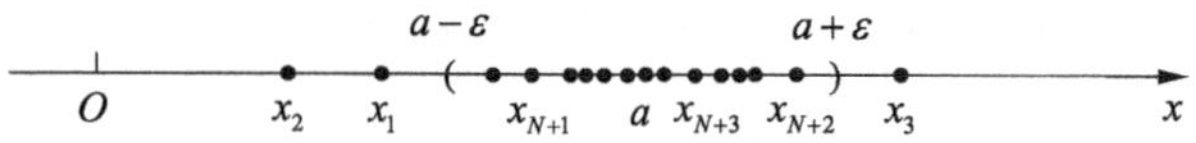

Figure 1-13

Example 2 Prove that the limit of the sequence
$$\frac{2}{1}, \frac{1}{2}, \frac{4}{3}, \cdots, \frac{n+(-1)^{n-1}}{n}, \cdots$$
is 1.

Proof $|x_n - 1| = \left|\dfrac{n+(-1)^{n-1}}{n} - 1\right| = \dfrac{1}{n}$. Given a number $\varepsilon > 0$, we want to find an integer $N > 0$ such that $\dfrac{1}{n} < \varepsilon$ for $n > N$.

Thus, we take $N = \left[\dfrac{1}{\varepsilon}\right]$. As $n > N$, we have
$$\left|\frac{n+(-1)^{n-1}}{n} - 1\right| = \frac{1}{n} < \varepsilon.$$

Hence, by the definition of limit of the sequence, we have
$$\lim_{n\to\infty} \frac{n+(-1)^{n-1}}{n} = 1.$$

Example 3 Let $x_n = \dfrac{2(n^2-1)}{n^2+n+1}$. Prove $\lim\limits_{n\to\infty} x_n = 2$.

Proof Since
$$\left|\frac{2(n^2-1)}{n^2+n+1} - 2\right| = \frac{2n+4}{n^2+n+1} < \frac{2n+4}{n^2} < \frac{2n+n}{n^2} = \frac{3}{n} \quad (\text{let } n > 4).$$

so $\left|\dfrac{2(n^2-1)}{n^2+n+1} - 2\right| < \varepsilon$ is equivalent to $\dfrac{3}{n} < \varepsilon$, namely, $n > \dfrac{3}{\varepsilon}$.

Thus, $\forall \varepsilon > 0$, we take $N = \max\left(\left[\dfrac{3}{\varepsilon}\right], 4\right)$. When $n > N$, we have
$$\left|\frac{2(n^2-1)}{n^2+n+1} - 2\right| < \varepsilon,$$

Therefore, by the definition of limit of the sequence, we have
$$\lim_{n\to\infty}\frac{2(n^2-1)}{n^2+n+1}=2.$$

Example 4 Let $|q|<1$, prove the limit of the sequence $1, q, q^2, \cdots, q^{n-1}, \cdots$ is 0.

Proof $\forall \varepsilon>0$ (let $\varepsilon<1$), since $|x_n-0|=|q^{n-1}-0|=|q|^{n-1}$, we want to find $N(\varepsilon)$, such that $|x_n-0|<\varepsilon$, that is
$$|q|^{n-1}<\varepsilon.$$
Take the natural logarithm of both sides, we have
$$(n-1)\ln|q|<\ln\varepsilon.$$
Because $|q|<1$, $\ln|q|<0$, the requirement is $n>1+\dfrac{\ln\varepsilon}{\ln|q|}$.

Hence for the given $\varepsilon>0$, we take $N=\left[1+\dfrac{\ln\varepsilon}{\ln|q|}\right]$. When $n>N$, we have
$$|x_n-0|=|q|^{n-1}<\varepsilon.$$
Therefore
$$\lim_{n\to\infty}x_n=\lim_{n\to\infty}q^{n-1}=0.$$

3) Properties of convergent sequence

Theorem 1 (Uniqueness) If the sequence $\{x_n\}$ is convergent, its limit is unique.

Proof We will prove the conclusion by method of contradiction. Suppose that $\lim_{n\to\infty}x_n=a$, $\lim_{n\to\infty}x_n=b$, and $a\neq b$. There is no harm in assuming that $a<b$, let $\varepsilon_0=\dfrac{b-a}{2}>0$.

Because $\lim_{n\to\infty}x_n=a$, $\exists N_1\in\mathbf{N}^+$, when $n>N_1$, we have
$$|x_n-a|<\varepsilon_0=\frac{b-a}{2}.$$
That is
$$a-\frac{b-a}{2}<x_n<a+\frac{b-a}{2}=\frac{b+a}{2}.$$

Again because $\lim_{n\to\infty}x_n=b$, $\exists N_2\in\mathbf{N}^+$, when $n>N_2$, we have
$$|x_n-b|<\varepsilon_0=\frac{b-a}{2}.$$
That is
$$\frac{b+a}{2}=b-\frac{b-a}{2}<x_n<b+\frac{b-a}{2}.$$

Take $N=\max\{N_1, N_2\}$. Then when $n>N$, we have both inequalities $x_n<\dfrac{b+a}{2}$ and $x_n>\dfrac{b+a}{2}$. This is impossible. Hence the limit of $\{x_n\}$ must be unique.

Theorem 2 (Boundedness) If the sequence $\{x_n\}$ is convergent, then $\{x_n\}$ is bounded.

Proof Suppose that $\{x_n\}$ is convergent to a, that is $\lim_{n\to\infty}x_n=a$. By the definition of the limit, for given $\varepsilon_0=1$, there exists a positive integer N. When $n>N$, we have

$$|x_n-a|<\varepsilon_0=1.$$
That is
$$|x_n|=|x_n-a+a|\leqslant|x_n-a|+|a|<1+|a|.$$
Let $M=\max\{|x_1|,|x_2|,\cdots,|x_N|,1+|a|\}$, for all $n\in N^+$, we have $|x_n-a|\leqslant M$. Therefore, the sequence $\{x_n\}$ is bounded.

It should be noted that boundedness is only a necessary condition of a convergent sequence, that is if a sequence is convergent, then it must be bounded, but a bounded sequence is not necessarily convergent. For example, the sequence
$$1,0,1,0,1,0,\cdots$$
is bounded, but it is not convergent.

Corollary 1 If the sequence $\{x_n\}$ is unbounded, then the sequence is divergent.

Theorem 3 (Preservation of sign) If the sequence $\{x_n\}$ is convergent to a, and $a>0$ (or $a<0$). Then there exists a positive integer N, when $n>N$, $x_n>0$ (or $x_n<0$).

Proof We only prove the case $a>0$. Since $\lim\limits_{n\to\infty}x_n=a$, for $\varepsilon_0=\dfrac{a}{2}>0$, there exists a positive integer N. When $n>N$, we have
$$|x_n-a|<\varepsilon_0=\dfrac{a}{2},$$
then we have
$$x_n>a-\dfrac{a}{2}=\dfrac{a}{2}>0.$$

Corollary 2 If there exists a positive integer N, for any $n>N$, such that $\lim\limits_{n\to\infty}x_n=a$ and $x_n\geqslant 0$ (or $x_n\leqslant 0$), then $a\geqslant 0$ (or $a\leqslant 0$).

Theorem 4 (Subsequence) If the sequence $\{x_{n_k}\}$ is convergent to a, then any subsequence $\{x_{n_k}\}$ of the sequence $\{x_n\}$ is convergent to a.

Thus it can be seen that if one subsequence of a sequence is divergent or there are two subsequences convergent to different limits, the sequence is divergent.

For example, since the two subsequences of the sequence $\{(-1)^{n+1}\}$,
$$1,1,1,1,\cdots \quad \text{and} \quad -1,-1,-1,-1,\cdots$$
Converge to 1 and -1 respectively, hence $\{(-1)^{n+1}\}$ is divergent.

1.2.2 Limit of function

In this section, we will extend the concepts and results obtained from sequences to functions. As we know, a sequence is a special function $f(n)$ defined on a number set N. Its independent variable n is a discrete variable and tends to positive infinity. However, for a function $f(x)$, its independent variable x changes continuously and may have various limits. This makes the situation more complex. We will mainly discuss two cases in the following, and other cases are similar:

(1) limit of a function $f(x)$ for independent variable x tending to a finite value x_0;

(2) limit of a function $f(x)$ for independent variable x tending to infinity ∞.

1) Limit of a function for independent variable tending to a finite value

If the independent variable x tends to a finite value x_0, the corresponding function value $f(x)$ approaches constant A, then A is called the limit of function $f(x)$ as $x \to x_0$. Similar to the definition of limit of sequence.

$f(x)$ which is infinitely close to A can be expressed by $|f(x)-A|<\varepsilon$, where ε is any given positive value. Because $f(x) \to A$ is achieved in the process of $x \to x_0$, for any given $\varepsilon>0$, $|f(x)-A|<\varepsilon$, doesn't hold for any x, but the x close enough to x_0, which can be written as $0<|x-x_0|<\delta$, where δ is a positive value. From the geometric point of view, x falls in a deleted δ-neighborhood of x_0, and the radius δ describes the fact that x may be arbitrarily close to x_0.

Example 5 Find the trend of change of the function $f(x)=\dfrac{x^2-4}{2(x-2)}$ as $x \to 2$.

Solution It is obvious that when $x \to 2$, $f(x)$ is infinitely close to 2 (Figure 1-14). Because as $x \neq 2$, $|f(x)-2|=\dfrac{1}{2}|(x-2)|$,

to make $|f(x)-2|<0.01$, then $0<|x-2|<0.02$;

to make $|f(x-2)|<10^{-8}$, then $0<|x-2|<2 \cdot 10^{-8}$;

....

Therefore, to make $|f(x)-2|<\varepsilon$, then $0<|x-2|<\delta$, that is $\delta=2\varepsilon$.

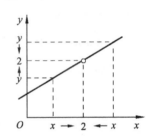

Figure 1-14

> **Definition 2** Suppose that function $f(x)$ is defined on $\overset{\circ}{U}(x_0, \delta)$, A is a constant. If $\forall \varepsilon>0$, $\exists \delta>0$, such that $|f(x)-A|<\varepsilon$, for all $0<|x-x_0|<\delta$, then A is called the limit of function $f(x)$ as $x \to x_0$, denoted by
> $$\lim_{x \to x_0} f(x)=A \quad \text{or} \quad f(x) \to A(x \to x_0).$$

If such a constant A doesn't exist, we say that when $x \to x_0$, the function $f(x)$ doesn't have a limit.

Definition 2 can be simply written as:
$\lim\limits_{x \to x_0} f(x)=A \Leftrightarrow \forall \varepsilon>0$, $\exists \delta>0$ such that $|f(x)-A|<\varepsilon$ for all x satisfying $0<|x-x_0|<\delta$.

Note $0<|x-x_0|<\delta$ in the definition implies that $x \neq x_0$. So as $x \to x_0$, the question that whether $f(x)$ has a limit, and what the limit is, has nothing to do with whether $f(x)$ is defined at x_0, and the value of $f(x)$ at x_0.

Geometric interpretation (Figure 1-15) of the definition is: for any given $\varepsilon>0$, we draw a strip region between the straight lines $y=A-\varepsilon$ and $y=A+\varepsilon$, and no matter how narrow the strip region is, there always exists a $\delta(\varepsilon)>0$ such that the graph of the function

$y=f(x)$ located in the deleted neighborhood $\overset{\circ}{U}(a,\delta)$ of x_0 is totally contained in that strip region.

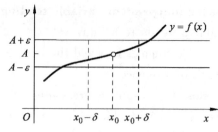

Figure 1-15

Example 6 Prove $\lim\limits_{x\to x_0} C = C$, where C is a constant.

Proof Since $|f(x)-A|=|C-C|=0$, $\forall \varepsilon>0$, we take any $\delta>0$, when $0<|x-x_0|<\delta$, we have
$$|f(x)-A|=|C-C|=0<\varepsilon.$$
Thus, $\lim\limits_{x\to x_0} C = C$.

Example 7 Prove $\lim\limits_{x\to x_0} x = x_0$.

Proof Since $|f(x)-A|=|x-x_0|$, $\forall \varepsilon>0$, we take $\delta=\varepsilon$, when $0<|x-x_0|<\delta=\varepsilon$, we have
$$|f(x)-A|=|x-x_0|<\varepsilon.$$
Thus, $\lim\limits_{x\to x_0} x = x_0$.

Example 8 Prove that $\lim\limits_{x\to 2} \dfrac{1}{3}(x+2) = \dfrac{4}{3}$.

Proof Since $|f(x)-A| = \left|\dfrac{1}{3}(x+2)-\dfrac{4}{3}\right| = \dfrac{1}{3}|x-2|$, to make $|f(x)-A|<\varepsilon$, we get
$$|x-2|<3\varepsilon.$$
Therefore, $\forall \varepsilon>0$, take $\delta=3\varepsilon$, when $0<|x-2|<\delta$, we have
$$|f(x)-2| = \left|\dfrac{1}{3}(x+2)-\dfrac{4}{3}\right| = \dfrac{1}{3}|x-2|<\varepsilon.$$
Thus,
$$\lim\limits_{x\to 2}\dfrac{1}{3}(x+2) = \dfrac{4}{3}.$$

Example 9 When $x_0>0$, prove $\lim\limits_{x\to x_0}\sqrt{x} = \sqrt{x_0}$.

Proof Since
$$|f(x)-A| = |\sqrt{x}-\sqrt{x_0}| = \left|\dfrac{x-x_0}{\sqrt{x}+\sqrt{x_0}}\right| \leqslant \dfrac{|x-x_0|}{\sqrt{x_0}},$$
for $|x-x_0|\leqslant x_0$, $|f(x)-A|<\varepsilon$ is equivalent to $\dfrac{|x-x_0|}{\sqrt{x_0}}<\varepsilon$.

Therefore $\forall \varepsilon>0$, we take $\delta=\min\{x_0, \sqrt{x_0}\varepsilon\}$. When $0<|x-x_0|<\delta$, we have

$$|\sqrt{x}-\sqrt{x_0}|<\varepsilon.$$
Thus
$$\lim_{x\to x_0}\sqrt{x}=\sqrt{x_0}.$$

Example 10 Prove that $\lim\limits_{x\to 0}\cos x=1$.

Proof Since $x\to 0$, there is no harm in assuming that $|x|<\dfrac{\pi}{2}$. Draw a unit circle whose radius is 1, shown in Figure 1-16. When $0<x<\dfrac{\pi}{2}$,

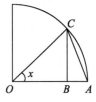

Figure 1-16

$$|\cos x-1|=|\overline{OB}-\overline{OA}|=\overline{BA}<\overline{AC}<\overset{\frown}{AC}=x.$$

Because $\cos(-x)=\cos x$, then when $0<|x|<\dfrac{\pi}{2}$, we have
$$|\cos x-1|<|x|.$$
Thus $\forall \varepsilon>0$, take $\delta=\varepsilon$, when $0<|x|<\delta$, we have $|\cos x-1|<\varepsilon$, that is
$$\lim_{x\to 0}\cos x=1.$$

The definition of limit of $f(x)$ as $x\to x_0$ contains two cases: x tends to x_0 only form the left side of x_0, that is $x\to x_0\,(x<x_0)$ denoted by $x\to x_0^-$; x tends to x_0 only form the right side of x_0, that is $x\to x_0\,(x>x_0)$ denoted by $x\to x_0^+$. Similarly, we can define the cases $f(x)\to A$ as $x\to x_0^+$ and $x\to x_0^-$.

> **Definition 3** If there exists a constant A, for $\forall \varepsilon>0$, $\exists \delta(\varepsilon)>0$ such that
> $$|f(x)-A|<\varepsilon$$
> for all x satisfying $x_0-\delta<x<x_0$, then A is called the left limit of $f(x)$ as $x\to x_0^-$, denoted by
> $$\lim_{x\to x_0^-}f(x)=A \quad \text{or} \quad f(x_0-0)=A.$$

> **Definition 4** If there exists a constant A, for $\forall \varepsilon>0$, $\exists \delta(\varepsilon)>0$ such that
> $$|f(x)-A|<\varepsilon$$
> for all x satisfying $x_0<x<x_0+\delta$, then A is called the right limit of $f(x)$ as $x\to x_0^+$, denoted by
> $$\lim_{x\to x_0^+}f(x)=A \quad \text{or} \quad f(x_0+0)=A.$$

The left limit and right limit are both called one-sided limits, to give a unified name. It is easy to prove the following conclusion.

> **Theorem 5** $\lim\limits_{x\to x_0}f(x)=A \Leftrightarrow \lim\limits_{x\to x_0^-}f(x)=\lim\limits_{x\to x_0^+}f(x)=A.$

By the relationship, we know that if one of the one-sided limits does not exist, or if both exist but the right limit and left limit are not equal, then the limit of $f(x)$ doesn't exist.

Example 11 Prove that $\lim\limits_{x\to 0}\dfrac{|x|}{x}$ does not exist.

Proof
$$\lim_{x\to 0^-}\dfrac{|x|}{x}=\lim_{x\to 0^-}\dfrac{-x}{x}=\lim_{x\to 0^-}(-1)=\lim_{x\to 0}(-1)=-1,$$

$$\lim_{x\to 0^+}\dfrac{|x|}{x}=\lim_{x\to 0^+}\dfrac{x}{x}=\lim_{x\to 0^+}1=1.$$

Since the left limit and the right limit are not equal, according to Theorem 5, $\lim\limits_{x\to 0}f(x)$ does not exist.

2) Limit of a function for independent variable tending to infinity

Definition 5 Suppose that $f(x)$ is defined when $|x|$ is greater than some positive number and A is a constant. If $\forall \varepsilon>0$, $\exists X>0$ such that $|f(x)-A|<\varepsilon$, for all $|x|>X$, then A is called the limit of the function $f(x)$ as x tends to infinity, denoted by
$$\lim_{x\to\infty}f(x)=A \quad \text{or} \quad f(x)\to A \text{ as } x\to\infty.$$

Definition 5 can also be simply described as
$$\lim_{x\to\infty}f(x)=A \Leftrightarrow \forall \varepsilon>0, \exists X>0, \text{ when } |x|>X, \text{ such that } |f(x)-A|<\varepsilon.$$

For the cases that $x>0$, x tends to infinity (denoted as $x\to +\infty$) and $x<0$, $|x|$ tends to infinity (denoted as $x\to -\infty$), we can define the limits of $f(x)$ as x tends to positive infinity and negative infinity.

$$\lim_{x\to +\infty}f(x)=A \Leftrightarrow \forall \varepsilon>0, \exists X>0, \text{ when } x>X, \text{ such that } |f(x)-A|<\varepsilon;$$

$$\lim_{x\to -\infty}f(x)=A \Leftrightarrow \forall \varepsilon>0, \exists X>0, \text{ when } x<-X, \text{ such that } |f(x)-A|<\varepsilon.$$

It is not difficult to prove that
$$\lim_{x\to\infty}f(x)=A \Leftrightarrow \lim_{x\to +\infty}f(x)=\lim_{x\to -\infty}f(x)=A.$$

The geometry meanings of $\lim\limits_{x\to\infty}f(x)=A$, $\lim\limits_{x\to +\infty}f(x)=A$ and $\lim\limits_{x\to -\infty}f(x)=A$ are relatively represented in Figure 1-17 a~c. Take $\lim\limits_{x\to\infty}f(x)=A$ for example, for any given $\varepsilon>0$, we can always find a point $X>0$ on the x-axis such that the graph of the function $y=f(x)$, which is on the right side of the line $x=X$ and on the left side of the line $x=-X$, is totally located in the strip. In this case, the line $y=A$ is called the horizontal asymptote of the function $y=f(x)$ (Figure 1-17a). Similarly, when $\lim\limits_{x\to +\infty}f(x)=A$ or $\lim\limits_{x\to -\infty}f(x)=A$, the straight line $y=A$ is also called the horizontal asymptote of the function $y=f(x)$ (Figure 1-17b~c).

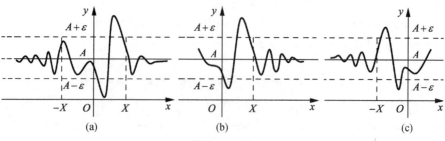

Figure 1-17

Example 12 Prove that $\lim\limits_{x\to\infty}\dfrac{1}{x}=0$.

Proof $\forall \varepsilon>0$, since $\left|\dfrac{1}{x}-0\right|=\dfrac{1}{|x|}$, in order to make $\left|\dfrac{1}{x}-0\right|<\varepsilon$, we get

$$\dfrac{1}{|x|}<\varepsilon, \text{ that is } |x|>\dfrac{1}{\varepsilon}.$$

Therefore, take $X=\dfrac{1}{\varepsilon}$, when $|x|>X$, we have

$$\left|\dfrac{1}{x}-0\right|=\dfrac{1}{|x|}<\varepsilon.$$

Hence
$$\lim\limits_{x\to\infty}\dfrac{1}{x}=0.$$

Here, the straight line $y=0$ is the horizontal asymptote of the figure of the function $y=\dfrac{1}{x}$.

Example 13 Prove that $\lim\limits_{x\to\infty}\dfrac{1-2x}{x+1}=-2$.

Proof $\forall \varepsilon>0$, since

$$\left|\dfrac{1-2x}{x+1}-(-2)\right|=\left|\dfrac{3}{x+1}\right|<\dfrac{3}{|x|-1}(\text{let } |x|>1).$$

In order to make $\left|\dfrac{1-2x}{x+1}-(-2)\right|<\varepsilon$, we obtain

$$\dfrac{3}{|x|-1}<\varepsilon, \text{ that is } |x|>\dfrac{3}{\varepsilon}+1.$$

Therefore, take $X=\dfrac{3}{\varepsilon}+1$, when $|x|>X$ (we also have $|x|>1$), then

$$\left|\dfrac{1-2x}{x+1}-(-2)\right|<\varepsilon.$$

Therefore
$$\lim\limits_{x\to\infty}\dfrac{1-2x}{x+1}=-2.$$

Here, the straight line $y=-2$ is the horizontal asymptote of the figure of the function $y=\dfrac{1-2x}{x+1}$.

3) Properties of function limit

No matter the case $x\to x_0$ or $x\to\infty$, the properties of function limit are similar to that of sequence limit. In the following we take the case $x\to x_0$ for example.

Theorem 6 (Uniqueness) If the limit of function $f(x)$ exists as $x\to x_0$, then the limit of $f(x)$ is unique. That is to say if $\lim\limits_{x\to x_0}f(x)=A$, $\lim\limits_{x\to x_0}f(x)=B$, then $A=B$.

Theorem 7 (Local boundedness) If $\lim\limits_{x\to x_0}f(x)$ exists, then $f(x)$ is bounded in a deleted neighborhood of x_0. That is $\exists M>0$, $\delta>0$ such that $|f(x)|\leqslant M$ for all x satisfying $0<|x-x_0|<\delta$.

Proof Suppose $\lim\limits_{x \to x_0} f(x) = A$, for $\varepsilon_0 = 1$, there exists $\delta > 0$, when $0 < |x - x_0| < \delta$, we have
$$|f(x) - A| < \varepsilon_0 = 1.$$
Thus, $\qquad |f(x)| \leqslant |A| + |f(x) - A| < |A| + 1.$

Take $M = |A| + 1$, when $0 < |x - x_0| < \delta$, we have $|f(x)| \leqslant M$, namely, $f(x)$ is bounded in $x \in \mathring{U}(x_0, \delta)$.

Theorem 8 (Local preservation of sign) If $\lim\limits_{x \to x_0} f(x) = A$ and $A > 0$ $(A < 0)$, then $\exists \delta > 0$ such that $f(x) > 0$ $(f(x) < 0)$ for all x satisfying that $0 < |x - x_0| < \delta$.

Proof Suppose that $\lim\limits_{x \to x_0} f(x) = A > 0$, for $\varepsilon = \dfrac{A}{2} > 0$, $\exists \delta > 0$ such that
$$|f(x) - A| < \varepsilon = \frac{A}{2},$$
then $\qquad A - \dfrac{A}{2} < f(x) < A + \dfrac{A}{2}.$

Thus, when $x \in \mathring{U}(x_0, \delta)$, $f(x) > \dfrac{A}{2} > 0$.

Similarly, we can prove the case of $A < 0$.

Corollary 1 If $\lim\limits_{x \to x_0} f(x) = A$ and $f(x) \geqslant 0 (f(x) \leqslant 0)$ for $x \in \mathring{U}(x_0, \delta)$, then $A \geqslant 0$ $(A \leqslant 0)$.

Corollary 2 If $\lim\limits_{x \to x_0} f(x) = A$, $\lim\limits_{x \to x_0} f(x) = B$, and $f(x) \geqslant g(x) (f(x) \leqslant g(x))$ for $x \in \mathring{U}(x_0, \delta)$, then $A \geqslant B (A \leqslant B)$.

Notice that Theorem 6 to Theorem 8 are also right for the cases $x \to x_0^-$, $x \to x_0^+$ and $x \to \infty$, $x \to +\infty$, $x \to -\infty$.

Exercises 1-2

1. Which of the following sequences are convergent? Which are divergent? For the convergent sequence, find its limit by observing the trend of changing.

(1) $x_n = \dfrac{10}{3n}$, $n = 1, 2, \cdots$;

(2) $x_n = (-1)^n \dfrac{1}{n+1}$, $n = 1, 2, \cdots$;

(3) $x_n = (-1)^n \dfrac{n}{n+1}$, $n = 1, 2, \cdots$;

(4) $x_n = \dfrac{2n-1}{3n+2}$, $n = 1, 2, \cdots$;

(5) $x_n = \dfrac{2^n - 1}{3^n}$, $n = 1, 2, \cdots$;

(6) $x_n = \dfrac{1 - (-1)^n}{n^3}$, $n = 1, 2, \cdots$;

(7) $x_n = 1 + \dfrac{1}{2^n}$, $n = 1, 2, \cdots$;

(8) $x_n = [(-1)^n + 1] \dfrac{n+1}{n}$, $n = 1, 2, \cdots$.

2. Prove the following limits by the definition of functions:

(1) $\lim\limits_{n\to\infty}\dfrac{1}{n^2+3}=0$;

(2) $\lim\limits_{n\to\infty}\dfrac{2n^2+3n+5}{n^2+5n+1}=2$;

(3) $\lim\limits_{n\to\infty}(\sqrt{n+1}-\sqrt{n})=0$;

(4) $\lim\limits_{n\to\infty}\left[\dfrac{1}{1\times 2}+\dfrac{1}{2\times 3}+\cdots+\dfrac{1}{(n-1)\times n}\right]=1$.

3. If $\lim\limits_{n\to\infty}x_n=A\neq 0$, prove $\lim\limits_{n\to\infty}\dfrac{1}{x_n}=\dfrac{1}{A}$.

4. For the function $f(x)$ as Figure 1-18, find the following limit. If the limit does not exist, instruct the reason.

(1) $\lim\limits_{x\to -1}f(x)$;

(2) $\lim\limits_{x\to -\frac{1}{2}}f(x)$;

(3) $\lim\limits_{x\to 0}f(x)$;

(4) $\lim\limits_{x\to 1}f(x)$.

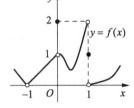

Figure 1-18

5. Prove the following limits by the definition of functions:

(1) $\lim\limits_{x\to 2}\dfrac{1+x}{2x}=\dfrac{3}{4}$;

(2) $\lim\limits_{x\to 4}\sqrt{x}=2$;

(3) $\lim\limits_{x\to -2}\dfrac{x^2-4}{x+2}=-4$;

(4) $\lim\limits_{x\to -\frac{1}{2}}\dfrac{1-4x^2}{2x+1}=2$.

6. Prove the following limits by the definition of functions:

(1) $\lim\limits_{x\to\infty}\dfrac{3x-1}{2x+1}=\dfrac{3}{2}$;

(2) $\lim\limits_{x\to +\infty}\dfrac{1}{\sqrt{x}}=0$.

7. Prove:

(1) The necessary and sufficient condition of $\lim\limits_{x\to\infty}f(x)=A$ is $\lim\limits_{x\to +\infty}f(x)=\lim\limits_{x\to -\infty}f(x)=A$.

(2) The necessary and sufficient condition of $\lim\limits_{x\to x_0}f(x)=A$ is $\lim\limits_{x\to x_0^+}f(x)=\lim\limits_{x\to x_0^-}f(x)=A$.

8. Find the left limit and right limit of $f(x)=\dfrac{x}{|x|}$ and $g(x)=\dfrac{x^2}{|x|}$ as $x\to 0$, and instruct whether the limit of them as $x\to 0$ exists or not.

9. If $\lim\limits_{x\to x_0}f(x)=A$, prove $\lim\limits_{x\to x_0}|f(x)|=|A|$. Give an example to illustrate that the inverse proposition does not hold.

1.3 Rules of limit operations

For some simple functions, limits may be seen by observation. Based on those simple functions, the question is how to extend our knowledge to find the limit for more complicated functions? The most fundamental method is by means of rational operations.

1.3.1 Rational operation rules

For the limit of sequence, we have the following conclusions.

Theorem 1 If $\lim\limits_{n\to\infty}x_n=a$, $\lim\limits_{n\to\infty}y_n=b$, then

(1) $\lim\limits_{n\to\infty}(x_n\pm y_n)=\lim\limits_{n\to\infty}x_n\pm\lim\limits_{n\to\infty}y_n=a\pm b$;

(2) $\lim\limits_{n\to\infty}(x_n\cdot y_n)=\lim\limits_{n\to\infty}x_n\cdot\lim\limits_{n\to\infty}y_n=a\cdot b$;

(3) If there is $b \neq 0$, then $\lim\limits_{n\to\infty} \dfrac{x_n}{y_n} = \dfrac{\lim\limits_{n\to\infty} x_n}{\lim\limits_{n\to\infty} y_n} = \dfrac{a}{b}$.

Proof Here we only prove (2). Leave the proof of (1) and (3) to readers.

$\forall \varepsilon > 0$, because
$$|x_n y_n - ab| = |(x_n y_n - a y_n) + (a y_n - ab)| \leqslant |x_n - a||y_n| + |a||y_n - b|,$$
in order to make $|x_n y_n - ab| < \varepsilon$, we have to make
$$|x_n - a||y_n| + |a||y_n - b| < \varepsilon.$$
$\{y_n\}$ is convergent, so $\{y_n\}$ is bounded. That is to say there exists $M > 0$, we have
$$|y_n| \leqslant M \quad (n = 1, 2, \cdots).$$
Moreover $\lim\limits_{n\to\infty} x_n = a$, $\lim\limits_{n\to\infty} y_n = b$, so for the above given $\varepsilon > 0$, there exists positive integers N_1, N_2, when $n > N_1$, we have $|x_n - a| < \dfrac{\varepsilon}{2M}$; when $n > N_2$, we have $|y_n - b| < \dfrac{\varepsilon}{2|a|+1}$.

Take $N = \max\{N_1, N_2\}$, when $n > N$, the inequations
$$|x_n - a| < \dfrac{\varepsilon}{2M}, \quad |y_n - b| < \dfrac{\varepsilon}{2|a|+1}$$
hold. Hence,
$$|x_n y_n - ab| \leqslant |x_n - a||y_n| + |a||y_n - b|$$
$$< \dfrac{\varepsilon}{2M} M + |a| \dfrac{\varepsilon}{2|a|+1} < \dfrac{\varepsilon}{2} + \dfrac{\varepsilon}{2} = \varepsilon.$$

Note If $\lim\limits_{n\to\infty} x_n$, $\lim\limits_{n\to\infty} y_n$ and $\lim\limits_{n\to\infty} z_n$ exist, we have
$$\lim_{n\to\infty}(x_n + y_n - z_n) = \lim_{n\to\infty} x_n + \lim_{n\to\infty} y_n - \lim_{n\to\infty} z_n;$$
$$\lim_{n\to\infty}(x_n \cdot y_n \cdot z_n) = \lim_{n\to\infty} x_n \cdot \lim_{n\to\infty} y_n \cdot \lim_{n\to\infty} z_n.$$

Corollary 1 If $\lim\limits_{n\to\infty} x_n = a$, then $\lim\limits_{n\to\infty}(kx_n) = k \lim\limits_{n\to\infty} x_n = ka$, where k is a constant.

Corollary 2 If $\lim\limits_{n\to\infty} x_n = a$, then $\lim\limits_{n\to\infty}(x_n)^m = (\lim\limits_{n\to\infty} x_n)^m = a^m$, where m is a constant.

Example 1 Find $\lim\limits_{n\to\infty} \dfrac{2n^3 - 5n + 1}{3n^3 + 4n^2 - 2n}$.

Solution Because the limits of the numerator and denominator do not exist, we can't use Theorem 1 directly. Divide the numerator and denominator by n^3, we get
$$x_n = \dfrac{2 - \dfrac{5}{n^2} + \dfrac{1}{n^3}}{3 + \dfrac{4}{n} - \dfrac{2}{n^2}},$$

Then
$$\lim_{n\to\infty} x_n = \dfrac{2 - 5\lim\limits_{n\to\infty}\dfrac{1}{n^2} + \lim\limits_{n\to\infty}\dfrac{1}{n^3}}{3 + 4\lim\limits_{n\to\infty}\dfrac{1}{n} - 2\lim\limits_{n\to\infty}\dfrac{1}{n^2}} = \dfrac{2}{3}.$$

For the limit of function, we have the same properties as following.

Theorem 2 If $\lim f(x) = A$, $\lim g(x) = B$, then
(1) $\lim[f(x) \pm g(x)] = \lim f(x) \pm \lim g(x) = A \pm B$;
(2) $\lim[f(x) \cdot g(x)] = \lim f(x) \cdot \lim g(x) = A \cdot B$;
(3) $B \neq 0$, $\lim \dfrac{f(x)}{g(x)} = \dfrac{\lim f(x)}{\lim g(x)} = \dfrac{A}{B}$.

Corollary 1 If $\lim f(x) = A$, then $\lim[k \cdot f(x)] = k \lim f(x) = kA$, where k is a constant.

Corollary 2 If $\lim f(x) = A$, then $\lim[f(x)]^m = [\lim f(x)]^m = A^m$, where m is a constant.

In the above theorems and corollaries, the change of the independent variable is not indicated under the symbol "lim", which means that the conclusion is true for $x \to x_0$, $x \to \infty$ and one-sided limits.

Example 2 Find $\lim\limits_{x \to 2}(3x^2 - 2x + 1)$.

Solution $\lim\limits_{x \to 2}(3x^2 - 2x + 1) = \lim\limits_{x \to 2} 3x^2 - \lim\limits_{x \to 2} 2x + \lim\limits_{x \to 2} 1$
$= 3(\lim\limits_{x \to 2} x)^2 - 2 \lim\limits_{x \to 2} x + 1 = 3 \cdot 2^2 - 2 \cdot 2 + 1 = 9.$

Example 3 Find $\lim\limits_{x \to 2} \dfrac{x^2 + 1}{x^3 - 5x^2}$.

Solution The limit of the denominator is not equal to 0, hence
$$\lim\limits_{x \to 2} \dfrac{x^2+1}{x^3-5x^2} = \dfrac{\lim\limits_{x \to 2}(x^2+1)}{\lim\limits_{x \to 2}(x^3-5x^2)} = \dfrac{(\lim\limits_{x \to 2} x)^2+1}{(\lim\limits_{x \to 2} x)^3 - 5(\lim\limits_{x \to 2} x)^2} = \dfrac{2^2+1}{2^3 - 5 \cdot 2^2} = -\dfrac{5}{12}.$$

For the two examples above, it can be found that when finding limits of polynomials or rational fraction functions as $x \to x_0$, we only need to substitute x by x_0. But don't do this when the denominator approaches zero.

In fact, assume the polynomial $f(x) = a_0 x^n + a_1 x^{n-1} + \cdots + a_n$, then
$$\lim\limits_{x \to x_0} f(x) = \lim\limits_{x \to x_0}(a_0 x^n + a_1 x^{n-1} + \cdots + a_n)$$
$$= a_0 (\lim\limits_{x \to x_0} x)^n + a_1 (\lim\limits_{x \to x_0} x)^{n-1} + \cdots + \lim\limits_{x \to x_0} a_n$$
$$= a_0 x_0^n + a_1 x_0^{n-1} + \cdots + a_n$$
$$= f(x_0).$$

Suppose that $R(x) = \dfrac{P(x)}{Q(x)}$ is a rational fraction function, where $P(x)$ and $Q(x)$ are both polynomials, then
$$\lim\limits_{x \to x_0} P(x) = P(x_0), \quad \lim\limits_{x \to x_0} Q(x) = Q(x_0).$$

If $Q(x_0) \neq 0$, then

$$\lim_{x \to x_0} R(x) = \lim_{x \to x_0} \frac{P(x)}{Q(x)} = \frac{\lim_{x \to x_0} P(x)}{\lim_{x \to x_0} Q(x)} = \frac{P(x_0)}{Q(x_0)} = R(x_0).$$

But we have to notice that: if $Q(x_0)=0$, the quotient rule for the limit of function cannot be used directly. We should consider another way.

Example 4 Find $\lim\limits_{x \to 1} \dfrac{(1-\sqrt{x})(1-\sqrt[3]{x})}{(1-x)^2}$.

Solution When $x \to 0$, the denominator tends to 0, the quotient rule can't be used directly. Here, we rationalize the numerator and then reduce the common factor. Hence

$$\lim_{x \to 1} \frac{(1-\sqrt{x})(1-\sqrt[3]{x})}{(1-x)^2} = \lim_{x \to 1} \frac{(1-x)(1-x)}{(1-x)^2(1+\sqrt{x})(1+\sqrt[3]{x}+\sqrt[3]{x^2})}$$

$$= \lim_{x \to 1} \frac{1}{(1+\sqrt{x})(1+\sqrt[3]{x}+\sqrt[3]{x^2})} = \frac{1}{6}.$$

Example 5 Find the following limit:

(1) $\lim\limits_{x \to \infty} \dfrac{1\,000x+2}{x^2+1}$; (2) $\lim\limits_{x \to \infty} \dfrac{2x^3-4x+x}{3x^3+2x^2-10}$.

Solution (1) Divide the numerator and denominator by x^2 and then take limit

$$\lim_{x \to \infty} \frac{1\,000x+2}{x^2+1} = \lim_{x \to \infty} \frac{\frac{1\,000}{x}+\frac{2}{x^2}}{1+\frac{1}{x^2}} = \frac{1\,000 \lim_{x \to \infty} \frac{1}{x} + 2\left(\lim_{x \to \infty} \frac{1}{x}\right)^2}{1+\left(\lim_{x \to \infty} \frac{1}{x}\right)^2} = \frac{0}{1} = 0.$$

(2) Divide the numerator and denominator by x^3 and then take limit

$$\lim_{x \to \infty} \frac{2x^3-4x+x}{3x^3+2x^2-10} = \lim_{x \to \infty} \frac{2-\frac{4}{x^2}+\frac{1}{x^3}}{3+\frac{2}{x}-\frac{10}{x^3}} = \frac{2}{3}.$$

1.3.2 Operation rule for composite function

> **Theorem 3** Suppose that $y=f[\varphi(x)]$ is composite function of $u=\varphi(x)$ and $y=f(u)$. The composition $f[\varphi(x)]$ is defined in a deleted neighborhood of x_0. If $\lim\limits_{x \to x_0}\varphi(x)=u_0$, $\lim\limits_{u \to u_0}f(u)=A$, and there exists $\delta_0>0$, when $x \in \overset{\circ}{U}(x_0, \delta_0)$, $\varphi(x) \neq u_0$, Then
>
> $$\lim_{x \to x_0} f[\varphi(x)] = \lim_{u \to u_0} f(u) = A.$$

Proof According to the definition of the function limit, we have to prove: $\forall \varepsilon>0$, $\exists \delta>0$, such that $|f[\varphi(x)]-A|<\varepsilon$ for $0<|x-x_0|<\delta$.

$\lim\limits_{u \to u_0} f(u)=A$, so for the above given $\varepsilon>0$, there exists $\eta>0$, when $0<|u-u_0|<\eta$, we have

$$|f(u)-A|<\varepsilon.$$

And because $\lim_{x \to x_0}\varphi(x)=u_0$, for the above $\eta>0$, there exists $\delta_1>0$, when $0<|x-x_0|<\delta_1$, we have
$$|\varphi(x)-u_0|<\eta.$$
According to the assumption, when $x\in \overset{\circ}{U}(x_0,\delta_0)$, $\varphi(x)\neq u_0$. Take $\delta=\min\{\delta_0,\delta_1\}$, then when $0<|x-x_0|<\delta$, we can get $0<|\varphi(x)-u_0|$ and $|\varphi(x)-u_0|<\eta$, namely $0<|\varphi(x)-u_0|<\eta$, then
$$|f[\varphi(x)]-A|<\varepsilon,$$
Therefore, $\lim_{x\to x_0} f[\varphi(x)]=A=\lim_{u\to u_0} f(u)$.

According to this theorem, by using replacement of variable $u=\varphi(x)$, we can find $\lim_{x\to x_0} f[\varphi(x)]$ by applying $\lim_{u\to u_0} f(u)$, that is $\lim_{x\to x_0} f[\varphi(x)]=\lim_{u\to u_0} f(u)=A$.

Example 6 Find $\lim_{x\to 1}\sqrt{\dfrac{1-x^3}{x-x^2}}$.

Solution $\lim_{x\to 1}\dfrac{1-x^3}{x-x^2}=\lim_{x\to 1}\dfrac{(1-x)(1+x+x^2)}{x(1-x)}=\lim_{x\to 1}\dfrac{(1+x+x^2)}{x}=3.$

let $u=\dfrac{1-x^3}{x-x^2}$, then we have
$$\lim_{x\to 1}\sqrt{\dfrac{1-x^3}{x-x^2}}=\lim_{u\to 3}\sqrt{u}=\sqrt{3}.$$

Exercises 1-3

1. Find the following limit:

(1) $\lim_{n\to\infty}\left(1+\dfrac{1}{2n}\right)\left(3-\dfrac{1}{4n}\right)$;   (2) $\lim_{n\to\infty}\dfrac{3n^2+n+5}{n^3+3n+1}$;

(3) $\lim_{n\to\infty}\dfrac{\sqrt{n}-8}{4n+1}$;   (4) $\lim_{n\to\infty}\dfrac{n(2n+1)(3n+2)}{5n^3}$;

(5) $\lim_{n\to\infty}\left(1+\dfrac{1}{2}+\dfrac{1}{2^2}+\cdots+\dfrac{1}{2^n}\right)$;   (6) $\lim_{n\to\infty}\dfrac{1+2+3+\cdots+n}{n^2}$.

2. Find the following limit:

(1) $\lim_{x\to 2}\dfrac{2x^2+3x+1}{3x-5}$;   (2) $\lim_{x\to\sqrt{2}}\dfrac{x^2-3}{x^2+1}$;

(3) $\lim_{x\to 1}\dfrac{x^2+2x-3}{x^3-1}$;   (4) $\lim_{x\to -1}\dfrac{x^2+6x+5}{1-x^2}$;

(5) $\lim_{x\to 0}\dfrac{5x^3+2x^2-x}{3x^2+2x}$;   (6) $\lim_{x\to 0}\dfrac{\sqrt{1+2x^2}-1}{x^2}$;

(7) $\lim_{x\to 1}\left(\dfrac{1}{1-x}-\dfrac{3}{1-x^3}\right)$;   (8) $\lim_{x\to 0}\dfrac{(1+x)^n-1-nx}{x^2}$ ($n\geqslant 2$, n is a positive integer).

3. Find the following limit:

(1) $\lim_{x\to\infty}\dfrac{2x+1}{x^2+2}$;   (2) $\lim_{x\to\infty}\dfrac{3x^2-1}{x^2-2x+3}$;

(3) $\lim\limits_{x\to+\infty}\dfrac{\sqrt{x^2+1}}{x+1}$; (4) $\lim\limits_{x\to+\infty}\sqrt{x}(\sqrt{x+1}-\sqrt{x})$.

4. Suppose $\lim\limits_{x\to\infty}\left(\dfrac{x^2+1}{x+1}\right)-ax-b=0$, find the values of a and b.

5. Find the value of a to make the limit of
$$f(x)=\begin{cases}-ax^2+x+1, & x\leqslant 1,\\ \dfrac{1-\sqrt{x}}{1-x}, & x>1,\end{cases}$$
exists as $x\to 1$.

6. In the following statements, which are right and which are wrong? If it is right, give reasons; if it is wrong, give one counter example.

(1) If $\lim\limits_{x\to x_0}f(x)$ exists and $\lim\limits_{x\to x_0}g(x)$ does not exist, then $\lim\limits_{x\to x_0}[f(x)+g(x)]$ doesn't exist;

(2) If $\lim\limits_{x\to x_0}f(x)$ exists and $\lim\limits_{x\to x_0}g(x)$ does not exist, then $\lim\limits_{x\to x_0}[f(x)\cdot g(x)]$ doesn't exist;

(3) If $\lim\limits_{x\to x_0}f(x)$ and $\lim\limits_{x\to x_0}g(x)$ don't exist, then $\lim\limits_{x\to x_0}[f(x)+g(x)]$ doesn't exist;

(4) If $\lim\limits_{n\to\infty}x_n$ and $\lim\limits_{n\to\infty}y_n$ exist, then $\lim\limits_{n\to\infty}x_ny_n$ doesn't exist.

7. Prove (1) and (3) in Theorem 1 in this section.

1.4 Principle of limit existence and two important limits

In this section, we introduce two principles of limit existence, and as applications, we discuss two important limits: $\lim\limits_{x\to 0}\dfrac{\sin x}{x}=1$ and $\lim\limits_{x\to\infty}\left(1+\dfrac{1}{x}\right)^x=e$.

1.4.1 Principle of limit existence

Principle 1 (The sandwich principle) Suppose that $\{x_n\}$, $\{y_n\}$ and $\{z_n\}$ are sequences such that

(1) $y_n\leqslant x_n\leqslant z_n\ (n=1,2,\cdots)$,

(2) $\lim\limits_{n\to\infty}y_n=a$, $\lim\limits_{n\to\infty}z_n=a$,

then the limit of the sequence $\{x_n\}$ exists and $\lim\limits_{n\to\infty}x_n=a$.

Proof $\forall \varepsilon>0$, since $\lim\limits_{n\to\infty}y_n=\lim\limits_{n\to\infty}z_n=a$, so $\exists N_1, N_2\in \mathbf{N}^+$, such that $|y_n-a|<\varepsilon$ for all $n>N_1$ and $|z_n-a|<\varepsilon$ for all $n>N_2$.

Take $N=\max\{N_1, N_2\}$, when $n>N$, we have $|y_n-a|<\varepsilon$ and $|z_n-a|<\varepsilon$, namely
$$a-\varepsilon<y_n<a+\varepsilon,\ a-\varepsilon<z_n<a+\varepsilon.$$
And because $y_n\leqslant x_n\leqslant z_n$, for all $n>N$, we have
$$a-\varepsilon<y_n\leqslant x_n\leqslant z_n<a+\varepsilon,$$
that is $|x_n-a|<\varepsilon$. Therefore $\lim\limits_{n\to\infty}x_n=a$.

Note From the proof, we can see condition (1) can be changed as: $\exists n_0 \in \mathbf{N}^+$, such that $y_n \leqslant x_n \leqslant z_n$ for all $n > n_0$.

This principle can be extended to the limit of functions.

Principle 1' Suppose that functions $f(x)$, $g(x)$ and $h(x)$ satisfy:

(1) when $x \in \mathring{U}(a, \delta)$ (or $|x| > X$), we have $g(x) \leqslant f(x) \leqslant h(x)$,

(2) $\lim\limits_{\substack{x \to x_0 \\ (x \to \infty)}} g(x) = A$, $\lim\limits_{\substack{x \to x_0 \\ (x \to \infty)}} h(x) = A$,

then $\lim\limits_{\substack{x \to x_0 \\ (x \to \infty)}} f(x) = A$.

Principle 1 and Principle 1' are called Squeeze Theorem and Sandwich Theorem.

Example 1 Find $\lim\limits_{n \to \infty} \left(\dfrac{1}{\sqrt{n^2+1}} + \dfrac{1}{\sqrt{n^2+2}} + \cdots + \dfrac{1}{\sqrt{n^2+2n}} \right)$.

Solution Since $\dfrac{2n}{\sqrt{n^2+2n}} < \dfrac{1}{\sqrt{n^2+1}} + \dfrac{1}{\sqrt{n^2+2}} + \cdots + \dfrac{1}{\sqrt{n^2+2n}} < \dfrac{2n}{\sqrt{n^2+1}}$, and

$$\lim_{n \to \infty} \dfrac{2n}{\sqrt{n^2+2n}} = \lim_{n \to \infty} \dfrac{2}{\sqrt{1+\dfrac{2}{n}}} = 2,$$

$$\lim_{n \to \infty} \dfrac{2n}{\sqrt{n^2+1}} = \lim_{n \to \infty} \dfrac{2}{\sqrt{1+\dfrac{1}{n^2}}} = 2,$$

according to Principle 1, we get

$$\lim_{n \to \infty} \left(\dfrac{1}{\sqrt{n^2+1}} + \dfrac{1}{\sqrt{n^2+2}} + \cdots + \dfrac{1}{\sqrt{n^2+2n}} \right) = 2.$$

Principle 2 A monotonous and bounded sequence must be convergent. That is to say, if

$$x_1 \leqslant x_2 \leqslant \cdots \leqslant x_n \leqslant x_{n+1} \leqslant \cdots \leqslant M,$$

or

$$x_1 \geqslant x_2 \geqslant \cdots \geqslant x_n \geqslant x_{n+1} \geqslant \cdots \geqslant m,$$

then the limit of the sequence exists.

The proof of the theorem is beyond the scope of this book. However the conclusion is apparent from the geometry. Suppose the sequence $\{x_n\}$ is monotone increasing. Then the corresponding points are arranged on the number axis from left to right successively as illustrated in Figure 1-19a. Since $\{x_n\}$ is bounded above, all the points x_n ($n=1, 2, \cdots$) can not go past some point M. Because $\{x_n\}$ has an infinite number of terms, those points x_n must accumulate forward to a point A as n increases. It is easy to recognize that A should be the limit of sequence $\{x_n\}$. The case for monotone decreasing and bounded sequence is shown in Figure 1-19b.

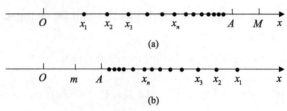

(a)

(b)

Figure 1-19

Principle 2 can also be described as, the monotone increasing sequence with upper bound, or the monotone decreasing sequence with lower bound which has the limit.

1.4.2 Two important limits

(1) $\lim\limits_{x \to 0} \dfrac{\sin x}{x} = 1$.

As an application of Principle $1'$, we prove this important limit. In the unit circle shown in Figure 1-20, assume the central angle $\angle AOC = x \left(0 < x < \dfrac{\pi}{2}\right)$. The tangent line at the point A intersects the prolongation line of OC at D, and $CB \perp OA$, we have

$$\overline{BC} = \sin x, \quad \overparen{AC} = x, \quad \overline{AD} = \tan x.$$
$$\text{area}(\triangle AOC) < \text{area}(\text{sector } AOC) < \text{area}(\triangle AOD),$$

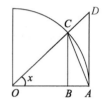

Figure 1-20

so $\dfrac{1}{2} \sin x < \dfrac{1}{2} x < \dfrac{1}{2} \tan x$, that is

$$\sin x < x < \tan x.$$

Divide each term by $\sin x$, we get $1 < \dfrac{x}{\sin x} < \dfrac{1}{\cos x}$, that is

$$\cos x < \dfrac{\sin x}{x} < 1.$$

Because $\cos x$, $\dfrac{\sin x}{x} (x \neq 0)$ are even functions, when $0 < |x| < \dfrac{\pi}{2}$, we have

$$\cos x < \dfrac{\sin x}{x} < 1.$$

And because $\lim\limits_{x \to 0} \cos x = 1$ (Example 10 in Section 1.2) and $\lim\limits_{x \to 0} 1 = 1$, according to Principle $1'$, we obtain $\lim\limits_{x \to 0} \dfrac{\sin x}{x} = 1$.

Example 2 Find $\lim\limits_{x \to 0} \dfrac{\tan x}{x}$.

Solution $\lim\limits_{x \to 0} \dfrac{\tan x}{x} = \lim\limits_{x \to 0} \left(\dfrac{\sin x}{x} \cdot \dfrac{1}{\cos x} \right) = \lim\limits_{x \to 0} \dfrac{\sin x}{x} \cdot \lim\limits_{x \to 0} \dfrac{1}{\cos x} = 1.$

Example 3 Find $\lim\limits_{x \to 0} \dfrac{x}{\arcsin x}$.

Solution Let $u = \arcsin x$, then $x = \sin u$, when $x \to 0$, $u \to 0$. So

$$\lim\limits_{x \to 0} \dfrac{x}{\arcsin x} = \lim\limits_{u \to 0} \dfrac{\sin u}{u} = 1.$$

Example 4 Find $\lim\limits_{x\to 0}\dfrac{1-\cos x}{x^2}$.

Solution $\lim\limits_{x\to 0}\dfrac{1-\cos x}{x^2}=\lim\limits_{x\to 0}\dfrac{2\sin^2\dfrac{x}{2}}{x^2}=\lim\limits_{x\to 0}\dfrac{2\sin^2\dfrac{x}{2}}{2^2\left(\dfrac{x}{2}\right)^2}=\dfrac{1}{2}\lim\limits_{x\to 0}\left[\dfrac{\sin\dfrac{x}{2}}{\dfrac{x}{2}}\right]^2=\dfrac{1}{2}.$

(2) $\lim\limits_{x\to\infty}\left(1+\dfrac{1}{x}\right)^x=\mathrm{e}.$

As an application of Principle 2, we prove another important limit. Firstly we consider the case that x is a positive integer n and approaches $+\infty$. For the sequence $\left\{\left(1+\dfrac{1}{n}\right)^n\right\}$ we will prove the sequence is monotone increasing. By the binomial theorem we know

$$x_n=\left(1+\dfrac{1}{n}\right)^n=1+C_n^1\dfrac{1}{n}+C_n^2\dfrac{1}{n^2}+\cdots+C_n^n\dfrac{1}{n^n}$$

$$=1+\dfrac{n}{1!}\cdot\dfrac{1}{n}+\dfrac{n(n-1)}{2!}\cdot\dfrac{1}{n^2}+\dfrac{n(n-1)(n-2)}{3!}\cdot\dfrac{1}{n^3}+\cdots+$$

$$\dfrac{n(n-1)(n-2)\cdots(n-n+1)}{n!}\cdot\dfrac{1}{n^n}$$

$$=1+1+\dfrac{1}{2!}\left(1-\dfrac{1}{n}\right)+\dfrac{1}{3!}\left(1-\dfrac{1}{n}\right)\left(1-\dfrac{2}{n}\right)+\cdots+$$

$$\dfrac{1}{n!}\left(1-\dfrac{1}{n}\right)\left(1-\dfrac{2}{n}\right)\cdots\left(1-\dfrac{n-1}{n}\right).$$

Similarly, we have

$$x_{n+1}=1+1+\dfrac{1}{2!}\left(1-\dfrac{1}{n+1}\right)+\dfrac{1}{3!}\left(1-\dfrac{1}{n+1}\right)\left(1-\dfrac{2}{n+1}\right)+\cdots+$$

$$\dfrac{1}{n!}\left(1-\dfrac{1}{n+1}\right)\left(1-\dfrac{2}{n+1}\right)\cdots\left(1-\dfrac{n-1}{n+1}\right)+$$

$$\dfrac{1}{(n+1)!}\left(1-\dfrac{1}{n+1}\right)\left(1-\dfrac{2}{n+1}\right)\cdots\left(1-\dfrac{n-1}{n+1}\right)\left(1-\dfrac{n}{n+1}\right).$$

Comparing the expansion of x_n and x_{n+1}, we can see that x_{n+1} has one more term (and is positive) than x_n, that is to say, except the first two terms, each term x_{n+1} is bigger than the corresponding term $x_n<x_{n+1}$. Hence, the sequence $\{x_n\}$ is monotone increasing.

Secondly, we will prove the sequence $\{x_n\}$ is upper bounded. According to the expression of x_n, we can get

$$x_n<1+1+\dfrac{1}{2!}+\dfrac{1}{3!}+\cdots+\dfrac{1}{n!}<1+1+\dfrac{1}{1\times 2}+\dfrac{1}{2\times 3}+\cdots+\dfrac{1}{(n-1)n}$$

$$=1+1+\left(1-\dfrac{1}{2}\right)+\left(\dfrac{1}{2}-\dfrac{1}{3}\right)+\cdots+\left(\dfrac{1}{n-1}-\dfrac{1}{n}\right)$$

$$=3-\dfrac{1}{n}<3.$$

thus $\{x_n\}$ has an upper bound.

According to Principle 2, the sequence $\{x_n\}$ is convergent, namely, $\lim\limits_{n\to\infty}\left(1+\dfrac{1}{n}\right)^n$ exists. The limit is denoted by e, thus,

$$\lim_{n\to\infty}\left(1+\frac{1}{n}\right)^n = e.$$

It can be proved that the limit of function $\left(1+\frac{1}{x}\right)^x$ exists and equal e as x approaches either $+\infty$ or $-\infty$, that is

$$\lim_{x\to\infty}\left(1+\frac{1}{x}\right)^x = e.$$

Let $t=\frac{1}{x}$, then as $x\to\infty$, $t\to 0$, so we have $\lim_{t\to 0}(1+t)^{\frac{1}{t}} = e$, namely,

$$\lim_{x\to 0}(1+x)^{\frac{1}{x}} = e.$$

It can be proved that e is an irrational number, and $e = 2.718\ 281\ 828\ 459\ 045\cdots$, which is a non-terminating decimal.

Example 5 Find $\lim\limits_{x\to\infty}\left(1-\frac{1}{x}\right)^x$.

Solution $\lim\limits_{x\to\infty}\left(1-\frac{1}{x}\right)^x = \lim\limits_{x\to\infty}\left[\left(1+\frac{1}{-x}\right)^{-x}\right]^{-1} = e^{-1}$.

Example 6 Find $\lim\limits_{x\to\infty}\left(\frac{x+3}{x+2}\right)^{2x}$.

Solution $\lim\limits_{x\to\infty}\left(\frac{x+3}{x+2}\right)^{2x} = \lim\limits_{x\to\infty}\left[\left(1+\frac{1}{x+2}\right)^x\right]^2 = \lim\limits_{x\to\infty}\left[\left(1+\frac{1}{x+2}\right)^{x+2-2}\right]^2$

$$= \lim_{x\to\infty}\left\{\left[\left(1+\frac{1}{x+2}\right)^{x+2}\right]^2\left(1+\frac{1}{x+2}\right)^{-4}\right\}$$

$$= \lim_{x\to\infty}\left[\left(1+\frac{1}{x+2}\right)^{x+2}\right]^2 \cdot \lim_{x\to\infty}\left(1+\frac{1}{x+2}\right)^{-4} = e^2 \cdot 1 = e^2.$$

Exercises 1-4

1. Find the following limit:

(1) $\lim\limits_{x\to 0}\dfrac{\sin 3x}{x}$;

(2) $\lim\limits_{x\to 0}\dfrac{\tan kx}{x}$ (k is a constant);

(3) $\lim\limits_{x\to 0}\dfrac{\sin 2x}{\sin 5x}$;

(4) $\lim\limits_{x\to 0}x\cot 2x$;

(5) $\lim\limits_{x\to 0}\dfrac{1-\cos 2x}{x\sin x}$;

(6) $\lim\limits_{n\to\infty}2^n\sin\dfrac{x}{2^{n-1}}$ (x is a non-zero constant);

(7) $\lim\limits_{n\to\infty}\dfrac{\sin\tan\dfrac{1}{n}}{\sin\dfrac{3}{n}}$;

(8) $\lim\limits_{x\to 1}(1-x)\tan\dfrac{\pi x}{2}$.

2. Find the following limit:

(1) $\lim\limits_{x\to 0}(1-2x)^{\frac{1}{x}}$;

(2) $\lim\limits_{x\to 0}\left(\dfrac{x+2}{2}\right)^{-\frac{1}{x}}$;

(3) $\lim\limits_{x\to\infty}\left(\dfrac{x-1}{x}\right)^{3x}$;

(4) $\lim\limits_{x\to\infty}\left(\dfrac{x-1}{x+1}\right)^x$.

3. Find the value of a to make the limit of

$$f(x)=\begin{cases} \dfrac{\sin ax}{x}, & x<0, \\ (1+2x)^{\frac{1}{x}}, & x>0, \end{cases}$$

exists as $x\to 0$.

4. By using Principle Ⅱ, prove the limit of the sequence $\sqrt{2}, \sqrt{2+\sqrt{2}}, \sqrt{2+\sqrt{2+\sqrt{2}}}, \cdots$ exists and find its limit.

5. Prove by the principle of limit existence.

(1) $\lim\limits_{n\to\infty}\left(\dfrac{1}{n^2+1}+\dfrac{1}{n^2+2}+\cdots+\dfrac{1}{n^2+n}\right)=0$;

(2) $\lim\limits_{n\to\infty}n\left(\dfrac{1}{n^2+\pi}+\dfrac{1}{n^2+2\pi}+\cdots+\dfrac{1}{n^2+n\pi}\right)=1$;

(3) $\lim\limits_{x\to\infty}\dfrac{[x]}{x}=1$.

1.5 Infinitesimal and infinity

Infinitesimal and infinity are two special limits, and play an important role in the limit theory. This section mainly discuss the concepts of infinitesimal and infinity, and comparison of infinitesimal order.

1.5.1 Infinitesimal

Definition 1 If the limit of a function $f(x)$ as $x\to x_0$ (or $x\to\infty$) is 0. Then the function $f(x)$ is called an infinitesimal quantity with respect to $x\to x_0$ (or $x\to\infty$).

For example, $x^2, 2x+x^3, 1-\cos x$ are infinitesimals; as $n\to\infty$ 时, $\dfrac{1}{n}, \dfrac{1+n}{n^2}$ are infinitesimals,

Note An infinitesimal is a variable, which tends to zero. It relates to the process of the change of the independent variable. For example, x^2 is an infinitesimal as $x\to 0$, while x^2 is not an infinitesimal as $x\to 1$. Besides, infinitesimal cannot be regarded as a number whose absolute value is quite small. Infinitesimal is a variable which approaches zero as the independent variable changes. Any small non-zero value is a constant, and doesn't tend to zero. However, zero is an infinitesimal, and also can be seen as a constant function, the limit of which is zero, whatever the independent variable changes. Zero is the only constant that can be seen as an infinitesimal.

Because infinitesimal is a variable whose limit is zero, infinitesimal has a close relationship with limit as follows.

Theorem 1 The necessary and sufficient condition for the limit of function $f(x)$ is A as $x\to x_0$ (or $x\to\infty$), $f(x)=A+\alpha(x)$, where $\alpha(x)$ is an infinitesimal.

Proof Necessity. Assume that $\lim\limits_{x \to x_0} f(x) = A$, then
$$\lim_{x \to x_0}[f(x) - A] = \lim_{x \to x_0} f(x) - A = A - A = 0,$$
that is $f(x) - A$ is an infinitesimal as $x \to x_0$. Let $\alpha(x) = f(x) - A$, then
$$f(x) = A + (f(x) - A) = A + \alpha(x),$$
that is $f(x)$ is the sum of the limit A and an infinitesimal $\alpha(x)$.

Sufficiency. If $f(x) = A + \alpha(x)$, where $\alpha(x)$ is an infinitesimal, then $\lim\limits_{x \to x_0} \alpha(x) = 0$, and
$$\lim_{x \to x_0} f(x) = A + \lim_{x \to x_0} \alpha(x) = A.$$

By means of the rational operation rules, it is easy to prove the following conclusions.

> **Theorem 2** Suppose the independent variables change in the same way,
> (1) a finite sum of infinitesimals is an infinitesimal;
> (2) a finite product of infinitesimals is an infinitesimal;
> (3) the product of an infinitesimal and a bounded functions is an infinitesimal.

Proof According to the operation rules of limit, (1) and (2) are easy to prove. Here we prove (3) under the condition $x \to x_0$.

Suppose function $u(x)$ is bounded in the deleted neighborhood $\overset{\circ}{U}(x_0, \delta_1)$ of x_0, then $\exists M > 0$, such that $|u(x)| \leqslant M$ for all $x \in \overset{\circ}{U}(x_0, \delta_1)$.

Let function $\alpha(x)$ is an infinitesimal as $x \to x_0$, then $\forall \varepsilon > 0, \exists \delta_2 > 0$, such that $|\alpha(x)| < \dfrac{\varepsilon}{M}$ for all $x \in \overset{\circ}{U}(x_0, \delta_2)$.

Here we take $\delta = \min\{\delta_1, \delta_2\}$, for all $x \in \overset{\circ}{U}(x_0, \delta)$, we get
$$|u(x)| \leqslant M \quad \text{and} \quad |\alpha(x)| < \frac{\varepsilon}{M}.$$
Thus
$$|\alpha(x) \cdot u(x)| = |\alpha(x)| \cdot |u(x)| < \frac{\varepsilon}{M} \cdot M = \varepsilon.$$
Therefore, $\alpha(x)u(x)$ is an infinitesimal as $x \to x_0$. That is to say the product of an infinitesimal and a bounded function is an infinitesimal.

Example 1 Suppose the independent variables vary in the same way, $\lim f(x)$ and $\lim g(x)$ exist. Prove
$$\lim[f(x) \cdot g(x)] = \lim f(x) \cdot \lim g(x).$$
Proof Suppose $\lim f(x) = A$, $\lim g(x) = B$. According to Theorem 1,
$$f(x) = A + \alpha(x), g(x) = B + \beta(x),$$
where $\alpha(x)$ and $\beta(x)$ are both infinitesimals. Thus
$$f(x)g(x) = AB + [A\beta(x) + B\alpha(x) + \alpha(x)\beta(x)].$$
According to Theorem 2, $A\beta(x) + B\alpha(x) + \alpha(x)\beta(x)$ is an infinitesimal. Therefore, according to Theorem 1, we have

$$\lim [f(x) \cdot g(x)] = AB = \lim f(x) \cdot \lim g(x).$$

Example 2 Find $\lim\limits_{x\to 0} x^2 \left(2\sin\dfrac{1}{x} + 3\cos\dfrac{1}{x}\right)$.

Solution Since $2\sin\dfrac{1}{x} + 3\cos\dfrac{1}{x}$ is a bounded function. As $x \to 0$, x^2 is an infinitesimal. According to the Theorem 2, we have

$$\lim\limits_{x\to 0} x^2 \left(2\sin\dfrac{1}{x} + 3\cos\dfrac{1}{x}\right) = 0.$$

Therefore,

as $n \to \infty$, the limit of $\dfrac{1}{n}\sin n$ is zero;

as $x \to \infty$, the limit of $\dfrac{1}{x}(2 + \sin x^2)$ is zero;

as $x \to x_0$, the limit of $(x - x_0)\cos\dfrac{1}{x - x_0}$ is zero.

1.5.2 Infinity

Definition 2 Suppose that we have a function $f: \mathring{U}(x_0) \to \mathbf{R}$. If $\forall M > 0$, $\exists \delta > 0$ (or $X > 0$) such that $|f(x)| > M$ for all $x \in \mathring{U}(x_0, \delta)$ (or $|x| > X$), then $f(x)$ is called an infinity as $x \to x_0$ (or $x \to \infty$), denoted as

$$\lim\limits_{x \to x_0} f(x) = \infty \text{ (or } \lim\limits_{x \to \infty} f(x) = \infty\text{)}.$$

Example 3 Prove $\lim\limits_{x \to 1}\dfrac{1}{x - 1} = \infty$ (Figure 1-21).

Proof $\forall M > 0$, we want to find $\delta > 0$ such that $|f(x)| = \left|\dfrac{1}{x-1}\right| = \dfrac{1}{|x-1|} > M$ for $0 < |x - 1| < \delta$. Since $\dfrac{1}{|x-1|} > M$, we have

$$|x - 1| < \dfrac{1}{M}.$$

Thus take $\delta = \dfrac{1}{M}$, for all $0 < |x - 1| < \delta$, we get

$$|f(x)| = \left|\dfrac{1}{x-1}\right| > M,$$

Therefore $\lim\limits_{x \to 1}\dfrac{1}{x-1} = \infty$.

In general, if $\lim\limits_{x \to x_0} f(x) = \infty$, then $x = x_0$ is called the vertical asymptote of the graph of the function $y = f(x)$. Obviously, as shown in Figure 1-21, the straight line $x = 1$ is the vertical asymptote of the graph of the function $f(x) = \dfrac{1}{x - 1}$.

In Definition 2, if we change $|f(x)| > M$ to $f(x) > M$ or $f(x) <$

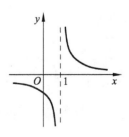

Figure 1-21

$-M$, we can get the definitions of
$$\lim_{\substack{x\to x_0\\(x\to\infty)}} f(x)=+\infty \text{ and } \lim_{\substack{x\to x_0\\(x\to\infty)}} f(x)=-\infty$$

Similarly, we also get the definitions of
$$\lim_{x\to x_0^-} f(x)=\infty(+\infty,-\infty),\ \lim_{x\to x_0^+} f(x)=\infty(+\infty,-\infty),$$
$$\lim_{x\to -\infty} f(x)=\infty(+\infty,-\infty),\ \lim_{x\to +\infty} f(x)=\infty(+\infty,-\infty)$$

etc..

For example, it is not difficult to find that
$$\lim_{x\to 2}\frac{1}{(2-x)^2}=+\infty;$$
$$\lim_{x\to 1^-}\frac{1}{1-x}=+\infty,\ \lim_{x\to 1^+}\frac{1}{1-x}=-\infty;$$
$$\lim_{x\to \frac{\pi}{2}^-}\tan x=+\infty,\ \lim_{x\to -\frac{\pi}{2}^+}\tan x=-\infty.$$

Note (1) Infinity is a variable, and means the limit doesn't exist. In the notation of infinity, although we use the symbol of limit, the limit does not exist.

(2) If $\lim_{x\to x_0} f(x)=\infty$, then $f(x)$ is unbounded in the deleted neighborhood of x_0. However, an unbounded function may not be an infinity. For instance, $f(x)=\frac{1}{x}\sin\frac{1}{x}$ is unbounded in the deleted neighborhood of point 0. However, as $x\to 0$, $f(x)$ does not approach infinity.

There is a close relationship between infinity and infinitesimal.

Theorem 3 Suppose that the independent variables in the following functions vary in the same way.

(1) If $f(x)$ is an infinity, then $\frac{1}{f(x)}$ is an infinitesimal;

(2) If $f(x)$ is an infinitesimal, and $f(x)\neq 0$, then $\frac{1}{f(x)}$ is an infinity.

Proof (1) Suppose $\lim_{x\to x_0} f(x)=\infty$.

$\forall \varepsilon>0$, then take $M=\frac{1}{\varepsilon}$, $\exists \delta>0$, when $0<|x-x_0|<\delta$, we have $|f(x)|>M=\frac{1}{\varepsilon}$, that is
$$\left|\frac{1}{f(x)}\right|<\varepsilon,$$

thus $\lim_{x\to x_0}\frac{1}{f(x)}=0$, namely, $\frac{1}{f(x)}$ is an infinitesimal.

(2) Suppose $\lim_{x\to x_0} f(x)=0$ and $f(x)\neq 0$.

$\forall M>0$, then take $\varepsilon=\frac{1}{M}$, $\exists \delta>0$, when $0<|x-x_0|<\delta$, we have $|f(x)|<\varepsilon=\frac{1}{M}$, thus,

$$\left|\frac{1}{f(x)}\right| > M,$$

namely, $\frac{1}{f(x)}$ is an infinity.

Similarly, we can prove the case $x \to \infty$.

Example 4 Find $\lim\limits_{x \to 1} \frac{x^5 - 3x + 5}{x - 1}$.

Solution Since $\lim\limits_{x \to 1} \frac{x - 1}{x^5 - 3x + 5} = \frac{0}{3} = 0$, thus $\lim\limits_{x \to 1} \frac{x^5 - 3x + 5}{x - 1} = \infty$.

Example 5 Find $\lim\limits_{x \to \infty} \frac{2x^4 - 3x + 10}{3x^3 + 5x^2 - 2x}$.

Solution Since

$$\lim_{x \to \infty} \frac{3x^3 + 5x^2 - 2x}{2x^4 - 3x + 10} = \lim_{x \to \infty} \frac{\frac{3}{x} + \frac{5}{x^2} - \frac{2}{x^4}}{2 - \frac{3}{x^3} + \frac{10}{x^4}} = 0,$$

thus
$$\lim_{x \to \infty} \frac{2x^4 - 3x + 10}{3x^3 + 5x^2 - 2x} = \infty.$$

Note From Example 5 in Section 1.3 and this Example 5, we find that when $a_0 \neq 0$, $b_0 \neq 0$, m and n are nonnegative integers, generally, we have

$$\lim_{x \to \infty} \frac{a_0 x^m + a_1 x^{m-1} + \cdots + a_{m-1} x + a_m}{b_0 x^n + b_1 x^{n-1} + \cdots + b_{n-1} x + b_n} = \begin{cases} \frac{a_0}{b_0}, & n = m, \\ 0, & n > m, \\ \infty, & n < m. \end{cases}$$

1.5.3 Comparison of infinitesimal order

We already knew that: the sum, difference and product of two infinitesimals are also infinitesimals. But the situation is different from the quotient of two infinitesimals. For example, as $x \to 0$, x, $2x$ and x^2 are all infinitesimals,

$$\lim_{x \to 0} \frac{x^2}{x} = 0, \quad \lim_{x \to 0} \frac{x}{x^2} = \infty, \quad \lim_{x \to 0} \frac{x}{2x} = \frac{1}{2}.$$

The different results show that the difference in the speed of infinitesimal tending to zero. The various circumstances of the limit of the ratio of two infinitesimals reflect the difference in the degree of "speed" which tends to be zero. As shown from the above limits, as x is much faster than x tends to zero, which in turn is much slower than zero and x and $2x$ are about zero.

> **Definition 3** Suppose that the independent variables change in the same way. Let α, β be two infinitesimals and $\beta \neq 0$.
>
> (1) If $\lim \frac{\alpha}{\beta} = 0$, then we say α is an infinitesimal of higher order than β, denoted as $\alpha = o(\beta)$;
>
> (2) If $\lim \frac{\alpha}{\beta} = \infty$, then we say α is an infinitesimal of lower order than β;

(3) If $\lim\frac{\alpha}{\beta}=c\neq 0$, then α and β are called infinitesimals of the same order;

(4) If $\lim\frac{\alpha}{\beta}=1$, then α and β are said to be equivalent, denoted as $\alpha\sim\beta$;

(5) If $\lim\frac{\alpha}{\beta^k}=c\neq 0(k>0)$, then α is called an infinitesimal of order k compared with β.

Obviously, the equivalent infinitesimals is a special case of infinitesimals of the same order, that is the case of $c=1$.

Here are some examples.

Since $\lim\limits_{x\to 0}\frac{\sin x}{x}=1$, so as $x\to 0$, $\sin x$ and x are equivalent infinitesimals, that is $\sin x\sim x$.

Since $\lim\limits_{x\to 0}\frac{\tan x}{x}=1$ (Example 2 in Section 1.4), so as $x\to 0$, $\tan x$ and x are equivalent infinitesimals, that is $\tan x\sim x$.

Since $\lim\limits_{x\to 0}\frac{1-\cos x}{x^2}=\frac{1}{2}$ (Example 4 in Section 1.4), so as $x\to 0$, $1-\cos x$ is an infinitesimal of order 2 compared with x, and $1-\cos x$ and x^2 are infinitesimals of the same order.

Since $\lim\limits_{x\to 0}\frac{1-\cos x}{\frac{1}{2}x^2}=1$, so as $x\to 0$, $1-\cos x$ and $\frac{1}{2}x^2$ are equivalent infinitesimals, that is $1-\cos x\sim\frac{1}{2}x^2$.

Since $\lim\limits_{x\to 0}\frac{\sin x^2}{x}=\lim\limits_{x\to 0}\frac{\sin^2 x}{x^2}\cdot\lim\limits_{x\to 0}x=0$, so as $x\to 0$, $\sin^2 x$ is an infinitesimal of higher order than x.

Since $\lim\limits_{x\to\infty}\frac{\frac{1}{x}}{\frac{1}{x^2}}=\infty$, so as $x\to\infty$, $\frac{1}{x}$ is an infinitesimal of lower order than $\frac{1}{x^2}$.

There exist two important conclusions for equivalent infinitesimals.

Theorem 4 Suppose that the independent variables change in the same way, the necessary and sufficient condition of α and β are equivalent infinitesimals which is $\beta=\alpha+o(\alpha)$.

Proof Suppose $\alpha\sim\beta$, then
$$\lim\frac{\beta-\alpha}{\alpha}=\lim\frac{\beta}{\alpha}-1=0,$$
therefore $\beta-\alpha=o(\alpha)$, that is $\beta=\alpha+o(\alpha)$.

On the contrary, let $\beta=\alpha+o(\alpha)$, then

$$\lim \frac{\beta}{\alpha} = \lim \frac{\alpha + o(\alpha)}{\alpha} = 1 + \lim \frac{o(\alpha)}{\alpha} = 1,$$

therefore $\alpha \sim \beta$.

Theorem 5 Suppose that the independent variables change in the same way, $\alpha \sim \alpha'$, $\beta \sim \beta'$, and $\lim \frac{\alpha'}{\beta'}$ exist. Then $\lim \frac{\alpha}{\beta} = \lim \frac{\alpha'}{\beta'}$.

Proof Since $\alpha \sim \alpha'$, $\beta \sim \beta'$, so $\lim \frac{\alpha'}{\alpha} = 1$ and $\lim \frac{\beta}{\beta'} = 1$ exist. Then we have

$$\lim \frac{\alpha}{\beta} = \lim \left(\frac{\alpha}{\alpha'} \cdot \frac{\alpha'}{\beta'} \cdot \frac{\beta'}{\beta} \right) = \lim \frac{\alpha}{\alpha'} \cdot \lim \frac{\alpha'}{\beta'} \cdot \lim \frac{\beta'}{\beta} = \lim \frac{\alpha'}{\beta'}.$$

Theorem 5 indicates that, when evaluating the limit of a ratio of two infinitesimals, we can replace the terms of the ratio by their equivalent infinitesimals. Getting limit with substitution of equivalent infinitesimal is a common, convenient and efficient way.

In particular, as $x \to 0$, the following infinitesimals are equivalent:

(1) $\sin x \sim \tan x \sim \arcsin x \sim \arctan x \sim x$;

(2) $1 - \cos x \sim \frac{1}{2} x^2$;

(3) $\ln(1+x) \sim x$;

(4) $e^x - 1 \sim x$;

(5) $(1+x)^\alpha - 1 \sim \alpha x$, where α is a non-zero constant.

Example 6 Find $\lim\limits_{x \to 0} \frac{\tan^2 x}{\sin 2x^2}$.

Solution We know that $\sin 2x^2 \sim 2x^2$, $\tan^2 x \sim x^2$ (since $\tan x \sim x$), as $x \to 0$. Hence

$$\lim_{x \to 0} \frac{\tan^2 x}{\sin 2x^2} = \lim_{x \to 0} \frac{x^2}{2x^2} = \frac{1}{2}.$$

Example 7 Find $\lim\limits_{x \to 0} \frac{a^x - 1}{x}$ $(a > 0, a \neq 1)$.

Solution As $x \to 0$, $a^x - 1 = e^{x \ln a} - 1 \sim x \ln a$, hence

$$\lim_{x \to 0} \frac{a^x - 1}{x} = \lim_{t \to 0} \frac{x \ln a}{x} = \ln a.$$

Example 8 Find $\lim\limits_{x \to 0} \frac{(1+x^2)^{\frac{2}{3}} - 1}{\cos x - 1}$.

Solution When $x \to 0$, $(1+x^2)^{\frac{2}{3}} - 1 \sim \frac{2}{3} x^2$, $\cos x - 1 \sim -\frac{1}{2} x^2$, the limit can be written as

$$\lim_{x \to 0} \frac{(1+x^2)^{\frac{2}{3}} - 1}{\cos x - 1} = \lim_{x \to 0} \frac{\frac{2}{3} x^2}{-\frac{1}{2} x^2} = -\frac{4}{3}.$$

Example 9 Find $\lim\limits_{x \to 0} \frac{\tan x - \sin x}{\sin^3 2x}$.

Solution $\lim\limits_{x\to 0}\dfrac{\tan x-\sin x}{\sin^3 2x}=\lim\limits_{x\to 0}\dfrac{\tan x(1-\cos x)}{\sin^3 2x}$.

Since $\sin 2x \sim 2x$, $\tan x \sim x$, $1-\cos x \sim \dfrac{1}{2}x^2$ as $x\to 0$. This yields:

$$\lim_{x\to 0}\frac{\tan x-\sin x}{\sin^3 2x}=\lim_{x\to 0}\frac{x\cdot\dfrac{1}{2}x^2}{(2x)^3}=\frac{1}{16}.$$

Exercises 1-5

1. Suppose $\varphi(x)=x$, as $x\to 0$, which of the following function is an infinitesimal of higher order than $\varphi(x)$? Which is an infinitesimal of lower order than $\varphi(x)$? Which are equivalent infinitesimals? And instruct reasons.

   (1) x^3+2x^2;
   (2) $x^5\sin x^3$;
   (3) $2x-\sin x$;
   (4) $x^{\frac{1}{3}}$;
   (5) $\tan 2x$;
   (6) $1-\cos\sqrt{x}$.

2. Prove by the definition, the function $y=\dfrac{1-x}{x}$ is infinity as $x\to 0$.

3. Whether the function $f(x)=x\cos x$ is bounded in $(-\infty,+\infty)$? Whether the function is infinity as $x\to\infty$? Why?

4. Find the asymptote of the figure of the function $f(x)=\dfrac{x^2+2x}{4-x^2}$.

5. By the feature of equivalent infinitesimal, find the following limit.

   (1) $\lim\limits_{x\to 0}\dfrac{\tan 3x}{\sin 2x}$;
   (2) $\lim\limits_{x\to 0}\dfrac{\sin x^m}{(\sin x)^n}\ (m,n\in\mathbf{N})$;
   (3) $\lim\limits_{x\to 0}\dfrac{\sqrt{1+2x^2}-1}{x\sin\dfrac{x}{2}}$;
   (4) $\lim\limits_{x\to 0}\dfrac{x^2\sin\dfrac{1}{2x}}{\arcsin 2x}$;
   (5) $\lim\limits_{x\to 0}\dfrac{\sin^3 x\cos x}{\sin 3x\tan^2 x}$;
   (6) $\lim\limits_{x\to 0}\dfrac{\tan x-\sin x}{\sin^3 x}$;
   (7) $\lim\limits_{x\to 0}\dfrac{1-\cos 2x}{\ln(1+x\sin x)}$;
   (8) $\lim\limits_{x\to a}\dfrac{e^x-e^a}{x-a}$.

1.6 Continuity of function and the properties

1.6.1 Continuity of function and discontinuity

1) Continuity of function

As we know, the main goal of calculus is to investigate the rules of change from a quantitative point of view. Usually the changes in a variable may be divided into two types, namely gradual changes and sudden changes. For example, normally, air temperature changes gradually with time, but when a large cold wave arrives in the winter, the air temperature may change suddenly with time. Again, normally the amplitude of oscillation of the earth's crust varies gradually with time, but sometimes the earth's crust will suddenly buckle when an earthquake occurs. The concepts of continuity and discontinuity will be introduced for describing these two kinds of changes in real situations.

Suppose that function $y=f(x)$ is defined in some neighborhood of x_0. When the independent variable changes from x_0 to x, the value of the function changes from $f(x_0)$ to $f(x)$. We call

$$\Delta x = x - x_0$$

the increment of the independent variable, and

$$\Delta y = f(x) - f(x_0) = f(x_0 + \Delta x) - f(x_0)$$

is called the increment of the function value, or simply the increment of the function (Figure 1-22).

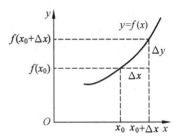

Figure 1-22

Definition 1 Let function $y=f(x)$ is defined in some neighborhood of x_0, if

$$\lim_{\Delta x \to 0} \Delta y = \lim_{\Delta x \to 0} [f(x_0 + \Delta x) - f(x_0)] = 0,$$

then the function $y=f(x)$ is said to be continuous at x_0.

Let $x = x_0 + \Delta x$, then $x \to x_0$ as $\Delta x \to 0$. Since

$$\Delta y = f(x_0 + \Delta x) - f(x_0) = f(x) - f(x_0),$$

that is

$$f(x) = f(x_0) + \Delta y.$$

So $\Delta y \to 0$ is equivalent to $f(x) \to f(x_0)$, thus $\lim_{\Delta x \to 0} \Delta y = 0$, $\lim_{x \to x_0} f(x) = f(x_0)$.

Therefore, the continuity of $y=f(x)$ at x_0 can be given as the definition below:

Definition 1' Let function $y=f(x)$ is defined in some neighborhood of x_0, if

$$\lim_{x \to x_0} f(x) = f(x_0),$$

then we say that the function $y=f(x)$ is continuous at x_0.

From Definition 1', we can see that if a function $f(x)$ is continuous at x_0, then it must satisfy the following three conditions.

(1) $f(x)$ is defined at x_0;

(2) $\lim_{x \to x_0} f(x)$ exists, that is, $\lim_{x \to x_0^-} f(x)$ and $\lim_{x \to x_0^+} f(x)$ both exist and are equal;

(3) $\lim_{x \to x_0} f(x) = f(x_0)$.

If at least one of the conditions of continuity is not fulfilled, then $f(x)$ is not continuous at x_0.

Similarly, we can define left-sided continuity and right-sided continuity by the following one-sided limits respectively.

Definition 2 If $\lim\limits_{x \to x_0^-} f(x) = f(x_0)$, that is $f(x_0-0) = f(x_0)$, then $f(x)$ is called left-sided continuity at x_0. If $\lim\limits_{x \to x_0^+} f(x) = f(x_0)$, that is $f(x_0+0) = f(x_0)$, the $f(x)$ is called right-sided continuity at x_0.

Obviously, the necessary and sufficient conditions for the continuity of $f(x)$ at x_0 are that $f(x)$ is left-sided continuity and right-sided continuity at x_0, that is

$$\lim_{x \to x_0^-} f(x) = f(x_0) = \lim_{x \to x_0^+} f(x).$$

If $f(x)$ is continuous at every point of an open interval (a,b), then $f(x)$ is said to be continuous in the interval (a,b). If $f(x)$ is continuous in the open interval (a, b) and is right-sided continuous at the left end point $x=a$, and is also left-sided continuous at the right end point $x=b$, then $f(x)$ is said to be continuous in the closed interval $[a, b]$. If $f(x)$ is continuous in an interval I, then $f(x)$ is called a continuous function in the interval I.

Example 1 Suppose a constant function $f(x) = C$, for any $x_0 \in (-\infty, +\infty)$, we have $\lim\limits_{x \to x_0} f(x) = C$, that is

$$\lim_{x \to x_0} f(x) = f(x_0),$$

thus $f(x)$ is continuous at x_0. Therefore, the constant function $y = C$ is continuous on $(-\infty, +\infty)$.

Example 2 Suppose the rational function $R(x) = \dfrac{P(x)}{Q(x)}$, where $P(x)$ and $Q(x)$ are polynomials. In the Section 1.3, we have pointed out that: when $Q(x_0) \neq 0$ (namely $R(x)$ has definition at x), we have

$$\lim_{x \to x_0} R(x) = R(x_0),$$

thus, $R(x)$ is continuous at x_0. Therefore, the rational function is continuous in its domain.

Example 3 Prove the function $y = \sin x$ is continuous on $(-\infty, +\infty)$.

Proof For any $x_0 \in (-\infty, +\infty)$, by a trigonometric identity, we have

$$\Delta y = \sin(x_0 + \Delta x) - \sin x_0 = 2\sin\frac{\Delta x}{2}\cos\left(x_0 + \frac{\Delta x}{2}\right).$$

As $\Delta x \to 0$, $\sin\dfrac{\Delta x}{2} \sim \dfrac{\Delta x}{2}$, thus $\sin\dfrac{\Delta x}{2} \to 0$. Since $\left|\cos\left(x_0 + \dfrac{\Delta x}{2}\right)\right| \leqslant 1$, according to the conclusion that the product of an infinitesimal and a bounded value is also an infinitesimal, we obtain

$$\lim_{\Delta x \to 0} \Delta y = \lim_{\Delta x \to 0} [\sin(x_0 + \Delta x) - \sin x_0] = 0.$$

Since x is an arbitrary point in the interval $(-\infty, +\infty)$, so $y = \sin x$ is continuous at x_0. Similarly, we can prove $y = \cos x$ is continuous on $(-\infty, +\infty)$.

Example 4 Discuss the continuity of the function

$$f(x)=\begin{cases}1-2x, & x\leqslant 0, \\ 1+2x^2, & 0<x\leqslant 1, \\ 4-\dfrac{x}{2}, & 1<x.\end{cases}$$

at $x=0$ and $x=1$.

Solution (1) At $x=0$,
$$\lim_{x\to 0^-}f(x)=\lim_{x\to 0^-}(1-2x)=1, \quad \lim_{x\to 0^+}f(x)=\lim_{x\to 0^+}(1+2x^2)=1.$$
So, $\lim_{x\to 0^-}f(x)=\lim_{x\to 0^+}f(x)=1$, that is
$$\lim_{x\to 0}f(x)=1.$$
And $f(0)=1-2\cdot 0=1$, thus,
$$\lim_{x\to 0}f(x)=f(0).$$
Therefore, $f(x)$ is not continuous at $x=0$.

(2) At $x=1$,
$$\lim_{x\to 1^-}f(x)=\lim_{x\to 1^-}(1+2x^2)=3,$$
$$\lim_{x\to 1^+}f(x)=\lim_{x\to 1^+}\left(4-\frac{x}{2}\right)=\frac{7}{2}.$$
Since $\lim_{x\to 1^-}f(x)\neq \lim_{x\to 1^+}f(x)$, then $\lim_{x\to 1}f(x)$ does not exist. So, $f(x)$ is not continuous at $x=1$.

2) Discontinuity of function

From the above discussion, we know that if function $f(x)$ is continuous at x_0, then it must satisfy three conditions. If at least one of the conditions of continuity is not fulfilled, that is

(1) $f(x)$ is not defined at x_0;

(2) If $\lim_{x\to x_0}f(x)$ does not exist;

(3) $\lim_{x\to x_0}f(x)\neq f(x_0)$ even though $\lim_{x\to x_0}f(x)$ exists, then x_0 is called a discontinuity of the function $f(x)$.

Generally, the discontinuity of the function are classified into two types.

(1) Discontinuity of the first kind.

If the left-side limit and right-side limit of function $y=f(x)$ at $x=x_0$ both exist, and x_0 is a discontinuity, then x_0 is called a discontinuity of the first kind of the function $y=f(x)$.

The discontinuity of the first kind is classified into two types. If both the left-side limit and the right-side limit of function $f(x)$ at x_0 exist, but
$$\lim_{x\to x_0^-}f(x)\neq \lim_{x\to x_0^+}f(x),$$
then x_0 is called a jump discontinuity of $f(x)$. If
$$\lim_{x\to x_0^-}f(x)=\lim_{x\to x_0^+}f(x),$$
that is $\lim_{x\to x_0}f(x)$ exists, but either $f(x)$ is not defined at x_0 or $\lim_{x\to x_0}f(x)\neq f(x_0)$, then x_0 is

called a removable discontinuity of $f(x)$.

(2) Discontinuity point of the second kind.

If at least one of the right-side limit and left-side limit of the function $f(x)$ does not exist at x_0, then x_0 is called discontinuity point of the second kind of the function $f(x)$.

Example 5 For function
$$f(x)=\begin{cases} \dfrac{x^2-4}{x-2}, & x\neq 2, \\ 2, & x=2, \end{cases}$$

since $\lim\limits_{x\to 2}f(x)=\lim\limits_{x\to 2}\dfrac{x^2-4}{x-2}=\lim\limits_{x\to 2}(x+2)=4$, while $f(2)=0$, so $x=2$ is a discontinuity of the function. Because $f(x)$ has a limit at $x=2$, so $x=2$ is a removable discontinuity of the function(Figure 1-23).

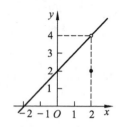

Figure 1-23

If we define $f(2)=4$, then
$$f(x)=\begin{cases} \dfrac{x^2-4}{x-2}, & x\neq 2, \\ 4, & x=2, \end{cases}$$

is continuous at $x=2$.

Example 6 The function $f(x)=\dfrac{\sin x}{x}$ is not defined at $x=0$ (Figure 1-24), so $x=0$ is a discontinuity of $f(x)$. Since $\lim\limits_{x\to 0}\dfrac{\sin x}{x}=1$, that is as $x\to 0$, the limit of $f(x)$ exists. Therefore, $x=0$ is a removable discontinuity of $f(x)$.

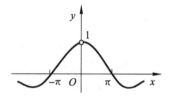

Figure 1-24

If we define $f(0)=1$, then
$$f(x)=\begin{cases} \dfrac{\sin x}{x}, & x\neq 0, \\ 1, & x=0, \end{cases}$$

is continuous at $x=0$.

If x_0 is a removable discontinuity of $f(x)$, because the limit $\lim\limits_{x\to x_0}f(x)$ exists, then we can change the definition of the functional value at x_0 or define the functional value at x_0, then the new function obtained is continuous at x_0.

Example 7 Let
$$f(x)=\begin{cases} x^2, & x\leqslant 0, \\ x+2, & x>0. \end{cases}$$
$$\lim\limits_{x\to 0^+}f(x)=\lim\limits_{x\to 0^+}(x+2)=2,$$
$$\lim\limits_{x\to 0^-}f(x)=\lim\limits_{x\to 0^-}x^2=0,$$

thus the right-side limit and the left-side limit exist, but are not equal (Figure 1-25), therefore $x=0$ is a jump discontinuity of $f(x)$.

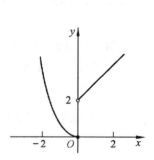

Figure 1-25

Example 8 Let $f(x)=\dfrac{1}{x-2}$, since

$$\lim_{x \to 2^-} f(x) = -\infty, \quad \lim_{x \to 2^+} f(x) = +\infty,$$

then both of the right-side limit and left-side limit of $f(x)$ at $x=2$ do not exist (Figure 1-26). So, $x=2$ is a discontinuity of the second kind of $f(x)$. Because

$$\lim_{x \to 2} f(x) = \lim_{x \to 2} \frac{1}{x-2} = \infty,$$

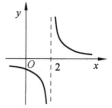

Figure 1-26

then $x=2$ is called an infinite discontinuity of $f(x)$.

Example 9 Suppose $f(x) = \sin \frac{1}{x}$, since the limit of $f(x)$ does not exist as $x \to 0$ (the left-side and the right-side limit don't exist). So $x=0$ is discontinuity of the second type.

Since the value of $f(x)$ often oscillate infinitely between -1 and $+1$ as x tends to 0. Thus $x=0$ is called an oscillating discontinuity(Figure 1-27).

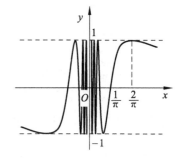

Figure 1-27

1.6.2 Operations on continuous functions and the continuity of elementary functions

1) Operations on continuous functions

The continuity of function is defined by the limit. According to the rational operation rules of limit, operations on continuous functions can be obtained.

Theorem 1 Suppose the functions $f(x)$ and $g(x)$ are both continuous at x_0, then $f(x) \pm g(x)$, $f(x) \cdot g(x)$ and $\frac{f(x)}{g(x)}(g(x_0) \neq 0)$ are also continuous at x_0.

Example 10 Since $\sin x$, $\cos x$ and the constant function are continuous on $(-\infty, +\infty)$, according to Theorem 1,

$$\tan x = \frac{\sin x}{\cos x}, \cot x = \frac{\cos x}{\sin x}, \sec x = \frac{1}{\cos x}, \csc x = \frac{1}{\sin x},$$

are continuous at their domains.

Theorem 2 Suppose the function $y = f[\varphi(x)]$ is composed from the function $y = f(u)$ and $u = \varphi(x)$. If $\lim_{x \to x_0} \varphi(x) = u_0$, and $f(u)$ is continuous at $u = u_0$, then we have

$$\lim_{x \to x_0} f[\varphi(x)] = \lim_{u \to u_0} f(u) = f(u_0).$$

According to Theorem 3 in Section 1.3 (the limit of the composite function) and the definition of continuity of function, we can get the conclusion of this theorem. Under the conditions of Theorem 2, we have

$$\lim_{x \to x_0} f[\varphi(x)] = f[\lim_{x \to x_0} \varphi(x)].$$

Example 11 Find the following limit:

(1) $\lim\limits_{x \to 0} \left(\dfrac{1-\cos x}{x^2} \right)^2$;

(2) $\lim\limits_{x \to 2} \sin\left(\pi \sqrt{\dfrac{x-2}{x^2-4}} \right)$.

Solution According to Theorem 2, we have

(1) $\lim\limits_{x \to 0} \left(\dfrac{1-\cos x}{x^2} \right)^2 = \left(\lim\limits_{x \to 0} \dfrac{1-\cos x}{x^2} \right)^2 = \left(\dfrac{1}{2} \right)^2 = \dfrac{1}{4}$.

(2) $\lim\limits_{x \to 2} \sin\left(\pi \sqrt{\dfrac{x-2}{x^2-4}} \right) = \sin\left[\lim\limits_{x \to 2} \left(\pi \sqrt{\dfrac{x-2}{x^2-4}} \right) \right] = \sin\left(\pi \sqrt{\lim\limits_{x \to 2} \dfrac{x-2}{x^2-4}} \right)$

$= \sin\left(\pi \sqrt{\lim\limits_{x \to 2} \dfrac{1}{x+2}} \right) = \sin \dfrac{\pi}{2} = 1$.

Theorem 3 Suppose the function $y = f[\varphi(x)]$ is composed from the functions $y = f(u)$ and $u = \varphi(x)$. If function $u = \varphi(x)$ is continuous at x_0, $\varphi(x_0) = u_0$, and $f(u)$ is continuous at u_0, then the composite function $f[\varphi(x)]$ is continuous at x_0.

Actually, according to Theorem 2, we have
$$\lim_{x \to x_0} f[\varphi(x)] = f[\lim_{x \to x_0} \varphi(x)] = f(u_0) = f[\varphi(x_0)],$$
therefore $y = f[\varphi(x)]$ is continuous at x_0.

Theorem 4 Suppose $y = f(x)$ is a monotone increasing (or decreasing) and continuous function in the interval I_x, then its inverse function $x = \varphi(y)$ is also monotone increasing (or decreasing) and is a continuous function in the set $I_y = \{y \mid y = f(x), x \in I_x\}$.

Theorem 4 is easily to proof from Figure 1-28. The proving process if omitted.

Example 12 Discuss the continuity of the inverse trigonometric functions $y = \arcsin x$ on $[-1, 1]$.

Solution $y = \sin x$ is monotone increasing and continuous on $I = \left[-\dfrac{\pi}{2}, \dfrac{\pi}{2} \right]$ and the range of the function is $I_y = [-1, 1]$. By Theorem 4, the inverse function $x = \arcsin y$ is continuous in I_y. That is to say, $y = \arcsin x$ is increasing and continuous on $[-1, 1]$.

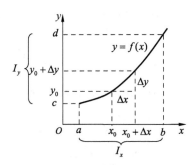

Figure 1-28

Similarly, $y = \arccos x$ is decreasing and continuous on $[-1, 1]$, $y = \arctan x$ is increasing and continuous in $(-\infty, +\infty)$, $y = \text{arccot}\, x$ is decreasing and continuous in $(-\infty, +\infty)$.

In one word, inverse trigonometric functions, such as $y = \arcsin x$, $y = \arccos x$, $y = \arctan x$ and $y = \text{arccot}\, x$ are all continuous in their domains.

2) Continuity of elementary functions

We have proved the continuity of constant function, trigonometric function and inverse trigonometric function in their domains.

Here we point out that, exponential function $y=a^x$ ($a>0$, $a\neq 1$) is continuous and monotone increasing in $(-\infty, +\infty)$. So, according to Theorem 4, its inverse function $y=\log_a x$ ($a>0$, $a\neq 1$) is also continuous in $(0, +\infty)$.

For the power function $y=x^\mu$ (μ is a real number), when $x>0$, we have
$$y=x^\mu=e^{\mu\ln x}.$$
Therefore, the power function $y=x^\mu$ can be regarded as a composite function formed by $y=e^u$ and $u=\mu\ln x$. Thus the power function $y=x^\mu$ is continuous in its domain of definition $(0, +\infty)$.

Summarizing the conclusions, all the basic elementary functions are continuous functions in their domains. According to the definition of the elementary functions, we can get the following conclusions.

Theorem 5 All the elementary functions are continuous in their domains.

Example 13 Find $\lim\limits_{x\to 0}\dfrac{\sqrt{1+\frac{x}{2}}-\sqrt{1-x}}{x}$.

Solution
$$\lim_{x\to 0}\dfrac{\sqrt{1+\frac{x}{2}}-\sqrt{1-x}}{x}=\lim_{x\to 0}\dfrac{\left(\sqrt{1+\frac{x}{2}}-\sqrt{1-x}\right)\left(\sqrt{1+\frac{x}{2}}+\sqrt{1-x}\right)}{x\left(\sqrt{1+\frac{x}{2}}+\sqrt{1-x}\right)}$$
$$=\lim_{x\to 0}\dfrac{\frac{3}{2}}{\sqrt{1+\frac{x}{2}}+\sqrt{1-x}}=\dfrac{\frac{3}{2}}{1+1}=\dfrac{3}{4}.$$

Example 14 Find $\lim\limits_{x\to\frac{\pi}{2}}\ln\left(1+\dfrac{x\sin 2x}{1-\cos x}\right)$.

Solution Since the logarithmic function $f(u)=\ln u$ is continuous. So, according to Theorem 2 in this section, we have
$$\lim_{x\to\frac{\pi}{2}}\ln\left(1+\dfrac{x\sin 2x}{1-\cos x}\right)=\ln\left(1+\lim_{x\to\frac{\pi}{2}}\dfrac{x\sin 2x}{1-\cos x}\right)$$
$$=\ln\left[1+\lim_{x\to\frac{\pi}{2}}\dfrac{x\cdot 2x}{\frac{1}{2}x^2}\right]=\ln 5.$$

Example 15 Find $\lim\limits_{x\to 0}\dfrac{\ln(1+x)}{x}$.

Solution $\lim\limits_{x\to 0}\dfrac{\ln(1+x)}{x}=\lim\limits_{x\to 0}\ln(1+x)^{\frac{1}{x}}=\ln\left[\lim\limits_{x\to 0}(1+x)^{\frac{1}{x}}\right]=\ln e=1.$

Example 16 Find $\lim\limits_{x\to 0}(1-2\sin x)^{\frac{1}{x}}$.

Solution $\lim\limits_{x\to 0}(1-2\sin x)^{\frac{1}{x}}=\lim\limits_{x\to 0}e^{\frac{1}{x}\ln(1-\sin 2x)}$
$$=e^{\lim\limits_{x\to 0}\frac{1}{x}\ln(1-\sin 2x)}=e^{\lim\limits_{x\to 0}\frac{1}{x}\cdot(-\sin 2x)}=e^{-2}.$$

1.6.3 Properties of continuous functions on a closed interval

In 1.6.1, we have learned the definition of continuity of the function on a closed interval. In the following, we will introduce the properties of continuous functions on a closed interval. These properties are very easily understood from geometric intuition but here omit the proofs because rigorous analyze proof are beyond the scope of this book.

Theorem 6 (Bounded and maximum-minimum theorem) Suppose that $f(x)$ is continuous on a closed interval $[a, b]$, then $f(x)$ must have maximum value M and minimum value m on $[a,b]$. That is, there exist at least two points $x_1, x_2 \in [a,b]$, such that for all $x \in [a,b]$, we have
$$f(x_1) = m \leqslant f(x) \leqslant M = f(x_2).$$

This theorem can be easily seen from the perspective of geometry (Figure 1-29).

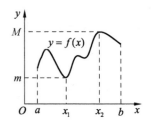

Figure 1-29

It must be emphasized that the conditions in Theorem 6 must be satisfied to ensure the conclusions.

(1) If the function is continuous on an open interval, the conclusion may not be established.

For example, although $y = \dfrac{1}{x}$ is continuous in $(0, 2)$, but it is unbounded in $(0, 2)$ (Figure 1-30). The maximum value and minimum value don't exist. The conclusion of theorem is not true.

(2) If the function is not continuous on a closed interval, the conclusion is not necessarily true. For example,
$$y = f(x) = \begin{cases} x+1, & -1 \leqslant x < 0, \\ 0, & x = 0, \\ x-1, & 0 < x \leqslant 1 \end{cases}$$
has a discontinuous point $x = 0$ in $[-1, -1]$, but the maximum value and minimum value don't exist in $[-1, -1]$ (Figure 1-31). The conclusion of this theorem is not true.

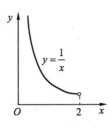

Figure 1-30

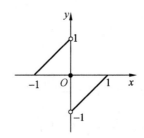

Figure 1-31

Corollary (Boundedness theorem) If the function $y = f(x)$ is continuous on a closed interval $[a, b]$, then it is bounded on $[a, b]$.

Proof Since $f(x)$ is continuous on $[a,b]$. According to Theorem 6, $f(x)$ have a maximum M and a minimum m on $[a, b]$. That is
$$m \leqslant f(x) \leqslant M, \quad a \leqslant x \leqslant b.$$
Take $K = \max\{|M|, |m|\}$, then for any $x \in [a,b]$, we have
$$|f(x)| \leqslant K.$$
Therefore, $y = f(x)$ is bounded on $[a,b]$.

Theorem 7 (Intermediate value theorem) Suppose that $f(x)$ is continuous on a closed interval $[a,b]$ and $f(a) \neq f(b)$, then for any given value λ between $f(a)$ and $f(b)$, there exists at least one point $\xi \in (a,b)$ such that $f(\xi) = \lambda$.

The correctness of the theorem can be easily understood from the perspective of geometry. As shown in Figure 1-32, for any given λ_1, λ_2 between $f(a)$ and $f(b)$, there exists $\xi_1 \in (a,b)$ such that $f(\xi_1) = \lambda_1$. And there exist ξ_2, ξ_3, ξ_4 on (a, b), such that
$$f(\xi_2) = f(\xi_3) = f(\xi_4) = \lambda_2.$$

Two useful corollaries are obtained from Intermediate value theorem.

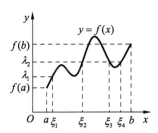

Figure 1-32

Corollary 1 (Zero point Theorem) Suppose that $f(x)$ is continuous on a closed interval $[a,b]$ and $f(a) \cdot f(b) < 0$, then there exists at least one point $\xi \in (a, b)$, such that $f(\xi) = 0$.

Actually, since $f(a) \cdot f(b) < 0$ and $\lambda = 0$ is a number between $f(a)$ and $f(b)$. According to Intermediate value theorem, there exists at least one point $\xi \in (a, b)$, such that
$$f(\xi) = 0.$$
This theorem may be used to prove the existence of a root of a given equation $f(x) = 0$ if f is continuous on a closed interval. Furthermore, it can also be used to find the approximate value of the root.

Corollary 2 If $f(x)$ is continuous on a closed interval $[a, b]$, then $f(x)$ can attain any value between its maximum M and its minimum m.

Actually, according to the Bounded and maximum-minimum theorem, there exist x_1, $x_2 \in [a, b]$, such that
$$f(x_1) = m, \quad f(x_2) = M.$$
Suppose that $m < M$, on the closed interval $[x_1, x_2]$ or $[x_2, x_1]$, Corollary 2 can be obtained by intermediate value theorem.

Example 17 Prove the equation $x^2 - x\ln x - 2 = 0$ has at least one root in $(1, 2)$.

Proof Let $f(x) = x^2 - x\ln x - 2$, $f(x)$ is continuous on $[1, 2]$, and
$$f(1) = 1^2 - 1\ln 1 - 2 = -1 < 0,$$
$$f(2) = 2^2 - 2\ln 2 - 2 = 2(1 - \ln 2) > 0,$$
that is $f(1) \cdot f(2) < 0$.

By zero point theorem, there exists at least one point ξ in $(1,2)$ such that $f(\xi) = 0$. That is, the equation $x^2 - x\ln x - 2 = 0$ has at least one root in $(1,2)$.

Example 18 Suppose the function $f(x)$ is continuous on $[a, b]$ and $a \leqslant f(x) \leqslant b$. Prove there exists at least on point $\xi \in (a,b)$ such that $f(\xi) = \xi$.

Proof Let $F(x) = f(x) - x$, obviously, $F(x)$ is continuous on $[a,b]$ and
$$F(a) = f(a) - a \geqslant 0,$$
$$F(b) = f(b) - b \leqslant 0.$$

If $F(a) = 0$ or $F(b) = 0$, then let $\xi = a$ or $\xi = b$, we have $F(\xi) = f(\xi) - \xi = 0$, that is $f(\xi) = \xi$.

If $F(a) > 0$ and $F(b) < 0$, that is $F(a) \cdot F(b) < 0$. By zero point theorem, there exists at least one point $\xi \in (a,b)$ such that $F(\xi) = f(\xi) - \xi = 0$, that is $f(\xi) = \xi$.

Therefore, there exists $\xi \in [a,b]$ such that $f(\xi) = \xi$.

Exercises 1-6

1. Judge the continuity of the following functions at the specified point:

(1) $f(x) = \begin{cases} x^3, & -1 \leqslant x \leqslant 1, \\ 1, & x > 1 \text{ or } x < -1, \end{cases}$ at $x = -1, x = 1$;

(2) $f(x) = \begin{cases} \dfrac{\sin x}{x}, & x < 0, \\ x^2 + 1, & x \geqslant 0, \end{cases}$ at $x = 0$.

2. Find the discontinuity of the following functions and point out their types. If it is a removable discontinuity, then change the definition of the function or define the functional value to make it continuous.

(1) $f(x) = \dfrac{x-2}{x^2-4}$;

(2) $f(x) = \dfrac{x^2-1}{x^2-3x+2}$;

(3) $f(x) = \cos^2 \dfrac{1}{x}$;

(4) $f(x) = \begin{cases} \dfrac{x^2}{3}, & -1 \leqslant x \leqslant 0, \\ 3-x, & 0 < x \leqslant 1; \end{cases}$

(5) $f(x) = e^{\frac{1}{x-1}}$;

(6) $f(x) = \dfrac{x}{\tan x}$.

3. Find constants a and b to make the following functions continuous:

(1) $f(x) = \begin{cases} e^x, & x \leqslant 0, \\ x+a, & x > 0; \end{cases}$

(2) $f(x) = \begin{cases} x^2 \sin x + b, & x < 0, \\ a, & x = 0, \\ (1+2x)^{\frac{1}{x}}, & x > 0; \end{cases}$

(3) $f(x)=\begin{cases} \dfrac{\ln(1-3x)}{bx}, & x<0, \\ 2, & x=0, \\ \dfrac{\sin ax}{x}, & x>0. \end{cases}$

4. Discuss the continuity of the function $f(x)=\lim\limits_{n\to\infty}\dfrac{x(1-x^{2n})}{1+x^{2n}}$. If there exist discontinuity, point out them and their types.

5. Find the following limit:

(1) $\lim\limits_{x\to 1}\cos\dfrac{x^2-1}{x-1}$;

(2) $\lim\limits_{x\to\frac{\pi}{6}}\sqrt{1+2\cos 2x}$;

(3) $\lim\limits_{x\to 1}\dfrac{\sqrt{5x-4}-\sqrt{x}}{x-1}$;

(4) $\lim\limits_{x\to 0}\dfrac{\sqrt{1+x^3}-1}{x^3}$;

(5) $\lim\limits_{x\to 0}\dfrac{\ln(x+a)-\ln a}{x}$;

(6) $\lim\limits_{x\to 0^+}[\ln(x+\sin 3x)-\ln x]$;

(7) $\lim\limits_{x\to\infty}\left(\dfrac{3+x}{6+x}\right)^{\frac{x-1}{2}}$;

(8) $\lim\limits_{x\to 0}(1+3\tan^2 x)^{\cot^2 x}$.

6. Prove the equation $x \cdot 2^x=1$ has at least one positive root, which is smaller than 1.

7. Prove the equation $x-a\sin x=b$ has at least one positive root, which is smaller than $a+b$, where the constants $a>0$, $b>0$.

8. Suppose the function $f(x)$ is continuous on $[0,1]$, $0\leqslant f(x)\leqslant 1$. Prove there exists at least one point ξ in $[0,1]$ such that $f(\xi)=\xi$.

9. Suppose the function $f(x)$ is continuous on $[a,b]$, $x_i\in[a,b]$, $i=1, 2, \cdots, n$, prove that there exists at least one point $\xi\in[a,b]$ such that $f(\xi)=\dfrac{f(x_1)+f(x_2)+\cdots+f(x_n)}{n}$.

Summary

1. Main contents

The main contents of this chapter can be divided into three parts: functions, limits and the continuity of functions, which are the basis of calculus theory.

(1) Concept of functions, properties of functions, inverse functions and composite functions, operations on functions and elementary functions.

(2) Definition of limit: for the limit of sequence, "ε-N" definition is given; for the limit of a function as independent variable tending to a finite value, "ε-δ" definition is given; and for the limit of a function as independent variable approaching infinity, "ε-X" definition is given. Because the trend of independent variables is different, limits include the following types:

$$\lim_{n\to\infty}x_n, \lim_{x\to\infty}f(x), \lim_{x\to+\infty}f(x), \lim_{x\to-\infty}f(x),$$

$$\lim_{x\to x_0}f(x), \lim_{x\to x_0^+}f(x), \lim_{x\to x_0^-}f(x).$$

(3) The properties of function limit: uniqueness, local boundedness, local preservation of sign.

(4) Rational operation rules and operation rule for composite function, principle of limit existence and two important limits: $\lim\limits_{x \to 0} \dfrac{\sin x}{x} = 1$ and $\lim\limits_{x \to 0}(1+x)^{\frac{1}{x}} = e$. These are the basic methods for distinguishing and finding the limit.

(5) Concepts of infinitesimal and infinity, the relationship between infinity and infinitesimal, properties of infinitesimal, comparison of infinitesimal order (infinitesimal of higher order, infinitesimal of lower order, infinitesimals of the same order, equivalent infinitesimals). Getting limit with substitution of equivalent infinitesimal is a common, convenient and efficient way.

(6) Concept of continuous function: $y = f(x)$ is continuous at $x_0 \Leftrightarrow \lim\limits_{\Delta x \to 0} \Delta y = 0 \Leftrightarrow \lim\limits_{x \to x_0} f(x) = f(x_0)$. The concepts of left-sided limit and right-side limit, discontinuity and their classifications, operations on continuous functions, continuity of inverse function and composite function, continuity of elementary functions.

(7) Properties of continuous functions on a closed interval: bounded and maximum-minimum theorems, boundedness theorem, intermediate value theorem and zero point theorem.

2. Basic requirements

(1) Understand the concept of functions, the representation of piecewise functions, properties of functions. Establish the function relationship of simple practical problems.

(2) Understand the concept of composite functions, operation rules of functions and concept of inverse functions.

(3) Grasp the properties of elementary functions, sketch their figures and understand the concept of elementary functions.

(4) Understand concepts of limit of sequence and limit of function (including left-sided limit and right-sided limit) and properties of the limit of functions.

(5) Skillfully use the rational operations of limit, two important limits, and operation rules for composite functions.

(6) Understand the concepts of infinitesimal and infinity, properties of infinitesimal, comparison of infinitesimal order. Be able to find limit with substitution of equivalent infinitesimal.

(7) Understand the concept of continuity of function (including left-sided continuity and right-sided continuity), find the types of the discontinuity points of functions.

(8) Understand the operation rules of continuous functions and continuity of elementary functions.

(9) Understand the properties of continuous functions a closed interval (bounded and maximum-minimum theorems, boundedness theorem, intermediate value theorem and the zero point theorem).

Quiz

1. Fill the blanks with "sufficient", "necessary" and "necessary and sufficient":

(1) The sequence $\{x_n\}$ is bounded is the _____ condition of that $\{x_n\}$ is convergent.

(2) $\lim\limits_{n\to\infty} x_{2n+1} = a$ is the _____ condition of $\lim\limits_{n\to\infty} x_n = a$, $\lim\limits_{n\to\infty} x_n = a$ is the _____ condition of $\lim\limits_{n\to\infty} x_{2n} = a$, $\lim\limits_{n\to\infty} x_{2n+1} = \lim\limits_{n\to\infty} x_{2n} = a$ is the _____ condition of $\lim\limits_{n\to\infty} x_n = a$.

(3) $f(x)$ is bounded in a deleted neighborhood of x_0 is the _____ condition of that $\lim\limits_{x\to x_0} f(x)$ exists. $\lim\limits_{x\to\infty} f(x)$ exists is the _____ condition of that "$\exists X > 0$ such that $f(x)$ is bounded in $(-\infty, -X) \cup (X, +\infty)$".

(4) $f(x)$ is boundless in a deleted neighborhood of x_0 is the _____ condition of $\lim\limits_{x\to x_0} f(x) = \infty$.

(5) $\lim\limits_{x\to x_0} f(x) = A$ is the _____ condition of $\lim\limits_{x\to x_0^-} f(x) = \lim\limits_{x\to x_0^+} f(x)$. $\lim\limits_{x\to x_0^-} f(x) = \lim\limits_{x\to x_0^+} f(x) = f(x_0)$ is the _____ condition of that $f(x)$ is continuous at x_0.

2. Find the correct answer and write the corresponding letter in the bracket.

(1) As $x \to 0$, ( ) and x^2 are equivalent infinitesimals, ( ) and x^2 are infinitesimals of the same order, ( ) is an infinitesimal of lower order than x^2.

A. $\tan^3 x$ B. $3x^3 + \sin x^2$ C. $1 - \cos\sqrt{x}$ D. $\sqrt{1+x^2} - 1$

(2) If $\lim\limits_{x\to x_0} f(x) = 0$, then ( ).

A. for any function $g(x)$, $\lim\limits_{x\to x_0} f(x)g(x) = 0$ holds

B. if $g(x)$ is a bounded function, $\lim\limits_{x\to x_0} f(x)g(x) = 0$ holds

C. only when $\lim\limits_{x\to x_0} g(x) = 0$, $\lim\limits_{x\to x_0} f(x)g(x) = 0$ holds

D. only when $g(x)$ is a constant, $\lim\limits_{x\to x_0} f(x)g(x) = 0$ holds

(3) Suppose that $f(x) = \dfrac{e^{\frac{1}{x}} - 1}{e^{\frac{1}{x}} + 1}$, then $x = 0$ is a _____ of $f(x)$.

A. removable discontinuity point   B. jump discontinuity point
C. discontinuity point of the second point   D. continuity point

3. Find the domains of the following functions:

(1) $y = \dfrac{\sqrt{x^2 - 3x - 10}}{x - 8}$;

(2) $y = \ln\dfrac{x - 2}{3 - x}$;

(3) $y = \arccos\dfrac{2x}{x^2 + 1}$;

(4) $y = \sqrt{\sin x} + \sqrt{16 - x^2}$.

4. Find the following limit:

(1) $\lim\limits_{n\to\infty} n(\sqrt{n^2 + 1} - n)$;

(2) $\lim\limits_{x\to 0} \tan 3x \csc 2x$;

(3) $\lim\limits_{x\to 0} \dfrac{1 - \cos x}{x \ln(1 + x)}$;

(4) $\lim\limits_{x\to\infty} \left(\dfrac{2x + 3}{2x + 1}\right)^{x+2}$;

(5) $\lim\limits_{x\to 0} \ln\left(\dfrac{x+\sin 3x}{\sin x}\right)$;

(6) $\lim\limits_{x\to 0} \dfrac{2\sin x - \sin 2x}{x^3}$.

5. Find the discontinuity of the functions and point out their types.

(1) $y = \dfrac{x}{\sin x}$;

(2) $f(x) = \begin{cases} e^{\frac{1}{x-1}}, & x > 0, \\ \ln(1+x), & -1 < x \leqslant 0; \end{cases}$

(3) $f(x) = \begin{cases} \dfrac{\sin x}{x}, & x < 0, \\ 0, & x = 0, \\ x\sin\dfrac{1}{x} + b, & x > 0. \end{cases}$

6. Suppose $f(x) = \begin{cases} \dfrac{\sin ax}{\ln(1+2x)}, & -\dfrac{1}{2} < x < 0, \\ 1, & x = 0, \\ x\sin\dfrac{1}{x^2} + 1, & x > 0. \end{cases}$

(1) Find the values of a and b to make $\lim\limits_{x\to 0} f(x)$ exist;

(2) Find the values of a and b to make $f(x)$ continuous at $x=0$.

7. A length of wire is cut into two sections, one is bent into a circle, the other one is bent into a square. Suppose that length of side of the square is x, the areas of the circle and the square is y. Find the function of y and x, and point out the domain.

8. Suppose $f(x)$ is continuous in $[a, b]$, and $a \leqslant f(x) \leqslant b$. Prove there exists at least one point $\xi \in [a, b]$ such that $f(\xi) = \xi$ (where $b > a$).

Exercises

1. Prove Dirichlet function $D(x) = \begin{cases} 1, & x \in Q, \\ 0, & x \notin Q \end{cases}$ is an even function, also a periodic function. Any non-zero rational number r is its period (so there is no minimum positive period).

2. Find the inverse functions of the following functions:

(1) $y = \arcsin(3^{x-1} - 2)$;

(2) $y = \begin{cases} e^x, & x \geqslant 0, \\ x+1, & x < 0. \end{cases}$

3. Suppose $f(x)$ is an even function, and is monotone increasing in $[0, +\infty)$. Solve the equation $f(x) = f\left(\dfrac{24}{x+10}\right)$.

4. Suppose $y = f(x)$ is an odd function, when $x > 0$, $f(x) = x(1-x)$. Find the expression of $f(x)$ when $x < 0$.

5. Find the following limit:

(1) $\lim\limits_{x\to 0} \dfrac{1-\sqrt{1-x}}{e^x - 1}$;

(2) $\lim\limits_{x\to +\infty} x(\sqrt{x^2+1} - \sqrt{x^2-2})$;

(3) $\lim_{x\to 0^+}(\cos\sqrt{x})^{\frac{\pi}{x}}$;

(4) $\lim_{x\to 0}\dfrac{\cos 3x-\cos 4x}{2x\sin x}$;

(5) $\lim_{x\to 0}(x+e^x)^{\frac{2}{x}}$;

(6) $\lim_{x\to 0}\dfrac{\sin x-\tan x}{(\sqrt[3]{1+x^2}-1)(\sqrt{1+\sin x}-1)}$.

6. Find the following limit:

(1) $\lim_{n\to\infty}\left(\cos\dfrac{x}{2}\cdot\cos\dfrac{x}{4}\cdot\cos\dfrac{x}{8}\cdot\cdots\cdot\cos\dfrac{x}{2^n}\right)$;

(2) $\lim_{x\to\infty}\sin^2(\pi\sqrt{n^2+n})$;

(3) $\lim_{n\to\infty}\left(\dfrac{1^2}{n^3+1^2}+\dfrac{2^2}{n^3+2^2}+\cdots+\dfrac{n^2}{n^3+n^2}\right)$;

(4) Let $x_n=\sqrt{6+\sqrt{6+\cdots+\sqrt{6}}}$ (the number of radical signs is n), find $\lim_{n\to\infty}x_n$.

7. Find the discontinuity points of the following functions and point out their types:

(1) $y=\dfrac{x^2-x}{|x|(x^2-1)}$;

(2) $y=\dfrac{1}{e^{\frac{x}{x-1}}-1}$;

(3) $y=\sqrt[3]{\dfrac{1-\cos\pi x}{4-x^2}}$;

(4) $y=(1+x)^{\tan\left(x-\frac{\pi}{4}\right)}$.

8. Suppose that $f(x_1+x_2)=f(x_1)\cdot f(x_2)$ holds for any x_1, x_2, and $f(x)$ is continuous at $x=0$. Prove that $f(x)$ is continuous in $(-\infty,+\infty)$.

9. Prove the cubic equation $x^3-3x^2+1=0$ has three real roots.

10. Suppose $f(x)$ is continuous on $[0,2a]$ $(a>0)$, and $f(2a)=f(0)$. Prove that there exists at least one point $\xi\in[0,a]$ such that $f(\xi)=f(\xi+a)$.

11. Prove the following limits by the definition of functions:

(1) $\lim_{x\to 3}\dfrac{x^2-x-6}{x-3}=5$;

(2) $\lim_{x\to 3}\sqrt{x}=\sqrt{3}$.

Chapter 2 Derivative and differential

> The main purpose of economic theory is to explain the relationship between economic variables. The relationship between economic variables is mainly how a variable's change affects other variables. For example, how will the interest rates be affected by the money supply? If the government expenditure increases by 1 million Yuan, will the total output increase or decrease? If so, how much will it increase or fall? When we use the form of a linear function to express this relationship, the effect of the change of a variable to another one will be expressed as the "slope" of the function. For the more general non-linear function, the impact of this change will be expressed as a function of the "derivative". Derivative and differential is the basic concepts of differential calculus. The derivative is change rate of a dependent variable of function with respect to the independent variable. Differential is the linear approximation of the function increment. In this chapter, we will define the derivative and differential of a function and learn how to calculate the them. At the same time, we will introduce the application of derivative and differential in economics.

2.1 Definition of derivatives

Since the Renaissance, the European industry, agriculture and seafaring have achieved large-scale development. Under the stimulation of vigorous development of capitalist productivity, natural science began to enter the comprehensive and breakthrough stage, which made the basic problem of calculus become the focus of public attention.

(1) Determine the velocity and acceleration of non-uniform motion, that is to say, study the rate of instantaneous change.

(2) The optical path design of the telescope needs to determine the normal line of any point on the lens surface, and transforms it into the tangent problem of any curve.

(3) To determine the maximum range of shells and seek the problem of the maximum and minimum function value which are involved in perihelion and aphelion of the planets orbit.

The archetype of these three practical problems can be attributed to the speed of the change of a function with respect to the independent variable's change in mathematics, namely the so-called change rate problem of a function. Newton started from the first issue, while Leibniz started from the second problem. They relatively gave the concept of derivative, and led to the birth of calculus.

2.1.1 Cited examples

1) The speed problem of a variable rectilinear motion

For uniform linear motion, use the distance to divide time, then we can obtain the speed of any time during this period. But for a variable rectilinear motion, we can only obtain the average speed of the time period by this method, so how to calculate the instantaneous speed at some time?

Assume there is a mass point M, which is doing a linear motion. The rule of its motion is $s=s(t)$. When the time t changes from t_0 to $t_0+\Delta t$, the position function s changes from to $s(t_0)$ to $s(t_0+\Delta t)$. That is to say, the position function has the increment $\Delta s = s(t_0+\Delta t)-s(t_0)$ (Figure 2-1)

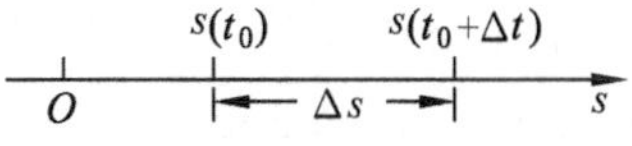

Figure 2-1

We call the ratio

$$\frac{\Delta s}{\Delta t}=\frac{s(t_0+\Delta t)-s(t_0)}{\Delta t}$$

of the increment Δs of the position function s and the increment Δt of the independent variable time t as the average speed of the mass point M in the period Δt, recorded as $\bar{v}$. That is

$$\bar{v}=\frac{\Delta s}{\Delta t}=\frac{s(t_0+\Delta t)-s(t_0)}{\Delta t}.$$

When the mass point M is doing a uniform linear motion, the average speed $\bar{v}$ is the motion speed of M at each time point during the period Δt. Also, it is the (instantaneous) speed of M at the time point $t=t_0$. When the mass point M is doing a variable linear motion, the average speed is the approximate value of the speed of the mass point M at the time point t_0. It is not the speed of M at the time point t_0. However, when $|\Delta t|$ is rather small, we can approximately take the average speed as the instantaneous speed at the time point t_0. And the smaller $|\Delta t|$ is, the better the extent is. When $\Delta t \to 0$, the limit of the average speed $\bar{v}=\frac{\Delta s}{\Delta t}$ is the instantaneous speed of the mass point M at the time point t_0, namely,

$$v=\lim_{\Delta t \to 0}\bar{v}=\lim_{\Delta t \to 0}\frac{\Delta s}{\Delta t}=\lim_{\Delta t \to 0}\frac{s(t_0+\Delta t)-s(t_0)}{\Delta t}. \tag{1}$$

2) The tangent slope problem of plane curve

In elementary mathematics, the tangent of a circle is defined as the line which has only one point of intersection with the circle. But for a general curve, this definition is obviously not appropriate. As shown in Figure 2-2, although the line l_1 only has one point of intersection with the given curve L, apparently we can not consider it contact with L, while although the line l_2 has one more points of intersection with L, it is tangent with L at the point A. Therefore, in higher mathematics, the tangent is defined as the limit position of the secant.

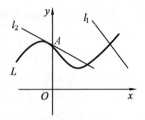

Figure 2-2

Assume the equation of the curve L is $y = f(x)$, as shown in Figure 2-3. Consider there are a point M_0 and a nearby point M on the curve L. Their coordinates are respectively $M_0(x_0, y_0)$ and $M(x_0 + \Delta x, y_0 + \Delta y)$ (in which Δx can be larger than 0, or smaller than 0, in this figure, we take Δx as lager than 0). Draw the secant $M_0 M$ of the curve L, which connects these two point. And then make the point M move along the curve L to M_0. At this time, the secant $M_0 M$ rotates along the curve L, centering on M_0. When $M \to M_0$, the limit position $M_0 T$ of the secant $M_0 M$

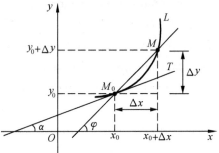

Figure 2-3

is called the tangent of the curve L at the point M_0. Here the meaning of the limit position is: only if the length of the segment $M_0 M$ is close to zero, $\angle MM_0 T$ is close to zero.

The key point of calculating the equation of the tangent $M_0 T$ of the curve $y = f(x)$ is to calculate its slope. Because the slope of the secant $M_0 M$ is

$$\bar{k} = \tan \varphi = \frac{\Delta y}{\Delta x} = \frac{f(x_0 + \Delta x) - f(x_0)}{\Delta x},$$

in which, φ is the dip angle of the secant $M_0 M$. When the point M approaches the point M_0 along the curve L, $\Delta x \to 0$. The limit value of the equation above is the slope of the tangent $M_0 T$, namely, the slope of the tangent $M_0 T$ is

$$k = \tan \alpha = \lim_{\Delta x \to 0} \frac{\Delta y}{\Delta x} = \lim_{\Delta x \to 0} \frac{f(x_0 + \Delta x) - f(x_0)}{\Delta x}, \tag{2}$$

in which α is the dip angle of the tangent $M_0 T$.

3) The marginal cost problem

In economics, we define the increase and decrease of total cost when the output increases or decreases one unit as the marginal cost, namely, the definition below:

Assume when the output of some product is x unit, the total cost is $C = C(x)$. If x changes from x_0 to $x_0 + \Delta x$, the increment of the total cost is

$$\Delta C = C(x_0 + \Delta x) - C(x_0).$$

At this time, the average change rate of the total cost is

$$\frac{\Delta C}{\Delta x} = \frac{C(x_0 + \Delta x) - C(x_0)}{\Delta x},$$

in which $\Delta x = 1$ or $\Delta x = -1$. We call it the marginal cost.

For example, assume the cost function of some product is

$$C(x) = 5\,000 + 13x + 30\sqrt{x},$$

in which x represents the output (unit: t), $C(x)$ represents the total cost when the output is x t (unit: Yuan). When the output is 400 t, add one more ton, namely when $\Delta x=1$, the change of the total cost is
$$\Delta C(x)=C(401)-C(400)=13.7495.$$
When the output is 400 t, add 1 t output, the added cost is
$$\left.\frac{\Delta C(x)}{\Delta x}\right|_{\substack{x=400\\ \Delta x=1}}=\frac{13.7495}{1}=13.7495.$$
When the output decreases one ton from 400 t, that is to say, when $\Delta x=-1$, the change of the total cost is
$$\Delta C(x)=C(399)-C(400)=-13.7505.$$
When the output is 400 t, decrease 1 t output, the decreased cost is
$$\left.\frac{\Delta C(x)}{\Delta x}\right|_{\substack{x=400\\ \Delta x=-1}}=\frac{13.7505}{-1}=13.7505.$$

From the examples above, we can know that when the output is $x_0=400$, add one ton ($\Delta x=1$) output, the marginal cost is 13.7495; decrease one ton ($\Delta x=-1$) output, the marginal cost is 13.7505. According to the definition of the marginal cost above, the marginal cost is not a certain value when the output is $x_0=400$ t. It is a defect in both theory and application, which need further perfection.

Notice the value of the independent variable x in the total cost function, according to the economic meaning, the output of the product is usually a positive integer. For example, the output unit of cars is "frame", the output unit of machines is "set", the output unit of clothes is "piece", etc.. They are positive integers. Therefore, the output x is a discrete variable. In economics, assume the output unit is detachable infinitely, then we can take the output x as a continuous variable. So we can introduce the limit method to express the marginal cost, that is to say,
$$\lim_{\Delta x\to 0}\frac{C(x_0+\Delta x)-C(x_0)}{\Delta x},$$
recorded as MC, that is
$$MC=\lim_{\Delta x\to 0}\frac{C(x_0+\Delta x)-C(x_0)}{\Delta x}.$$
According to the relation of the existence of limit and the infinitesimal,
$$\frac{C(x_0+\Delta x)-C(x_0)}{\Delta x}=MC+\alpha.$$
in which $\lim_{\Delta x\to 0}\alpha=0$, when $|\Delta x|$ is small, we shall have
$$\frac{C(x_0+\Delta x)-C(x_0)}{\Delta x}\approx MC.$$
When the product increases $|\Delta x|=1$, with respect to the total output of the product, it is quite a small change. So when $|\Delta x|=1$, the equation above is right. The error can also meet the need of practical problems. It shows that we can use
$$\lim_{\Delta x\to 0}\frac{C(x_0+\Delta x)-C(x_0)}{\Delta x}$$

to approximately represent the marginal cost when the output is x_0. As shown in the above example, when the output is $x_0 = 400$, the marginal cost is approximately:

$$\lim_{\Delta x \to 0} \frac{C(400+\Delta x)-C(400)}{\Delta x}$$

$$= \lim_{\Delta x \to 0} \frac{5\,000+13\times(400+\Delta x)+30\sqrt{400+\Delta x}-(5\,000+13\times 400+30\sqrt{400})}{\Delta x}$$

$$= 13+30\times \lim_{\Delta x \to 0} \frac{\sqrt{400+\Delta x}-20}{\Delta x}$$

$$= 13.75.$$

The error is 0.05. In economics, it is quite a small number, which can be totally ignored. Therefore, modern economics define the marginal cost as

$$\lim_{\Delta x \to 0} \frac{C(x_0+\Delta x)-C(x_0)}{\Delta x}. \tag{3}$$

It not only overcomes the defects in the original definition of marginal cost, but also makes the calculation of marginal cost more simple.

There are more similar examples in practical problems. For example, the linear density ρ of fine rods in physics, the current intensity I in electricity, the specific heat c in thermodynamics and the angular velocity ω in rigid body rotation, etc..

Although the above problems have different definitions, the mathematical model for solving them is the same, namely to calculate the limit of the ratio of the increment of the function and the independent variable when the increment of the independent approaches zero. Ignoring all the concrete definitions of these problems, only consider their mathematical structure, then we can abstractly have the concept of derivative.

2.1.2 Definition of derivatives

1) Derivatives at a point

Definition Suppose that the function $y=f(x)$ is defined in a neighborhood of x_0. And when x has an increment Δx at x_0 ($\Delta x \neq 0$, $x_0+\Delta x$ is still in that neighborhood), respectively the function y has an increment $\Delta y = f(x_0+\Delta x)-f(x_0)$. If when $\Delta x \to 0$, the limit of the ratio $\frac{\Delta y}{\Delta x}$ of Δy and Δx exists, we call the function $f(x)$ is derivative at the point x_0, and the value of it is called the derivative of the function $y=f(x)$ at the point x_0, recorded as

$$f'(x_0) = \lim_{\Delta x \to 0} \frac{\Delta y}{\Delta x} = \lim_{\Delta x \to 0} \frac{f(x_0+\Delta x)-f(x_0)}{\Delta x}, \tag{4}$$

or

$$y'\bigg|_{x=x_0}, \quad \frac{\mathrm{d}y}{\mathrm{d}x}\bigg|_{x=x_0} \quad \text{or} \quad \frac{\mathrm{d}f(x)}{\mathrm{d}x}\bigg|_{x=x_0}.$$

If the limit (4) does not exist, then $f(x)$ said to be non-derivable at x_0. If the limit

(4) is infinite, $f(x)$ is also non-derivable at x_0, but for convenience, we often say that the derivative of $f(x)$ at x_0 is infinite and write recorded as $f'(x_0)=\infty$.

2) Some explanations for the definition of derivatives

(1) The derivative of a function at the point x_0 is the change rate of the variable at the point x_0. It shows the speed of the change of the variable with respect to the independent variable.

(2) Equation (4) has some common equivalent forms:

Let $\Delta x = h$, then we have

$$f'(x_0) = \lim_{h \to 0} \frac{f(x_0+h)-f(x_0)}{h}. \tag{5}$$

Let $x = x_0 + \Delta x$, then $\Delta x = x - x_0$. When $\Delta x \to 0$, $x \to x_0$. Therefore, the Formula (4) can be also written in the following form

$$f'(x_0) = \lim_{x \to x_0} \frac{f(x)-f(x_0)}{x-x_0}. \tag{6}$$

And

$$\frac{\Delta y}{\Delta x} = \frac{f(x)-f(x_0)}{x-x_0}$$

is also called the difference quotient. So the derivative is the limit of the difference quotient.

(3) Calculating $f'(x_0)$ from the definition of derivative:

Step 1 Write down expression of the increment of the function

$$\Delta y = f(x_0+\Delta x)-f(x_0).$$

Step 2 Expand and simplify the difference quotient

$$\frac{\Delta y}{\Delta x} = \frac{f(x_0+\Delta x)-f(x_0)}{\Delta x}.$$

Step 3 Find $f'(x_0)$ by evaluating the limit

$$f'(x_0) = \lim_{\Delta x \to 0} \frac{\Delta y}{\Delta x} = \lim_{\Delta x \to 0} \frac{f(x_0+\Delta x)-f(x_0)}{\Delta x}.$$

3) One-sided derivative

Derivative is the limit of the ratio $\frac{\Delta y}{\Delta x}$ when $\Delta x \to 0$. Compared with the concepts of the left and right limits, naturally, if $\lim\limits_{\Delta x \to 0^-} \frac{\Delta y}{\Delta x}$ and $\lim\limits_{\Delta x \to 0^+} \frac{\Delta y}{\Delta x}$ both exist, then we respectively call them the left-hand and the right-hand derivative of $f(x)$ at x_0, denoted as $f'_-(x_0)$ and $f'_+(x_0)$, namely,

$$f'_-(x_0) = \lim_{\Delta x \to 0^-} \frac{f(x_0+\Delta x)-f(x_0)}{\Delta x}, \tag{7}$$

$$f'_+(x_0) = \lim_{\Delta x \to 0^+} \frac{f(x_0+\Delta x)-f(x_0)}{\Delta x}. \tag{8}$$

It is easy to prove the following resule.

Theorem The necessary and sufficient condition for the function $y=f(x)$ to be derivable at x_0 is that both left and are right derivatives of $f(x)$ at x_0 exist and are equal, that is $f'_-(x_0)=f'_+(x_0)$.

The concept of one-sided derivative can be used to calculate the derivative of the piecewise function at the piecewise point.

If the function $f(x)$ is derivable at every point of the open interval (a,b), then $f(x)$ is derivable on (a,b). If $f(x)$ is derivable on (a,b), and $f'_+(a)$, $f'_-(b)$ both exist, then $f(x)$ is called derivable on the closed interval $[a, b]$.

4) Derived function

If the function $y=f(x)$ is derivable at every point of the interval I, that is, for every point x of I, exists an unique corresponding derivative $f'(x)$. Therefore, the derivative $f'(x)$ is a new function defined on I, that is $f'(x): I \to R$, and is called the derived function of $f(x)$ on I, denoted by $f'(x)$ or $\dfrac{\mathrm{d}y}{\mathrm{d}x}$. That is to say,

$$f'(x)=\lim_{\Delta x \to 0}\frac{f(x+\Delta x)-f(x)}{\Delta x}. \tag{9}$$

Compared to the Equation (4) and the Equation (9), we can know that the value of the derived function $f'(x)$ at x_0 is the derivative $f'(x_0)$ of $f(x)$ at x_0, namely

$$f'(x_0)=f'(x)\Big|_{x=x_0}.$$

Example 1 The equation of the free-fall motion is $s=\dfrac{1}{2}gt^2$. Find

(1) the average speed of the falling subject during the time period from t_0 to t, and the average speed when $t_0=10$ s, $\Delta t=0.1$ s.

(2) the instantaneous speed of the falling subject when $t=t_0$, and the instantaneous speed at $t_0=10$ s.

Solution (1) When t changes from $t=t_0$ to $t=t_0+\Delta t$, the distance of the falling subject is

$$\Delta s=\frac{1}{2}g(t_0+\Delta t)^2-\frac{1}{2}gt_0^2.$$

The average speed in the time period $\Delta t=(t_0+\Delta t)-t_0$ is

$$\bar{v}=\frac{\Delta s}{\Delta t}=\frac{\frac{1}{2}g[(t_0+\Delta t)^2-t_0^2]}{\Delta t}=g\left(t_0+\frac{1}{2}\Delta t\right).$$

When $t_0=10$ s, $\Delta t=0.1$ s 时,$\bar{v}=g(10+0.05)=10.05\ g$ m/s.

(2) The instantaneous speed of the falling subject when $t=t_0$ is

$$v=\lim_{\Delta t \to 0}\bar{v}=\lim_{\Delta t \to 0}\left(t_0+\frac{1}{2}\Delta t\right)=gt_0.$$

When $t_0=10$ s, the instantaneous speed of the falling subject is $v=10g$ m/s.

Example 2 Discuss the derivability of the function $y=f(x)=|x|$ at $x=0$.

Solution $f(x)=|x|=\begin{cases} x, & x \geqslant 0, \\ -x, & x<0. \end{cases}$

$$f'_+(0) = \lim_{x \to 0^+} \frac{f(x)-f(0)}{x-0} = \lim_{x \to 0^+} \frac{x}{x} = 1,$$

$$f'_-(0) = \lim_{x \to 0^-} \frac{f(x)-f(0)}{x-0} = \lim_{x \to 0^-} \frac{-x}{x} = -1.$$

Because $f'_+(0) \neq f'_-(0)$, then $f(x)$ is not derivative at $x=0$. As shown in Figure 2-4, the figure of $f(x)$ has a shape point at $x=0$.

Find the derivatives of some functions by means of the definition of the derivative.

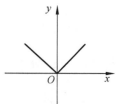

Example 3 Calculate the derivative of the function $y=C$ (C is a constant).

Figure 2-4

Solution $y' = \lim_{\Delta x \to 0} \frac{C-C}{\Delta x} = 0$, that is $(C)'=0$.

Example 4 Calculate the derivative of the function $y=\frac{1}{x}$ ($x \neq 0$) at the point $-1, \frac{1}{2}, 1$.

Solution $y' = \lim_{\Delta x \to 0} \frac{\frac{1}{x+\Delta x} - \frac{1}{x}}{\Delta x} = \lim_{\Delta x \to 0} \frac{-1}{x(x+\Delta x)} = -\frac{1}{x^2}$. That is,

$$\left(\frac{1}{x}\right)' = -\frac{1}{x^2}.$$

Therefore, $y'\big|_{x=-1} = -1$, $y'\big|_{x=\frac{1}{2}} = -4$, $y'\big|_{x=1} = -1$.

Example 5 Calculate the derivative of the function $y=x^n$ at x_0 (n is a positive integer).

Solution $y' = \lim_{\Delta x \to 0} \frac{(x_0+\Delta x)^n - x_0^n}{\Delta x}$

$$= \lim_{\Delta x \to 0} \frac{\Delta x[(x_0+\Delta x)^{n-1} + (x_0+\Delta x)^{n-2} x_0 + \cdots + x_0^{n-1}]}{\Delta x} = nx_0^{n-1}.$$

According to the arbitrary of x_0, for any $x \in (-\infty, +\infty)$, we have

$$(x^n)' = nx^{n-1}.$$

We can prove that, when $x \neq 0$, and n is any real number, the equation above is still right. If when $n=-1$, $\left(\frac{1}{x}\right)' = (-1)x^{-2} = -\frac{1}{x^2}$. This is the result of the Example 4.

2.1.3 Geometric interpretation of derivative

Suppose that the function f is derivable at x_0, then from cited Example 2 in 2.1.1, we know that the tangent of the curve $y=f(x)$ at the point $(x_0, f(x_0))$ exists and the derivative $f'(x_0)$ represents the slope of the tangent line M_0T to the plane curve $y=f(x)$ at the point $M_0(x_0, f(x_0))$ (Figure 2-3), namely,

$$k = \tan \alpha = \lim_{\Delta x \to 0} \frac{\Delta y}{\Delta x} = f'(x_0).$$

We can thus have the equation of the tangent to the graph of $y=f(x)$ at the point

$(x_0, f(x_0))$
$$y - f(x_0) = f'(x_0)(x - x_0). \tag{10}$$

The line that is perpendicular to the tangent line is called the normal line at that point. The equation of the normal line to the curve at the point $(x_0, f(x_0))$ is

$$y - f(x_0) = -\frac{1}{f'(x_0)}(x - x_0) \quad (f'(x_0) \neq 0). \tag{11}$$

Example 6 Find equations of the tangent line and the normal line to the curve $y = \frac{1}{x^2}$ at the point $M_0\left(2, \frac{1}{4}\right)$.

Solution $y' = -2x^{-3} = -\frac{2}{x^3}$, $y'\big|_{x=2} = -\frac{1}{4}$.

The tangent equation at the point M_0 is
$$y - \frac{1}{4} = -\frac{1}{4}(x - 2).$$
That is
$$x + 4y - 3 = 0.$$

The equation of the normal line at the point M_0 is
$$y - \frac{1}{4} = 4(x - 2).$$
That is
$$16x - 4y - 31 = 0.$$

Example 7 Find equation of the tangent line to the curve $y = \sqrt[3]{x}$ at the point $O(0,0)$.

Solution $f'(0) = \lim_{\Delta x \to 0} \frac{f(0 + \Delta x) - f(0)}{\Delta x}$

$$= \lim_{\Delta x \to 0} \frac{\sqrt[3]{\Delta x} - 0}{\Delta x}$$

$$= \lim_{\Delta x \to 0} \frac{1}{\sqrt[3]{(\Delta x)^2}} = +\infty.$$

We can see that $y = \sqrt[3]{x}$ is not derivative at $x = 0$ or the derivative is ∞. The slope of the tangent of the curve at $O(0,0)$ is ∞, namely the tangent of the curve $y = \sqrt[3]{x}$ at $O(0,0)$ is perpendicular to the x-axis, which is the y-axis. So the tangent equation is $x = 0$.

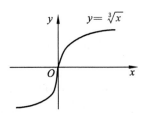

Figure 2-5

Therefore, the equation of the normal line of the curve $y = \sqrt[3]{x}$ is $y = 0$ at $O(0,0)$, which is the x-axis (Figure 2-5).

Example 8 Find equation of the tangent line to the curve $y = x^3 - 3x^2 + 5$ which is perpendicular to the line $x + 9y - 1 = 0$.

Solution Because the slope of the tangent of the curve $y = x^3 - 3x^2 + 5$ at any point is $y' = 3x^2 - 6x$. The slope of the line $x + 9y - 1 = 0$ is $-\frac{1}{9}$. According to the question, we can learn that $3x^2 - 6x = 9$, $x_1 = -1$, $x_2 = 3$.

So the point of contact is $(-1, 1)$ or $(3, 5)$, and the slopes of the tangent

are respectively
$$k_1 = y'\Big|_{x=-1} = 9, \quad k_2 = y'\Big|_{x=3} = 9.$$

The two tangent equations are
$$y - 1 = 9(x+1), \quad y - 5 = 9(x-3).$$
namely, $y - 9x - 10 = 0, y - 9x + 22 = 0$.

2.1.4 Use the unit to explain derivative

Assume $S = f(t)$ is a position function near a fixed point at the time t. The unit of S is m, and the unit of t is s. Then $\dfrac{\mathrm{d}S}{\mathrm{d}t} = f'(2) = 10$ m/s means that the subject is moving at the speed of 10 m/s when $t = 2$ s. If the subject keeps moving 1 s at this speed (from $t = 2$ to $t = 3$), it will move 10 m approximately. Generally, we have the following relations:

(1) The unit of the derivative of the function is the quotient of the unit of the variable divides the unit of the independent variable.

(2) If the derivative of a function does not change drastically near a point, then the derivative of the function at this point is approximately equal to the change of the function when the independent variable changes one unit.

Example 9 The cost of building a house of an area of A m² is $C = f(A)$ (unit: Yuan). What is the practical meaning of the function $f'(A)$?

Solution $f'(A) = \dfrac{\mathrm{d}C}{\mathrm{d}A}$ is the cost divided by the area. So it can be measured by "Yuan/m²". If $\mathrm{d}C$ is the change of the cost when the increment of the area is $\mathrm{d}A$ m². Then $\dfrac{\mathrm{d}C}{\mathrm{d}A}$ is the additional cost per square meter. Therefore, if a house of about A m² is designed to build, then $f'(A)$ is the increased area when the area is increased by one square meter.

Example 10 The cost of T t mining copper ore from a mine is $C = f(T)$ Yuan. Then what is the meaning of $f'(2\,000) = 1\,000$?

Solution
$$f'(2\,000) = \dfrac{\mathrm{d}C}{\mathrm{d}T}\Big|_{T=2\,000}.$$

Because the unit of C is Yuan, and the unit of T is t, then the unit of $\dfrac{\mathrm{d}C}{\mathrm{d}T}$ is "Yuan/t".

So $\dfrac{\mathrm{d}C}{\mathrm{d}T}\Big|_{T=2\,000} = 1\,000$ indicates that when mining 2 000 t ore from the mine, the cost of mining extra 1 t ore is approximately 1 000 Yuan. In other words, the cost of mining the 2001st t ore is about 1 000 Yuan.

Example 11 Known the water in the pipe is flowing at the speed of 10 m³/s. Try to use a derivative of some function to explain the speed.

Solution In fact, the water speed of 10 m³/s may be achieved by water flowing very slowly through a large pipe, and may also be achieved by water flowing very rapidly through to a narrow pipe. If inspecting the unit—m³/s, you will find that it is the change rate of some amount calculating by cubic meter. And cubic meter is the measuring unit of volume, so the unit is the change rate of volume. Imagine all the flowing water flows into a local water tank, and assume $V(t)$ represents the volume of a water tank at the time , then we can have the change rate of $V(t)$ is 10, that is

$$V'(t) = \frac{dV}{dt} = 10.$$

2.1.5 Relationship between derivability and continuity

If the function $f(x)$ is derivable at x_0, that is to say $f'(x_0)$ exists and is finite. According to $f'(x_0) = \lim\limits_{\Delta x \to 0} \frac{\Delta y}{\Delta x}$, we have $\frac{\Delta y}{\Delta x} = f'(x_0) + \alpha$, where $\lim\limits_{\Delta x \to 0} \alpha = 0$, or

$$\Delta y = f'(x_0)\Delta x + \alpha \Delta x. \tag{12}$$

Therefore, $\lim\limits_{\Delta x \to 0} \Delta y = \lim\limits_{\Delta x \to 0}[f'(x_0)\Delta x + \alpha \Delta x] = 0.$

That is to say, $f(x)$ is continuous at x_0. We have the following conclusions:

If the function $f(x)$ is derivable at x_0, then $f(x)$ is continuous at x_0.

The converse of the above conclusion is not followed, that means, the continuity of the function $f(x)$ at one point is only a necessary condition for its derivability at the point, but is not a sufficient condition. According to Example 2, the function $y = |x|$ is continuous at $x = 0$, but is not derivable at $x = 0$. Again according to Example 7, the function $y = \sqrt[3]{x}$ is continuous at $x = 0$, but is not derivable at $x = 0$.

Example 12 Discuss the derivability of the function

$$f(x) = \begin{cases} x\sin\dfrac{1}{x}, & x \neq 0, \\ 0, & x = 0, \end{cases}$$

at $x = 0$.

Solution $\lim\limits_{x \to 0} f(x) = \lim\limits_{x \to 0}\left(x\sin\dfrac{1}{x}\right) = 0 = f(0)$, so $f(x)$ is continuous at $x = 0$. While

$$\lim_{\Delta x \to 0}\frac{\Delta y}{\Delta x} = \lim_{\Delta x \to 0}\frac{f(0+\Delta x) - f(0)}{\Delta x}$$

$$= \lim_{\Delta x \to 0}\frac{\Delta x \sin\dfrac{1}{\Delta x} - 0}{\Delta x} = \lim_{\Delta x \to 0}\sin\dfrac{1}{\Delta x}$$

does not exist, then $f(x)$ is not derivative at $x = 0$.

Example 13 Assume the function

$$f(x) = \begin{cases} a, & x < 0, \\ x^2 + 1, & 0 \leqslant x < 1. \end{cases}$$

What is the value of a when $f(x)$ is derivative?

Solution Because when $x < 0$, $f(x) = a$ is derivable at any point. When $0 \leqslant x < 1$, $f(x) = x^2 + 1$ is also derivable at any point. Therefore, we only have to discuss the value of a when $f(x)$ is derivable at $x = 0$.

At $x = 0$, because

$$f'_-(0) = \lim_{x \to 0^-} \frac{f(x)-f(0)}{x-0} = \lim_{x \to 0^-} \frac{a-1}{x},$$

$$f'_+(0) = \lim_{x \to 0^+} \frac{f(x)-f(0)}{x-0} = \lim_{x \to 0^+} \frac{x^2+1-1}{x} = 0.$$

Try to make $f(x)$ be derivative at $x=0$, we must have $\lim\limits_{x \to 0^-} \dfrac{a-1}{x} = 0$. Then we have $a=1$.

So when $a=1$, $f(x)$ is a derived function.

Example 14 Determine the constants a and b, such that the function

$$f(x) = \begin{cases} 2e^x + a, & x < 0, \\ x^2 + bx + 1, & x \geq 0, \end{cases}$$

(1) is continuous at $x=0$;

(2) is derivable at $x=0$.

Solution (1) If $f(x)$ is continuous at $x=0$, then we have

$$\lim_{x \to 0^-} f(x) = \lim_{x \to 0^+} f(x) = f(0) = 1,$$

that is $2+a=1$, and we have $a=-1$.

So when $a=-1$, b is any real number, $f(x)$ is continuous at $x=0$.

(2) If $f(x)$ is derivable at $x=0$, then we have

$$f'_-(0) = \lim_{x \to 0^-} \frac{f(x)-f(0)}{x-0} = \lim_{x \to 0^-} \frac{2e^x+a-1}{x} = \lim_{x \to 0^-} \frac{2e^x-2}{x} = 2 \text{(because } a=-1\text{)},$$

$$f'_+(0) = \lim_{x \to 0^+} \frac{f(x)-f(0)}{x-0} = \lim_{x \to 0^+} \frac{x^2+bx+1-1}{x} = b.$$

According to $f'_+(0) = f'_-(0)$, we have $b=2$. Thus when $a=-1$, $b=2$, $f(x)$ is derivable at $x=0$.

2.1.6 The application of derivative in natural science

1) Grating chromatic dispersion

Grating instrument is used for determining wavelength of spectral line in optics. Every wavelength λ of the incident light cast in grating has a certain corresponding deflection angle θ, so the deflection angle is a function of wavelength λ. When the wavelength changes from λ to $\lambda + \Delta\lambda$, the average change rate of deflection is

$$\frac{\Delta\theta}{\Delta\lambda} = \frac{\theta(\lambda+\Delta\lambda) - \theta(\lambda)}{\Delta\lambda}.$$

The change rate of the deflection at the wavelength λ is

$$\frac{d\theta}{d\lambda} = \lim_{\Delta\lambda \to 0} \frac{\Delta\theta}{\Delta\lambda},$$

which is called the grating chromatic dispersion.

2) Compression and compressibility of gas

The volume V of a gas with constant temperature varies with the pressure p, and when the pressure increases from p to $p+\Delta p$, the corresponding volume increments are

$$\Delta V = V(p+\Delta p) - V(p),$$

and the average volume growth rate is

$$\frac{\Delta V}{\Delta p} = \frac{V(p+\Delta p)-V(p)}{\Delta p}.$$

When $\Delta p > 0$, apparently we have $\Delta V \leqslant 0$. The above ratio is not positive, so that the absolute value reflects the average compression rate of the volume to the pressure.

$$\frac{dV}{dp} = \lim_{\Delta p \to 0} \frac{\Delta V}{\Delta p}$$

is the compression rate of the volume at which the pressure is p. In thermodynamics

$$\beta = -\frac{V'(p)}{V}$$

is defined as the coefficient of isothermal compressibility for a gas, which reflects the volume compression rate of a unit volume of gas when the pressure is p.

3) Growth rate of biological population

Assume $N = N(t)$ represents the total number of a biological populations at the time t. Then the average growth rate of the population in this interval from t to $t+\Delta t$ is defined as

$$\frac{\Delta N}{\Delta t} = \frac{N(t+\Delta t)-N(t)}{\Delta t}.$$

In order to study the changing law of the number of individuals in the population, we shall have the rate of change at any given time N, namely,

$$\frac{dN}{dt} = \lim_{\Delta t \to 0} \frac{\Delta N}{\Delta t}.$$

In biology, it is called the rate of growth of a population.

Of course, the total number of individuals $N(t)$ can only be positive integers, which is not a continuous function. So it is not derivative. But because the majority of population reproduction is a generating, with its large number, when the time interval Δt is quite small, the number of individuals caused by the birth and death of the populations with respect to the change of the total numbers is also small. Therefore $N(t)$ can approximate be taken as the continuous derived function. In fact, the growth rate $N'(t)$ has played an important role in the study of species growth and related issues.

Exercises 2-1

1. An object falls from the height of 400 m, and the height from the ground when it falls at the time point t is (unit: s) $h = -16t^2 + 400$ (m).

   (1) Find the average speed of falling objects in the first 4 s.

   (2) Find the instantaneous speed of the object falling at the 4th s.

2. Find the equation of the tangent to the curve at the specified point.

   (1) $y = x^3 + 1$ at the point $(1,2)$;

   (2) $y = \sqrt{x}$ at the point $(1, 1)$.

3. Let the object rotate by a fixed axle. The angular speed of the object rotation is the ratio of the rotate angle and time consuming t. If the rotation is constant speed, then rotate angle ω is the function of t: $\omega = \omega(t)$. If the rotation is not constant speed, try to find the

expression of the angular speed.

4. Find the derivatives of the following functions from the definition of derivative:

(1) $f(x)=\dfrac{1}{x^2}$, $x_0=2$;

(2) $f(x)=\dfrac{1}{\sqrt{x}}$, $x_0=2$;

(3) $f(x)=\begin{cases} x^2\sin\dfrac{1}{x}, & x\neq 0, \\ 0, & x=0, \end{cases}$ $x_0=0$.

5. Discuss the derivability of the following functions at $x=0$.

(1) $y=\begin{cases} x, & x\geqslant 0, \\ x^3, & x<0; \end{cases}$

(2) $y=\begin{cases} x, & x\geqslant 0, \\ \sin x, & x<0; \end{cases}$

(3) $y=\begin{cases} e^x-1, & x\geqslant 0, \\ x-x^2, & x<0. \end{cases}$

6. Determine the constants a and b, such that the function $f(x)=\begin{cases} x^2+a, & x\geqslant 1, \\ bx+1, & x<1 \end{cases}$ is derivable at $x=1$?

7. Assume $f(x)=\begin{cases} x^a\sin\dfrac{1}{x}, & x\neq 0, \\ 0, & x=0. \end{cases}$ In order to make $f(x)$,

(1) continuous; (2) derivable; (3) continuous and derivable, at $x=0$, what value should a and b take?

8. Assume $f'(x_0)$ exist, find the following limits:

(1) $\lim\limits_{\Delta x\to 0}\dfrac{f(x_0-\Delta x)-f(x_0)}{\Delta x}$;

(2) $\lim\limits_{h\to 0}\dfrac{f(x_0)-f(x_0-2h)}{h}$;

(3) $\lim\limits_{h\to 0}\dfrac{f(x_0+2h)-f(x_0-3h)}{h}$.

9. At which point on the curve $y=x^3$ is the tangent line parallel to the line $y-12x-1=0$? At which point on the curve is the normal line parallel to the line $y+12x-1=0$?

10. Assume the curve L: $y=\sin x$, $x\in(-\infty,+\infty)$.

(1) Find the tangent equation and the normal equation of the curve L at the original point.

(2) Find the parallel tangent equation of the curve L and the coordinates of the points of intersection with L.

11. t min after the water is released from the tank, the remaining water in the tank is expressed by $Q(t)=200(30-t)^2$ (unit: L). Then how fast is the flow of water when 10 min is just out of date? What is the average outflow rate of water at the beginning of 10 min?

2.2 Rules of finding derivatives

To determine the change rate of the function—derivative, is a common problem often encountered in theoretical research and practical application. But it is difficult to find derivatives of more complicated functions by using the definition of derivative, sometimes even is not feasible. In this section, we will propose some fundamental rules of derivation, so that we can find the derivatives of elementary function much more easily instead of using the definition of the derivative.

2.2.1 Derivative formulas for several fundamental elementary functions

We apply the definition of derivative given in 2.1.2 to calculate the derivatives of the fundamental elementary functions $\sin x$, $\cos x$, $\log_a x$ and a^x.

1) $y = \sin x$

$$\frac{dy}{dx} = \lim_{\Delta x \to 0} \frac{\Delta y}{\Delta x} = \lim_{\Delta x \to 0} \frac{\sin(x+\Delta x)-\sin x}{\Delta x}$$

$$= \lim_{\Delta x \to 0} \frac{2\cos\left(x+\frac{\Delta x}{2}\right)\sin\frac{\Delta x}{2}}{\Delta x}$$

$$= \lim_{\Delta x \to 0} \cos\left(x+\frac{\Delta x}{2}\right) \frac{\sin\frac{\Delta x}{2}}{\frac{\Delta x}{2}} = \cos x.$$

Therefore, $(\sin x)' = \cos x \quad (-\infty < x < +\infty)$.

2) $y = \cos x$

$$\frac{dy}{dx} = \lim_{\Delta x \to 0} \frac{\Delta y}{\Delta x} = \lim_{\Delta x \to 0} \frac{\cos(x+\Delta x)-\cos x}{\Delta x}$$

$$= \lim_{\Delta x \to 0} \frac{-2\sin\left(x+\frac{\Delta x}{2}\right)\sin\frac{\Delta x}{2}}{\Delta x}$$

$$= -\lim_{\Delta x \to 0} \sin\left(x+\frac{\Delta x}{2}\right) \frac{\sin\frac{\Delta x}{2}}{\frac{\Delta x}{2}} = -\sin x.$$

Therefore, $(\cos x)' = -\sin x \quad (-\infty < x < +\infty)$.

3) $y = \log_a x \quad (a > 0, a \neq 1)$

$$\frac{dy}{dx} = \lim_{\Delta x \to 0} \frac{\log_a(x+\Delta x) - \log_a x}{\Delta x} = \lim_{\Delta x \to 0} \log_a \left(1+\frac{\Delta x}{x}\right)^{\frac{1}{\Delta x}}$$

$$= \frac{1}{x}\log_a e = \frac{1}{x \ln a}.$$

Therefore, $(\log_a x)' = \frac{1}{x \ln a}$.

Specially, when $a = e$, $(\ln x)' = \frac{1}{x}$.

4) $y = a^x$ $(a>0, a \neq 1)$

$$\frac{dy}{dx} = \lim_{\Delta x \to 0} \frac{a^{x+\Delta x} - a^x}{\Delta x} = \lim_{\Delta x \to 0} a^x \frac{a^{\Delta x} - 1}{\Delta x}$$

$$= a^x \lim_{\Delta x \to 0} \frac{\Delta x \cdot \ln a}{\Delta x} = a^x \ln a.$$

Therefore, $(a^x)' = a^x \ln a.$

Specially, when $a = e$, $(e^x)' = e^x$.

It is obviously that, although e is a irrational number, choosing it to be the bottom of the exponential function and logarithmic function can make these calculations easier.

2.2.2 Derivation rules of rational operations

Theorem 1 Suppose that the functions $u(x)$, $v(x)$ are derivable at x, then $u(x) \pm v(x)$, $u(x) \cdot v(x)$, $\frac{u(x)}{v(x)}(v(x) \neq 0)$ are all derivable at x, and

(1) $[u(x) \pm v(x)]' = u'(x) \pm v'(x)$;

(2) $[u(x) \cdot v(x)]' = u'(x)v(x) + u(x)v'(x)$;

(3) $\left[\frac{u(x)}{v(x)}\right]' = \frac{u'(x)v(x) - u(x)v'(x)}{v^2(x)}$ $(v(x) \neq 0)$.

Proof (1) Assume $y = u(x) \pm v(x)$, then

$$\frac{dy}{dx} = \lim_{\Delta x \to 0} \frac{\Delta y}{\Delta x} = \lim_{\Delta x \to 0} \frac{[u(x+\Delta x) \pm v(x+\Delta x)] - [u(x) \pm v(x)]}{\Delta x}$$

$$= \lim_{\Delta x \to 0} \left[\frac{u(x+\Delta x) - u(x)}{\Delta x} \pm \frac{v(x+\Delta x) - v(x)}{\Delta x}\right]$$

$$= \lim_{\Delta x \to 0} \left(\frac{\Delta u}{\Delta x} \pm \frac{\Delta v}{\Delta x}\right)$$

$$= u'(x) \pm v'(x).$$

So, $[u(x) \pm v(x)]' = u'(x) \pm v'(x).$

(2) Assume $y = u(x)v(x)$, then

$$\frac{dy}{dx} = \lim_{\Delta x \to 0} \frac{\Delta y}{\Delta x} = \lim_{\Delta x \to 0} \frac{u(x+\Delta x)v(x+\Delta x) - u(x)v(x)}{\Delta x}$$

$$= \lim_{\Delta x \to 0} \frac{[u(x+\Delta x)v(x+\Delta x) - u(x)v(x+\Delta x)] + [u(x)v(x+\Delta x) - u(x)v(x)]}{\Delta x}$$

$$= \lim_{\Delta x \to 0} \left[\frac{\Delta u}{\Delta x} v(x+\Delta x) + u(x) \frac{\Delta v}{\Delta x}\right] = \frac{du}{dx} v(x) + u(x) \frac{dv}{dx},$$

Therefore, $[u(x)v(x)]' = u'(x)v(x) + u(x)v'(x).$

(3) Assume $y = \frac{u(x)}{v(x)}$ $(v(x) \neq 0)$, then

$$\frac{dy}{dx} = \lim_{\Delta x \to 0} \frac{\Delta y}{\Delta x} = \lim_{\Delta x \to 0} \frac{\frac{u(x+\Delta x)}{v(x+\Delta x)} - \frac{u(x)}{v(x)}}{\Delta x} = \lim_{\Delta x \to 0} \frac{u(x+\Delta x)v(x) - u(x)v(x+\Delta x)}{v(x)v(x+\Delta x) \cdot \Delta x}$$

$$= \lim_{\Delta x \to 0} \frac{u(x+\Delta x)v(x) - u(x)v(x) + u(x)v(x) - u(x)v(x+\Delta x)}{v(x)v(x+\Delta x) \cdot \Delta x}$$

$$= \lim_{\Delta x \to 0} \frac{1}{v(x)v(x+\Delta x)} \left[\frac{\Delta u}{\Delta x} v(x) - \frac{\Delta v}{\Delta x} u(x) \right] = \frac{u'(x)v(x) - u(x)v'(x)}{v^2(x)}.$$

That is
$$\left[\frac{u(x)}{v(x)} \right]' = \frac{u'(x)v(x) - u(x)v'(x)}{v^2(x)}.$$

Corollary Assume $u_1(x), u_2(x), \cdots, u_m(x)$ are derivative at the point x, then we have

(1) $[u_1(x) + u_2(x) + \cdots + u_m(x)]' = u_1'(x) + u_2'(x) + \cdots + u_m'(x)$;

(2) $[Cu_i(x)]' = Cu_i'(x)$ $(i=1,2,\cdots,m)$, C is a constant number;

(3) $[u_1(x)u_2(x)\cdots u_m(x)]' = u_1'(x)u_2(x)\cdots u_m(x) + u_1(x)u_2'(x)\cdots u_m(x) + \cdots + u_1(x)\cdots u_{m-1}(x)u_m'(x)$.

Example 1 Find the derivatives of $\cot x$, $\sec x$, $\csc x$.

Solution Using the Quotient Rule, we get
$$(\tan x)' = \left(\frac{\sin x}{\cos x} \right)' = \frac{(\sin x)' \cos x - \sin x (\cos x)'}{\cos^2 x}$$
$$= \frac{\cos^2 x + \sin^2 x}{\cos^2 x} = \frac{1}{\cos^2 x} = \sec^2 x.$$

Similarly, we have
$$(\cot x)' = -\frac{1}{\sin^2 x} = -\csc^2 x;$$
$$(\sec x)' = \left(\frac{1}{\cos x} \right)' = \frac{1' \cos x - 1(\cos x)'}{\cos^2 x}$$
$$= \frac{\sin x}{\cos^2 x} = \sec x \tan x;$$
$$(\csc x)' = -\csc x \cot x.$$

Example 2 Find the derivative of $y = 6a^x + 3\log_a x + x^{10} + \sin \frac{\pi}{12}$.

Solution $y' = 6a^x \ln a + \frac{3}{x \ln a} + 10x^9$.

Example 3 Find the derivative of $y = xe^x \sin x$.

Solution $y' = x' e^x \sin x + x(e^x)' \sin x + xe^x (\sin x)'$
$= e^x \sin x + xe^x \sin x + xe^x \cos x$
$= e^x (\sin x + x\sin x + x\cos x)$.

Example 4 Find the derivative of $y = (x^n + \cos x) \sin x$.

Solution $y' = (x^n + \cos x)' \sin x + (x^n + \cos x)(\sin x)'$
$= (nx^{n-1} - \sin x) \sin x + (x^n + \cos x) \cos x$
$= x^{n-1}(n\sin x + x\cos x) + \cos 2x$.

Example 5 Find the derivative of $y = \frac{e^x + \sin x}{\sqrt{x}}$.

Solution $y' = \frac{\sqrt{x}(e^x + \sin x)' - (e^x + \sin x)(\sqrt{x})'}{(\sqrt{x})^2}$

$$= \frac{\sqrt{x}(e^x+\cos x)-(e^x+\sin x)\frac{1}{2\sqrt{x}}}{x}$$

$$= \frac{(2x-1)e^x+2x\cos x-\sin x}{2x\sqrt{x}}.$$

Example 6 (**The genetic data and its sensitivity to changes**)   In general, when the smaller change of x caused the larger change of the function value $f(x)$, then we call the function is relatively sensitive to the change of x. The derivative $f'(x)$ is the measurement method of this sensitivity. Austria geneticist Mendel planted pea and other plants in the garden, finding the genetic law, separation law and free combination rule. His research showed that if p (a number between 0 and 1) is the frequency of the gene which can make the pea skin smooth (dominant gene), while $1-p$ is the frequency of the gene which can make the pea skin wrinkle, then the proportion of the smooth skin peas in the next generation is

$$y=2p(1-p)+p^2=2p-p^2.$$

Then
$$\frac{\mathrm{d}y}{\mathrm{d}p}=2-2p.$$

In Figure 2-6, when p is quite small, y is more sensitive to the effect of change of p than when p is large. Indeed, the derivative figure in Figure 2-7 confirms the fact that the figure indicates that $\frac{\mathrm{d}y}{\mathrm{d}p}$ is close to 2 when p is near 0, and $\frac{\mathrm{d}y}{\mathrm{d}p}$ is close to 0 when p is near 1.

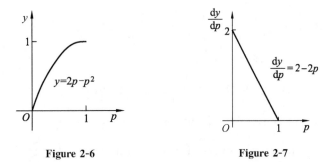

Figure 2-6   Figure 2-7

The meaning of the derivative in genetics is: introduce more positive genes to highly invisible groups (such as high frequency of skin wrinkling pea groups) and the highly advantageous groups. The former has more significant influence than the latter on the increase of the dominant gene in the next generation.

2.2.3 Derivative of inverse functions

For any given function, generally speaking, the inverse function may not exist. But if $f(x)$ is a single valued monotone continuous function on an interval I, then its inverse function $x=\varphi(y)$ on the corresponding interval I' is also a single valued, monotone continuous function.

Theorem 2 (The derivation rule of inverse function) Suppose the function $x=\varphi(y)$ is a strictly monotone continuous function on an interval I. If $x=\varphi(y)$ is derivable at y, and $\varphi'(y)\neq 0$, then its inverse function $y=\varphi^{-1}(x)=f(x)$ is also derivable at the corresponding $x\in I'$, and

$$f'(x)=\frac{1}{\varphi'(y)}. \qquad (1)$$

Proof By the existence theorem of inverse functions, the inverse function $y=\varphi^{-1}(x)=f(x)$ is also a strictly monotone continuous function on the corresponding interval I', thus

$$\Delta y=f(x+\Delta x)-f(x)\neq 0.$$

And because of the continuity of $x=\varphi(y)$, when $\Delta y\to 0$, then $\Delta x\to 0$. Therefore,

$$\frac{dy}{dx}=\lim_{\Delta x\to 0}\frac{\Delta y}{\Delta x}=\lim_{\Delta x\to 0}\frac{1}{\frac{\Delta x}{\Delta y}}=\frac{1}{\lim_{\Delta y\to 0}\frac{\Delta x}{\Delta y}}=\frac{1}{\frac{dx}{dy}}.$$

That is
$$f'(x)=\frac{1}{\varphi'(y)}.$$

Now we find the derivatives of the inverse trigonometric functions.

Example 7 Find derivative of the inverse trigonometric function $y=\arcsin x$.

Solution Because $y=\arcsin x$ is the inverse function of $x=\varphi(y)=\sin y$, $y\in \left[-\frac{\pi}{2},\frac{\pi}{2}\right]$. According to Theorem 2, we have

$$y'=(\arcsin x)'=\frac{1}{(\sin y)'}=\frac{1}{\cos y}=\frac{1}{\sqrt{1-\sin^2 y}}=\frac{1}{\sqrt{1-x^2}} \quad (-1<x<1).$$

Similarly, we have

$$(\arccos x)'=-\frac{1}{\sqrt{1-x^2}} \quad (-1<x<1).$$

Example 8 Find the derivative of the function $y=\arctan x$.

Solution $y=\arctan x$ is the inverse function of $x=\tan y$ $\left(-\frac{\pi}{2}<y<\frac{\pi}{2}\right)$. According to Theorem 2, we have

$$y'=(\arctan x)'=\frac{1}{(\tan y)'}=\frac{1}{\sec^2 y}=\frac{1}{1+\tan^2 y}=\frac{1}{1+x^2}.$$

Similarly, we have

$$(\text{arccot } x)'=-\frac{1}{1+x^2}.$$

2.2.4 Derivation rules of composite functions

Theorem 3 (The chain rule) If $u=\varphi(x)$ is derivable at x, and $y=f(u)$ is derivable at $u=\varphi(x)$, then the composition function $y=f[\varphi(x)]$ is derivable at x, and its derivative is

$$\frac{\mathrm{d}y}{\mathrm{d}x} = f'(u) \cdot \varphi'(x). \tag{2}$$

That is,
$$\frac{\mathrm{d}y}{\mathrm{d}x} = \frac{\mathrm{d}y}{\mathrm{d}u} \cdot \frac{\mathrm{d}u}{\mathrm{d}x}.$$

Proof Because $y = f(u)$ is derivable at u, that is
$$\lim_{\Delta u \to 0} \frac{\Delta y}{\Delta u} = f'(u).$$

Then we have
$$\frac{\Delta y}{\Delta u} = f'(u) + \alpha,$$

where $\lim_{\Delta u \to 0} \alpha = 0$. If $\Delta u \neq 0$, $\Delta y = f'(u)\Delta u + \alpha \Delta u.$ (3)

When $\Delta u = 0$, $\Delta y = f(u + \Delta u) - f(u) = 0$, the Equation (3) dividing by $\Delta x \neq 0$, we have
$$\frac{\Delta y}{\Delta x} = f'(u)\frac{\Delta u}{\Delta x} + \alpha \frac{\Delta u}{\Delta x}.$$

Therefore,
$$\frac{\mathrm{d}y}{\mathrm{d}x} = \lim_{\Delta x \to 0}\left[f'(u)\frac{\Delta u}{\Delta x} + \alpha \frac{\Delta u}{\Delta x}\right]. \tag{4}$$

When $\Delta x \to 0$, $\Delta u \to 0$,
$$\lim_{\Delta x \to 0}\left[f'(u)\frac{\Delta u}{\Delta x}\right] = f'(u)\frac{\mathrm{d}u}{\mathrm{d}x}.$$

Substitute it into Equation (4), we have
$$\frac{\mathrm{d}y}{\mathrm{d}x} = f'(u)\frac{\mathrm{d}u}{\mathrm{d}x} = f'(u)\varphi'(x).$$

Example 9 Find the derivative of $y = \left(2x - \dfrac{3}{x}\right)^{10}$.

Solution Assume $y = u^{10}$, $u = 2x - \dfrac{3}{x}$, then
$$\frac{\mathrm{d}y}{\mathrm{d}x} = \frac{\mathrm{d}y}{\mathrm{d}u}\frac{\mathrm{d}u}{\mathrm{d}x} = \frac{\mathrm{d}}{\mathrm{d}u}(u^{10}) \cdot \frac{\mathrm{d}}{\mathrm{d}x}\left(2x - \frac{3}{x}\right)$$
$$= 10u^9\left(2 + \frac{3}{x^2}\right) = 10\left(2 + \frac{3}{x^2}\right)\left(2x - \frac{3}{x}\right)^9.$$

Example 10 Find the derivative of $y = x^\mu$.

Solution
$$y = x^\mu = \mathrm{e}^{\mu \ln x}.$$

Assume
$$y = \mathrm{e}^u, \quad u = \mu \ln x,$$

thus,
$$\frac{\mathrm{d}y}{\mathrm{d}x} = (\mathrm{e}^u)'u_x' = \mathrm{e}^u \frac{\mu}{x} = \mathrm{e}^{\mu \ln x}\frac{\mu}{x} = x^\mu \cdot \frac{\mu}{x} = \mu x^{\mu - 1}.$$

That is to say, for any real number μ, we have
$$(x^\mu)' = \mu x^{\mu - 1}.$$

Example 11 Find the derivative of the hyperbolic functions $\operatorname{sh} x$, $\operatorname{ch} x$, $\operatorname{th} x$, $\operatorname{cth} x$.

Solution
$$(\operatorname{sh} x)' = \left(\frac{\mathrm{e}^x - \mathrm{e}^{-x}}{2}\right)' = \frac{\mathrm{e}^x + \mathrm{e}^{-x}}{2} = \operatorname{ch} x;$$
$$(\operatorname{ch} x)' = \left(\frac{\mathrm{e}^x + \mathrm{e}^{-x}}{2}\right)' = \frac{\mathrm{e}^x - \mathrm{e}^{-x}}{2} = \operatorname{sh} x;$$

$$(\text{th } x)' = \left(\frac{\text{sh } x}{\text{ch } x}\right)' = \frac{(\text{sh } x)' \text{ch } x - \text{sh } x (\text{ch } x)'}{\text{ch}^2 x}$$

$$= \frac{\text{ch}^2 x - \text{sh}^2 x}{\text{ch}^2 x} = \frac{1}{\text{ch}^2 x};$$

$$(\text{cth } x)' = \left(\frac{\text{ch } x}{\text{sh } x}\right)' = \frac{\text{sh}^2 x - \text{ch}^2 x}{\text{sh}^2 x} = -\frac{1}{\text{sh}^2 x}.$$

Example 12 Find the derivative of the inverse hyperbolic sine function $y = \text{arsh } x$.

Solution Sinse $y = \text{arsh } x$ is the inverse function of $x = \text{sh } y$,

$$y' = (\text{arsh } x)' = \frac{1}{(\text{sh } y)'} = \frac{1}{\text{ch } y} = \frac{1}{\sqrt{1 + \text{sh}^2 y}} = \frac{1}{\sqrt{1 + x^2}}.$$

Similarly, we have

$$(\text{arch } x)' = \frac{1}{\sqrt{x^2 - 1}}$$

and

$$(\text{arth } x)' = \frac{1}{1 - x^2}.$$

Example 13 Find the derivative of $y = \ln(x + \sqrt{1 + x^2})$.

Solution

$$y' = \frac{1}{x + \sqrt{1+x^2}} (x + \sqrt{1+x^2})' = \frac{1}{x + \sqrt{1+x^2}} \left[1 + \frac{1}{2\sqrt{1+x^2}} (1+x^2)' \right]$$

$$= \frac{1}{x + \sqrt{1+x^2}} \left[1 + \frac{x}{\sqrt{1+x^2}} \right] = \frac{\sqrt{1+x^2} + x}{(x + \sqrt{1+x^2})\sqrt{1+x^2}} = \frac{1}{\sqrt{1+x^2}}.$$

According to Chapter 1, we have

$$y = \text{arsh } x = \ln(x + \sqrt{1 + x^2}).$$

The result is exactly the same as Example 12.

Example 14 Find the derivative of $y = \ln|x|$.

Solution When $x > 0$, $y' = (\ln x)' = \frac{1}{x}$.

When $x < 0$, $y' = [\ln(-x)]' = \frac{1}{-x}(-x)' = \frac{1}{x}$.

Therefore, only if $x \neq 0$, we always have

$$(\ln|x|)' = \frac{1}{x}.$$

For convenience in application, we make a list of the derivatives for the fundamental elementary functions:

(1) $(C)' = 0$ (C is a constant number).

(2) $(x^\mu)' = \mu x^{\mu-1}$ ($\mu \in \mathbf{R}$).

(3) $(a^x)' = a^x \ln a$ ($a > 0$, $a \neq 1$); specially, $(e^x)' = e^x$.

(4) $(\log_a x)' = \frac{1}{x \ln a}$ ($a > 0$, $a \neq 1$); specially, $(\ln x)' = \frac{1}{x}$, $(\ln|x|)' = \frac{1}{x}$.

(5) $(\sin x)' = \cos x$.

(6) $(\cos x)' = -\sin x$.

(7) $(\tan x)' = \sec^2 x$.

(8) $(\cot x)' = -\csc^2 x$.

(9) $(\sec x)' = \sec x \tan x$.

(10) $(\csc x)' = -\csc x \cot x$.

(11) $(\arcsin x)' = \dfrac{1}{\sqrt{1-x^2}}$.

(12) $(\arccos x)' = -\dfrac{1}{\sqrt{1-x^2}}$.

(13) $(\arctan x)' = \dfrac{1}{1+x^2}$.

(14) $(\operatorname{arccot} x)' = -\dfrac{1}{1+x^2}$.

(15) $(\operatorname{sh} x)' = \operatorname{ch} x$.

(16) $(\operatorname{ch} x)' = \operatorname{sh} x$.

(17) $(\operatorname{th} x)' = \dfrac{1}{\operatorname{ch}^2 x}$.

(18) $(\operatorname{arsh} x)' = \dfrac{1}{\sqrt{1+x^2}}$.

(19) $(\operatorname{arch} x)' = \dfrac{1}{\sqrt{x^2-1}}$.

(20) $(\operatorname{arth} x)' = \dfrac{1}{1-x^2}$.

Example 15 Find the derivatives of the following functions:

(1) $y = \dfrac{1}{2}\arctan\dfrac{2x}{1-x^2}$;

(2) $y = 3^{\sin(1-2x)} + \dfrac{1}{2^x}$;

(3) $y = \dfrac{x}{2}\sqrt{a^2-x^2} + \dfrac{a^2}{2}\arcsin\dfrac{x}{a}$ $(a>0)$;

(4) $y = x\arctan x - xe^x + \dfrac{\ln x}{\sqrt{x}}$.

Solution (1) $y' = \dfrac{1}{2}\dfrac{1}{1+\left(\dfrac{2x}{1-x^2}\right)^2}\left(\dfrac{2x}{1-x^2}\right)'$

$= \dfrac{1}{2} \cdot \dfrac{(1-x^2)^2}{(1-x^2)^2+4x^2} \cdot \dfrac{2(1-x^2)-2x \cdot (-2x)}{(1-x^2)^2} = \dfrac{1}{1+x^2}$.

(2) $y' = [3^{\sin(1-2x)} \ln 3][\sin(1-2x)]' + \dfrac{-1}{(2^x)^2} \cdot (2^x)'$

$= [3^{\sin(1-2x)} \ln 3]\cos(1-2x) \cdot (-2) + \dfrac{-1}{(2^x)^2} \cdot 2^x \ln 2$

$= -2\ln 3 \cdot 3^{\sin(1-2x)} \cos(1-2x) - \dfrac{\ln 2}{2^x}$.

(3) $y' = \left(\dfrac{x}{2}\sqrt{a^2-x^2}\right)' + \left(\dfrac{a^2}{2}\arcsin\dfrac{x}{a}\right)'$

$= \left(\dfrac{x}{2}\right)'\sqrt{a^2-x^2} + \dfrac{x}{2}(\sqrt{a^2-x^2})' + \dfrac{a^2}{2}\left(\arcsin\dfrac{x}{a}\right)'$

$$= \frac{1}{2}\sqrt{a^2-x^2} + \frac{x}{2} \cdot \frac{1}{2} \frac{(a^2-x^2)'}{\sqrt{a^2-x^2}} + \frac{a^2}{2} \frac{\left(\frac{x}{a}\right)'}{\sqrt{1-\left(\frac{x}{a}\right)^2}}$$

$$= \frac{1}{2}\sqrt{a^2-x^2} - \frac{1}{2} \frac{x^2}{\sqrt{a^2-x^2}} + \frac{a^2}{2\sqrt{a^2-x^2}} = \sqrt{a^2-x^2}.$$

(4) $y' = \arctan x + \dfrac{x}{1+x^2} - xe^x - e^x + \dfrac{\frac{1}{x}\sqrt{x} - \ln x \cdot \frac{1}{2\sqrt{x}}}{(\sqrt{x})^2}$

$= \arctan x + \dfrac{x}{1+x^2} - (x+1)e^x + \dfrac{2-\ln x}{2x\sqrt{x}}.$

Example 16 Find the derivatives of the following functions:

(1) $y = \ln \dfrac{\sqrt{x^2+1}}{\sqrt[3]{x-2}}$ $(x>2)$;

(2) $y = \sqrt{x+\sqrt{x+\sqrt{x}}}$;

(3) $y = \log_x \sin x$ $(x>0, x \neq 1)$;

(4) $y = x^{a^a} + a^{x^a} + a^{a^x}$ $(a>0)$.

Solution (1) Since $y = \dfrac{1}{2}\ln(x^2+1) - \dfrac{1}{3}\ln(x-2)$, then we have

$y' = \dfrac{1}{2} \cdot \dfrac{1}{x^2+1}(x^2+1)' - \dfrac{1}{3} \cdot \dfrac{1}{x-2}(x-2)'$

$= \dfrac{1}{2} \cdot \dfrac{1}{x^2+1} \cdot 2x - \dfrac{1}{3} \cdot \dfrac{1}{x-2} = \dfrac{x}{x^2+1} - \dfrac{1}{3(x-2)}.$

(2) $y' = \dfrac{1}{2\sqrt{x+\sqrt{x+\sqrt{x}}}}(x+\sqrt{x+\sqrt{x}})'$

$= \dfrac{1}{2\sqrt{x+\sqrt{x+\sqrt{x}}}} \left[1 + \dfrac{1}{2\sqrt{x+\sqrt{x}}}(x+\sqrt{x})'\right]$

$= \dfrac{1}{2\sqrt{x+\sqrt{x+\sqrt{x}}}} \left[1 + \dfrac{1}{2\sqrt{x+\sqrt{x}}}\left(1+\dfrac{1}{2\sqrt{x}}\right)\right]$

$= \dfrac{4\sqrt{x^2+x\sqrt{x}}+2\sqrt{x}+1}{8\sqrt{x+\sqrt{x+\sqrt{x}}} \cdot \sqrt{x^2+x\sqrt{x}}}.$

(3) Because the base of the logarithmic function is variable, the function can be transformed into

$$y = \dfrac{\ln \sin x}{\ln x}.$$

Therefore,

$$y' = \dfrac{\cot x \ln x - \dfrac{1}{x}\ln \sin x}{\ln^2 x} = \dfrac{x\cos x \ln x - \sin x \ln \sin x}{x \sin x \ln^2 x}.$$

(4) $y' = a^a x^{a^a - 1} + a^{x^a} \cdot \ln a \cdot (x^a)' + a^{a^x} \cdot \ln a \cdot (a^x)'$

$= a^a x^{a^a - 1} + ax^{a-1} a^{x^a} \ln a + a^x a^{a^x} \ln^2 a.$

Example 17 Assume $f(x) = \begin{cases} x, & x<0, \\ \ln(1+x), & x \geq 0. \end{cases}$ Find $f'(x)$.

Solution When calculating the derivative of the piecewise function, the derivatives on each continuous interval can be found by the general derivative rules, but the derivatives at the piecewise points only be obtained by the definition of derivative.

When $x<0$, $f'(x)=1$. When $x>0$,
$$f'(x)=[\ln(1+x)]'=\frac{1}{1+x}(1+x)'=\frac{1}{1+x}.$$

When $x=0$,
$$f'_-(0)=\lim_{x\to 0^-}\frac{x-\ln(1+0)}{x}=1,$$
$$f'_+(0)=\lim_{x\to 0^+}\frac{\ln(1+x)-\ln(1+0)}{x}=1,$$

thus $f'(0)=1$.

Therefore,
$$f'(x)=\begin{cases} 1, & x\leqslant 0, \\ \dfrac{1}{1+x}, & x>0. \end{cases}$$

Example 18 If $f(u)$ is derivable, calculate the derivative of the function $y=f(\sec x)$.

Solution $y'=[f(\sec x)]'=f'(\sec x)\cdot(\sec x)'=f'(\sec x)\sec x\tan x.$

Exercises 2-2

1. Find the derivatives of the following functions:

   (1) $y=2x^3-\dfrac{1}{x^2}+5x+3$;

   (2) $y=x^2\cos x$;

   (3) $y=\dfrac{1}{\sqrt[3]{x}}+\dfrac{\pi}{2}$;

   (4) $y=3^x+x^4+e^2$;

   (5) $y=\dfrac{1}{x+\sin x}$;

   (6) $y=x\ln x+\dfrac{\ln x}{x}$;

   (7) $y=\left(\dfrac{1}{x}-x^2\right)\left(x-\dfrac{1}{x^2}\right)$;

   (8) $y=\dfrac{x}{\sqrt{4-x^2}}$;

   (9) $y=\dfrac{1}{1+\sqrt{t}}-\dfrac{1}{1-\sqrt{t}}$;

   (10) $y=e^x(\sin x-\cos x)$;

   (11) $y=\dfrac{\sin x+\cos x}{2^x}$;

   (12) $y=\dfrac{\sin x-\cos x}{\tan x}$.

2. Find the derivatives of the following functions at given points:

   (1) $s=\ln(1+a^{-2t})$, find $s'(0)$;

   (2) $y=\dfrac{1-\sqrt{x}}{1+\sqrt{x}}$, find $y'\big|_{x=4}$;

   (3) $y=\dfrac{\cos x}{2x-1}$, find $y'\big|_{x=\frac{\pi}{2}}$.

3. Given the function $f(x)=(ax+b)\sin x+(cx+d)\cos x$. In order to make $f'(x)=x\cos x$, what value should a, b, c, d take?

4. The line passing through the point $(0,-3)$ and the point $(5,-2)$ is tangent to the

curve $y=\dfrac{c}{x+1}$. Find the constant c.

5. Find the derivatives of the following functions:

(1) $y=\arcsin x^2 - xe^{x^2}$;

(2) $y=x\arccos\dfrac{x}{2}$;

(3) $y=\text{arccot}\sqrt{x^3-2x}$;

(4) $y=\dfrac{1}{4}\ln\dfrac{1+x}{1-x}+\dfrac{1}{2}\arctan x$.

6. Find the derivatives of the following functions:

(1) $y=\text{ch}(\text{sh } x)$;

(2) $y=(\text{sh } x)e^{\text{ch } x}$;

(3) $y=\text{th}(1-x^2)$;

(4) $y=\arctan\text{th } x$;

(5) $y=\ln\text{ch } x+\dfrac{1}{2\text{ch}^2 x}$.

7. Suppose $\varphi(x)$ is a derivable function, find the derivatives of the following functions:

(1) $f(x)=\varphi(x^2)$;

(2) $f(x)=\varphi(e^x)\cdot e^{\varphi(x)}$;

(3) $f(x)=x\varphi\left(\dfrac{1}{x}\right)$;

(4) $f(x)=\varphi[\varphi(x)]$.

8. Try to prove:

(1) The derivative of a derivable odd function is an even function. The derivative of a derivable even function is an odd function.

(2) The derivative of a derivable periodic function is still a periodic function, and the periods are the same.

2.3 Higher-order derivatives

2.3.1 The concept of higher order derivative

We know that the velocity of a body at time t is the derivative of the displacement function $s=s(t)$ at t, that is, $v(t)=s'(t)$. Moreover, the acceleration $a(t)$ is the derivative of the velocity function $v=v(t)$ with respect to time t, hence

$$a(t)=\lim_{\Delta t\to 0}\dfrac{\Delta v}{\Delta t}=v'(t)=\dfrac{\text{d}}{\text{d}t}\left(\dfrac{\text{d}s}{\text{d}t}\right).$$

We call $a(t)=\dfrac{\text{d}}{\text{d}t}\left(\dfrac{\text{d}s}{\text{d}t}\right)$ the second order derivative of $s(t)$ with respect to t, denoted as

$$a(t)=s''(t) \quad \text{or} \quad \dfrac{\text{d}^2 s}{\text{d}t^2}.$$

Suppose that the function $f(x)$ is derivable on interval I. If its derivative $f'(x)$ is also derivable on interval I, then $f(x)$ is called second derivable on I and the derivative of $f'(x)$ is called the second derivative of $f(x)$ and denoted by y'', $f''(x)$, $\dfrac{\text{d}^2 y}{\text{d}x^2}$ or $\dfrac{\text{d}^2 f}{\text{d}x^2}$, namely

$$y''=\dfrac{\text{d}y'}{\text{d}x}=\lim_{\Delta x\to 0}\dfrac{f'(x+\Delta x)-f'(x)}{\Delta x}.$$

Similarly, the derivative of the second order derivative of $f(x)$ is called the third derivative of $f(x)$. The derivative of the third derivative of $f(x)$ is called the fourth derivative of $f(x)$ …… In general, the derivative of the $n-1$ order derivative of $f(x)$ is call the n order derivative of $f(x)$. They are respectively denoted as

$$y''', f'''(x), \frac{d^3 y}{dx^3}, \frac{d^3 f}{dx^3}; y^{(4)}, f^{(4)}(x), \frac{d^4 y}{dx^4}, \frac{d^4 f}{dx^4}; \cdots; y^{(n)}, f^{(n)}(x), \frac{d^n y}{dx^n}, \frac{d^n f}{dx^n}.$$

The second order derivative and derivatives of order higher than second order are all called higher order derivatives. Commonly, $f'(x)$ is said to be the first order derivative of $f(x)$, and $f(x)$ itself is said to be the zero order derivative of $f(x)$ denoted as $f(x) = f^{(0)}(x)$.

Example 1 Find the nth order derivative of the exponential function $y = a^x$ ($a > 0$, $a \neq 1$).

Solution $y' = a^x \ln a$, $y'' = a^x \ln^2 a$, $y''' = a^x \ln^3 a$, $\cdots$, $y^{(n)} = a^x \ln^n a$, that is
$$(a^x)^{(n)} = a^x \ln^n a. \tag{1}$$

Specially, when $a = e$,
$$(e^x)^{(n)} = e^x. \tag{2}$$

Example 2 Find the nth order derivative of the sine function $y = \sin x$.

Solution
$$y' = \cos x = \sin\left(x + \frac{\pi}{2}\right),$$

$$y'' = \cos\left(x + \frac{\pi}{2}\right) = \sin\left(x + 2 \cdot \frac{\pi}{2}\right),$$

$$y''' = \cos\left(x + 2 \cdot \frac{\pi}{2}\right) = \sin\left(x + 3 \cdot \frac{\pi}{2}\right).$$

Generally, by means of mathematical induction, we can obtain
$$y^{(n)} = (\sin x)^{(n)} = \sin\left(x + n \cdot \frac{\pi}{2}\right). \tag{3}$$

Similarly we get
$$(\cos x)^{(n)} = \cos\left(x + n \cdot \frac{\pi}{2}\right). \tag{4}$$

Example 3 Find the nth order derivative of the logarithmic function $y = \ln(1+x)$.

Solution $y' = \dfrac{1}{1+x}$, $y'' = \dfrac{-1}{(1+x)^2}$, $y''' = \dfrac{(-1) \cdot (-2)}{(1+x)^3} = \dfrac{(-1)^2 2!}{(1+x)^3}$.

Generally, we have
$$y^{(n)} = (-1)^{n-1} \frac{(n-1)!}{(1+x)^n}.$$

That is
$$[\ln(1+x)]^{(n)} = (-1)^{n-1} \frac{(n-1)!}{(1+x)^n}. \tag{5}$$

Example 4 Find the nth order derivative of the power function $y = x^\alpha$ (α is a real number).

Solution
$$y' = \alpha x^{\alpha-1},$$
$$y'' = \alpha(\alpha-1) x^{\alpha-2},$$
$$y''' = \alpha(\alpha-1)(\alpha-2) x^{\alpha-3}.$$

Generally, we have
$$y^{(n)} = \alpha(\alpha-1)(\alpha-2) \cdots (\alpha-n+1) x^{\alpha-n},$$

that is
$$(x^\alpha)^{(n)} = \alpha(\alpha-1)(\alpha-2) \cdots (\alpha-n+1) x^{\alpha-n}. \tag{6}$$

Specially, when $\alpha = n$,
$$(x^n)^{(n)} = n(n-1)(n-2)\cdots 2 \cdot 1 = n!,$$
$$(x^n)^{(n+1)} = 0.$$

Example 5 Assume $y = xf(1-\cos x)$, $f''(x)$ both exist. Find $\dfrac{d^2 y}{dx^2}$.

Solution
$$\frac{dy}{dx} = f(1-\cos x) + xf'(1-\cos x) \cdot (1-\cos x)'$$
$$= f(1-\cos x) + x\sin x f'(1-\cos x).$$
$$\frac{d^2 y}{dx^2} = f'(1-\cos x) \cdot (1-\cos x)' + (x\sin x)' f'(1-\cos x) + x\sin x f''(1-\cos x)(1-\cos x)'$$
$$= \sin x f'(1-\cos x) + (\sin x + x\cos x) f'(1-\cos x) + x\sin^2 x f''(1-\cos x)$$
$$= (2\sin x + x\cos x) f'(1-\cos x) + x\sin^2 x f''(1-\cos x).$$

2.3.2 The linear property and Leibniz formula

Suppose that the functions $u(x)$, $v(x)$ are both derivable of order n. Then we have
$$(u \pm v)^{(n)} = u^{(n)} \pm v^{(n)}, \tag{7}$$
$$(Cu)^{(n)} = Cu^{(n)} \ (C \text{ is a constant number}). \tag{8}$$

The situation of $(u \cdot v)^{(n)}$ is complex.
$$(u \cdot v)' = u'v + uv',$$
$$(u \cdot v)'' = u''v + 2u'v' + uv'',$$
$$(u \cdot v)''' = u'''v + 3u''v' + 3u'v'' + uv'''.$$

It is obvious that, if we consider the orders of derivation as the exponent of a power function, then the right side of the equation is the binomial expansion formula. Therefore, we obtain the following formula by according to the mathematical induction.
$$(u \cdot v)^{(n)} = u^{(n)}v + C_n^1 u^{(n-1)}v' + C_n^2 u^{(n-2)}v'' + \cdots + C_n^k u^{(n-k)}v^{(k)} + \cdots + uv^{(n)},$$
that is
$$(u \cdot v)^{(n)} = \sum_{k=0}^{n} C_n^k u^{(n-k)} v^{(k)}, \tag{9}$$
where $C_n^k = \dfrac{n(n-1)\cdots(n-k+1)}{k!} = \dfrac{n!}{k!(n-k)!}$, Equation (9) is called the Leibniz formula.

By Leibniz formula, we can easily calculate the higher derivative.

Example 6 Find the tenth order derivative of the function $y = x^3 e^x$.

Solution Denote $u = e^x$, $v = x^3$ then
$$u^{(k)} = (e^x)^{(k)} = e^x, \ k = 0, 1, \cdots, 10,$$
$$v^{(k)} = (x^3)^{(k)} = 0, \ k = 4, 5, \cdots, 10.$$

By means of the Leibniz formula, we have
$$(u \cdot v)^{(10)} = (e^x)^{(10)} x^3 + C_{10}^1 (e^x)^{(9)} (x^3)' + C_{10}^2 (e^x)^{(8)} (x^3)'' + C_{10}^3 (e^x)^{(7)} (x^3)'''.$$

Therefore,
$$(x^3 e^x)^{(10)} = x^3 e^x + 30x^2 e^x + 45 \times 6x e^x + 720 e^x = (x^3 + 30x^2 + 270x + 720) e^x.$$

Example 7 Find the 50th order derivative of function $y = x^2 \sin x$.

Solution Denote $u = \sin x$, $v = x^2$, then

$$u^{(k)} = (\sin x)^{(k)} = \sin\left(x + k\frac{\pi}{2}\right), \quad k = 0, 1, 2, \cdots, 50,$$

$$v^{(k)} = (x^2)^{(k)} = 0, \quad k = 3, 4, \cdots, 50.$$

By using Leibniz formula, we obtain

$$y^{(50)} = (x^2 \sin x)^{(50)} = (\sin x)^{(50)} x^2 + C_{50}^1 (\sin x)^{(49)} \cdot 2x + C_{50}^2 (\sin x)^{(48)} \cdot 2$$

$$= x^2 \sin\left(x + 50 \times \frac{\pi}{2}\right) + 50 \times 2x \sin\left(x + 49 \times \frac{\pi}{2}\right) + 50 \times 49 \sin\left(x + 48 \times \frac{\pi}{2}\right)$$

$$= -x^2 \sin x + 100x \cos x + 2\ 450 \sin x.$$

Exercises 2-3

1. Find the second order derivatives of the following functions:
   (1) $y = xe^{-x^2}$;
   (2) $y = x^2 \ln x$;
   (3) $y = (1 + x^2) \arctan x$;
   (4) $y = \dfrac{e^x}{x}$;
   (5) $y = \ln(x + \sqrt{1 + x^2})$;
   (6) $y = e^{2x} \sin(2x + 1)$.

2. Find the third order derivatives of the following functions:
   (1) $y = x^2 \sin 2x$;
   (2) $y = (x^2 + a^2) \arctan \dfrac{x}{a}$.

3. Find the nth order derivatives of the following functions $y^{(n)}$:
   (1) $y = \dfrac{2(x+1)}{x^2 + 2x - 3}$;
   (2) $y = \ln \dfrac{a + bx}{a - bx}$.

4. Suppose that $f''(x)$ exists. Find the second order derivatives of the following functions:
   (1) $y = f(\sin^2 x)$;
   (2) $y = \ln[f(x^2)]$.

5. Assume $x = \varphi(y)$ is the inverse function of $y = f(x)$. $f'(x) \neq 0$, $f'''(x)$ both exist. Prove
   (1) $\varphi''(y) = -\dfrac{f''(x)}{[f'(x)]^3}$;
   (2) $\varphi'''(y) = \dfrac{3[f''(x)]^2 - f'(x) f'''(x)}{[f'(x)]^5}$.

6. Find the derivatives of given order:
   (1) $y = x^2 e^{3x}$, $y^{(10)}$;
   (2) $y = e^x \cos x$, $y^{(4n)}$;
   (3) $y = x \operatorname{sh} x$, $y^{(100)}$.

7. Prove $(\sin^4 x + \cos^4 x)^{(n)} = 4^{n-1} \cos\left(4x + \dfrac{n\pi}{2}\right)$.

2.4 Derivation of implicit functions and parametric equations

2.4.1 Derivation of implicit functions

In applied problems we often need to investigate a class of functions in which the dependent variable y and the independent variable x is determined by an equation of the form $F(x, y) = 0$. If there exists a function $y = f(x)$ defined on interval I, such that $F(x, f(x)) \equiv 0$, then $y = f(x)$ is called an implicit function defined by the equation $F(x, y) = 0$.

Some functions are defined implicitly by a relation between x and y, such as $e^{xy} + xy +$

$x+y-1=0$ and $\sin(xy)+\ln(x+y^2)+1=0$. It is not easy to solve these equations for y explicitly as a function of x.

Fortunately we don't need to solve an equation for y in terms of x in order to find the derivative of y. Instead we can use the method of implicit function derivation.

Suppose that $y=f(x)$ is a function defined implicitly by an equation $F(x,y)=0$ on the interval I. If we substitute $y=f(x)$ into the equation $F(x,y)=0$, then we have an identity $F(x,f(x))\equiv 0$ for all $x \in I$. We regard $F(x,f(x))$ as a composite function and take the derivative of both sides of the identity $F(x, f(x))\equiv 0$ with respect to x. Using the derivation rule for composite functions, we obtain the derivative $\dfrac{dy}{dx}$.

Note Implicit differentiation steps:

Step 1 Differentiable both sides of the equation with respect to x, treating y as a differentiable function of x;

Step 2 Collect the terms with $\dfrac{dy}{dx}$ on one side of the equation;

Step 3 Factor out $\dfrac{dy}{dx}$;

Step 4 Solve for $\dfrac{dy}{dx}$.

Example 1 Suppose that the implicit function $y=y(x)$ is defined by the equation
$$e^{xy}+xy+x+y-1=0.$$
Find $y', y'\big|_{x=0}$.

Solution Take the derivative of both sides of the equation $e^{xy}+xy+x+y-1=0$ with respect to x and obtain
$$e^{xy}(y+xy')+y+xy'+1+y'=0,$$
$$(xe^{xy}+x+1)y'=-(ye^{xy}+y+1),$$
thus,
$$y'=-\frac{ye^{xy}+y+1}{xe^{xy}+x+1}.$$

When $x=0$, $y=0$, we have $y'\big|_{x=0}=-1$.

Example 2 Suppose that the function $y=y(x)$ is defined by the equation
$$\sin(xy)+\ln(x+y^2)-1=0.$$
Find y'.

Solution Take the derivative of both sides of the equation $\sin(xy)+\ln(x+y^2)-1=0$ with respect to x. We obtain
$$\cos(xy)(y+xy')+\frac{1}{x+y^2}(1+2yy')=0,$$
$$\left[x\cos(xy)+\frac{2y}{x+y^2}\right]y'=-\left[y\cos(xy)+\frac{1}{x+y^2}\right],$$

that is,
$$y' = -\frac{y\cos(xy) + \frac{1}{x+y^2}}{x\cos(xy) + \frac{2y}{x+y^2}} = -\frac{(x+y^2)y\cos(xy) + 1}{(x+y^2)x\cos(xy) + 2y}.$$

Example 3 Find the equation of tangent line of the curve $xy + \ln y = 1$ at the point $M(1,1)$.

Solution Take the derivative of both sides of the equation $xy + \ln y = 1$ with respect to x, we obtain
$$y + xy' + \frac{1}{y}y' = 0,$$
that is,
$$y' = -\frac{y^2}{xy+1}.$$

At the point $M(1,1)$,
$$y'\bigg|_{\substack{x=1\\y=1}} = -\frac{1^2}{1\times 1+1} = -\frac{1}{2}.$$

So, the equation of the tangent line at the point $M(1,1)$ is $y-1 = -\frac{1}{2}(x-1)$, namely,
$$x + 2y - 3 = 0.$$

Example 4 Assume the function $y = y(x)$ is defined by the equation $e^x - e^y = xy$. Find $y''\big|_{x=0}$.

Solution Take the derivative of both sides of the equation with respect to x,
$$e^x - e^y y' = y + xy'.$$
Therefore,
$$y' = \frac{e^x - y}{e^y + x}.$$

From the given equation, we get $y = 0$ when $x = 0$. Substitute $x = 0$ and $y = 0$ into the equation y', we get $y'\big|_{x=0} = 1$. To find the second order derivative of the implicit function, take the derivatives of both sides of the above equation $e^x - e^y y' = y + xy'$ again with respect to x. Notice that y' is also a function of x, we get
$$y'' = \left(\frac{e^x - y}{e^y + x}\right)' = \frac{(e^x - y')(e^y + x) - (e^x - y)(e^y y' + 1)}{(e^y + x)^2}.$$

Substitute $x = 0$, $y = 0$ and $y'\big|_{x=0} = 1$ into the function y'', we obtain
$$y''\big|_{x=0} = -2.$$

2.4.2 Logarithmic derivation method

So far, we no longer use the definition of derivative for calculating the derivative of the function, but use the basic derivation rules and derivative formulas. However, for some complicated explicit function, $y = [\varphi(x)]^{\psi(x)}$ ($\varphi(x) > 0$), we can't find the derivative only from derivation rules. For this kind of function, we first take logarithm on both sides and then use the method of implicit derivation to find its derivative. This method is called the logarithmic derivation method.

The following examples are used to illustrate this method.

Example 5 Find the derivative of the function $y=x^x\ (x>0)$.

Solution Take logarithm on both sides and we have
$$\ln y = x\ln x.$$
Take derivatives of both sides with respect to x, then we obtain
$$\frac{1}{y}y' = \ln x + x \cdot \frac{1}{x} = 1 + \ln x,$$
therefore,
$$y' = y(1+\ln x) = x^x(1+\ln x).$$

Example 6 Let $(\cos y)^x = (\sin x)^y$, find y'.

Solution Take logarithm on both sides
$$x\ln\cos y = y\ln\sin x.$$
Take derivations of both sides of the equation with respect to x, we have
$$\ln\cos y - x\frac{\sin y}{\cos y}y' = y'\ln\sin x + y\frac{\cos x}{\sin x}.$$
That is
$$y' = \frac{\ln\cos y - y\cot x}{x\tan y + \ln\sin x}.$$

Example 7 Find the derivative of $y = \sqrt[4]{\dfrac{x(x-1)}{\sqrt[3]{x-2}}}$.

Solution Take logarithm on both sides (assume $x>2$), we have
$$\ln y = \frac{1}{4}\left[\ln x + \ln(x-1) - \frac{1}{3}\ln(x-2)\right].$$
Take derivations of both sides of the equation will respect to x, we obtain
$$\frac{1}{y}y' = \frac{1}{4}\left(\frac{1}{x} + \frac{1}{x-1} - \frac{1}{3}\cdot\frac{1}{x-2}\right),$$
therefore,
$$y' = \frac{1}{4}y\left(\frac{1}{x} + \frac{1}{x-1} - \frac{1}{3}\cdot\frac{1}{x-2}\right)$$
$$= \frac{1}{4}\sqrt[4]{\frac{x(x-1)}{\sqrt[3]{x-2}}}\left[\frac{1}{x} + \frac{1}{x-1} - \frac{1}{3(x-2)}\right].$$

2.4.3 Derivation of parametric equations

In practical problems, the dependent variable and independent variable are defined by the third parameter. That is to say, the function $y=f(x)$ where x and y are defined by the parametric equations
$$\begin{cases} x=\varphi(t), \\ y=\psi(t), \end{cases} \alpha \leqslant t \leqslant \beta,$$
where t is the parameter variable. For example, if the frictional damping of air is omitted, the locus of motion of a projectile may be represented by a parametric equation:
$$\begin{cases} x=v_1 t, \\ y=v_2 t - \dfrac{1}{2}gt^2, \end{cases} \tag{1}$$
where, v_1, v_2 are respectively the horizontal and vertical initial velocity of the projectile, g denotes the gravitational acceleration, t represents the time, and x, y are the abscissa and

ordinate of the projectile in the coordinate plane (Figure 2-8).

Eliminate the parameter t from Equation (1), we get
$$y = \frac{v_2}{v_1}x - \frac{g}{2v_1^2}x^2.$$

Figure 2-8

It is called the function defined by the parametric equations.

Generally, the function determined by the parametric equations
$$\begin{cases} x = \varphi(t), \\ y = \psi(t), \end{cases} \alpha \leqslant t \leqslant \beta, \qquad (2)$$
is called the function defined by the parametric equations.

If the parametric equations are very complicated, we can't get an expression of the function $y = f(x)$ by means of elimination. In this case, how to find the derivative of parametric equations?

Assume in Equation (2), $x = \varphi(t)$, $y = \psi(t)$ are both derivable. $\varphi'(t) \neq 0$ and $x = \varphi(t)$ has monotonous inverse function $t = \varphi^{-1}(x)$. Then, we have
$$y = \psi[\varphi^{-1}(x)] = f(x).$$
According to the chain rule, we have
$$\frac{dy}{dx} = \frac{dy}{dt} \cdot \frac{dt}{dx}.$$
According to the derivative rule of inverse function, we get
$$\frac{dt}{dx} = \frac{1}{\frac{dx}{dt}} = \frac{1}{\varphi'(t)}.$$

Therefore,
$$\frac{dy}{dx} = \frac{dy}{dt} \cdot \frac{1}{\frac{dx}{dt}} = \frac{\psi'(t)}{\varphi'(t)}. \qquad (3)$$

If $x = \varphi(t)$, $y = \psi(t)$ have second order derivatives, according to the derivation rule of the parametric equations, since $\frac{dy}{dx} = \frac{\psi'(t)}{\varphi'(t)}$ is a function of t, thus $\frac{d^2y}{dx^2}$ can also be taken as the derivative of the parametric equations. Then we have,
$$\frac{d^2y}{dx^2} = \frac{\frac{d}{dt}\left(\frac{dy}{dx}\right)}{\frac{dx}{dt}} = \frac{\frac{d}{dt}\left[\frac{\psi'(t)}{\varphi'(t)}\right]}{\varphi'(t)} = \frac{\varphi'(t)\psi''(t) - \psi'(t)\varphi''(t)}{\varphi'^2(t)} \cdot \frac{1}{\varphi'(t)},$$

that is,
$$\frac{d^2y}{dx^2} = \frac{\varphi'(t)\psi''(t) - \psi'(t)\varphi''(t)}{\varphi'^3(t)}. \qquad (4)$$

Similarly, if $\varphi(t)$, $\psi(t)$ have the third derivatives, we can get
$$\frac{d^3y}{dx^3} = \frac{\frac{d}{dt}\left(\frac{d^2y}{dx^2}\right)}{\frac{dx}{dt}}.$$

Example 8 Find the equation of the tangent line of the ellipse $\begin{cases} x = a\cos t, \\ y = b\sin t, \end{cases} 0 \leqslant t \leqslant 2\pi$

at the point $t=\dfrac{\pi}{4}$ (Figure 2-9).

Solution When $t=\dfrac{\pi}{4}$, the coordinate (x_0, y_0) of the point M_0 is

$$x_0=a\cos\dfrac{\pi}{4}=\dfrac{a}{\sqrt{2}},\quad y_0=b\sin\dfrac{\pi}{4}=\dfrac{b}{\sqrt{2}}.$$

Figure 2-9

The slope of the tangent line at M_0 is

$$\left.\dfrac{dy}{dx}\right|_{t=\frac{\pi}{4}}=\left.\dfrac{(b\sin t)'}{(a\cos t)'}\right|_{t=\frac{\pi}{4}}=\left.\dfrac{b\cos t}{-a\sin t}\right|_{t=\frac{\pi}{4}}=-\dfrac{b}{a}.$$

Therefore, the tangent line is

$$y-\dfrac{b}{\sqrt{2}}=-\dfrac{b}{a}\left(x-\dfrac{a}{\sqrt{2}}\right),$$

that is

$$bx+ay-\sqrt{2}ab=0.$$

Example 9 Suppose that a circle of radius is a is rolling along the x-axis. A is a fixed point on the circumference. At the very beginning, the point A is at the origin O. Assume the parameter is θ (Figure 2-10). Then the trajectory equation of the point A is

$$\begin{cases}x=a(\theta-\sin\theta),\\ y=a(1-\cos\theta).\end{cases}$$

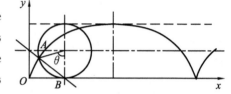

Figure 2-10

The curve is called the cycloid. Prove the normal line of the cycloid at $A(x,y)$ pass through the tangent point B of the circle and x-axis.

Proof The slope of the tangent line of cycloid at $A(x, y)$ is

$$\dfrac{dy}{dx}=\dfrac{[a(1-\cos\theta)]'}{[a(\theta-\sin\theta)]'}=\dfrac{\sin\theta}{1-\cos\theta}.$$

Thus, the slope of the normal line is $-\dfrac{1-\cos\theta}{\sin\theta}$ and the equation is

$$y-a(1-\cos\theta)=-\dfrac{1-\cos\theta}{\sin\theta}[x-a(\theta-\sin\theta)].$$

Let $y=0$, we obtain,

$$x=a(\theta-\sin\theta)+a(1-\cos\theta)\cdot\dfrac{\sin\theta}{1-\cos\theta}=a\theta.$$

The abscissa of the point B is $OB=\overset{\frown}{AB}=a\theta$. So the normal line of the cycloid at $A(x,y)$ passes through B.

Example 10 Assume that the function $y=y(x)$ is defined by the parametric equations $\begin{cases}x=\ln(1+t^2),\\ y=\arctan t.\end{cases}$ Find $\dfrac{d^2y}{dx^2},\dfrac{d^3y}{dx^3}$.

Solution

$$\dfrac{dy}{dx}=\dfrac{y'(t)}{x'(t)}=\dfrac{\dfrac{1}{1+t^2}}{\dfrac{2t}{1+t^2}}=\dfrac{1}{2t},$$

$$\frac{d^2 y}{dx^2} = \frac{d}{dt}\left(\frac{dy}{dx}\right) \cdot \frac{dt}{dx} = \frac{\left(\frac{1}{2t}\right)'}{[\ln(1+t^2)]'} = \frac{-\frac{1}{2t^2}}{\frac{2t}{1+t^2}} = -\frac{1+t^2}{4t^3},$$

$$\frac{d^3 y}{dx^3} = \frac{d}{dt}\left(\frac{d^2 y}{dx^2}\right) \cdot \frac{dt}{dx} = \frac{-\frac{1}{4}\left(\frac{1}{t^3}+\frac{1}{t}\right)'}{[\ln(1+t^2)]'}$$

$$= -\frac{\frac{1}{4}\left(\frac{-3}{t^4}-\frac{1}{t^2}\right)}{\frac{2t}{1+t^2}} = \frac{(1+t^2)(3+t^2)}{8t^5} = \frac{3+4t^2+t^4}{8t^5}.$$

Example 11 Let $\begin{cases} x=te^t, \\ e^t+e^y=2. \end{cases}$ Find $\left.\dfrac{d^2 y}{dx^2}\right|_{t=0}$.

Solution Since $x'_t = e^t(t+1)$ and $e^t + e^y y'_t = 0$, we have $y'_t = -e^{t-y}$. Then

$$\frac{dy}{dx} = \frac{y'_t}{x'_t} = \frac{-e^{t-y}}{e^t(1+t)} = \frac{-e^{-y}}{1+t},$$

$$\frac{d^2 y}{dx^2} = \left(-\frac{e^{-y}}{1+t}\right)'_t \cdot \frac{1}{x'_t} = \frac{-1[-e^{-y}y'_t(1+t)-e^{-y}]}{(1+t)^2 \cdot e^t(t+1)}.$$

When $t=0$, $x=0$, $y=0$, we get

$$\left.x'_t\right|_{t=0} = \left.e^t(t+1)\right|_{t=0} = 1,$$

$$\left.y'_t\right|_{t=0} = \left.-e^{t-y}\right|_{t=0} = -1.$$

Therefore, $\left.\dfrac{d^2 y}{dx^2}\right|_{t=0} = \dfrac{-1[-1(-1) \cdot 1 - 1]}{1^2 \cdot 1 \cdot 1} = 0.$

2.4.4 Derivative of the function expressed by the polar equation

If the polar equation of the given curve is

$$r = r(\theta). \tag{5}$$

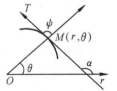

Figure 2-11

Using the relation between the rectangular coordinates and polar coordinates $x = r\cos\theta$, $y = r\sin\theta$, we get the parametric equation of curve (5) is

$$\begin{cases} x = r(\theta)\cos\theta, \\ y = r(\theta)\sin\theta, \end{cases}$$

where the polar angle is the parameter θ (Figure 2-11).

The slope of the tangent line to the curve is

$$\tan\alpha = \frac{dy}{dx} = \frac{y'(\theta)}{x'(\theta)} = \frac{r'(\theta)\sin\theta + r(\theta)\cos\theta}{r'(\theta)\cos\theta - r(\theta)\sin\theta} = \frac{r'(\theta)\tan\theta + r(\theta)}{r'(\theta) - r(\theta)\tan\theta}. \tag{6}$$

Example 12 Find the tangent and normal line equations of the curve $r = a\sin 2\theta$ (called the four-leaf rose line, Figure 2-12) at $\theta = \dfrac{\pi}{4}$.

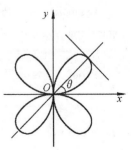

Figure 2-12

Solution Turn the polar equation of the curve into the parametric equations

$$\begin{cases} x = r(\theta)\cos\theta = a\sin 2\theta\cos\theta, \\ y = r(\theta)\sin\theta = a\sin 2\theta\sin\theta, \end{cases}$$

where the polar angle θ is the parameter. Then the slope of the tangent line is

$$k_1 = \frac{dy}{dx}\bigg|_{\theta=\frac{\pi}{4}} = \frac{2a\cos 2\theta\sin\theta + a\sin 2\theta\cos\theta}{2a\cos 2\theta\cos\theta - a\sin 2\theta\sin\theta}\bigg|_{\theta=\frac{\pi}{4}} = -1.$$

Thus the slope of the normal line is

$$k_2 = -\frac{1}{k_1} = 1.$$

When $\theta = \frac{\pi}{4}$, the coordinates of the point of tangency are

$$x = \frac{a}{\sqrt{2}},\ y = \frac{a}{\sqrt{2}}.$$

Thus the equation of the tangent line is

$$y - \frac{a}{\sqrt{2}} = -1\left(x - \frac{a}{\sqrt{2}}\right).$$

That is,
$$x + y = \sqrt{2}a.$$

The equation of the normal line is

$$y - \frac{a}{\sqrt{2}} = x - \frac{a}{\sqrt{2}},$$

that is,
$$x - y = 0.$$

Assume the angle of the tangent MT of the point $M(r,\theta)$ on the curve (5) and the line OM between the cut point and the pole is $\psi(0 \leqslant \psi \leqslant \pi)$. α is the angle of the polar axis OA making counterclockwise rotation to the tangent MT. Then we have $\psi = \alpha - \theta$ (Figure 2-11). And we have

$$\tan\psi = \tan(\alpha - \theta) = \frac{\tan\alpha - \tan\theta}{1 + \tan\alpha\tan\theta}.$$

Substitute $\tan\alpha = \frac{dy}{dx}$ into Equation (6) and the above equation, we have

$$\tan\psi = \frac{r(\theta)}{r'(\theta)}. \tag{7}$$

Therefore, it is convenient to use the angle ψ to determine the position of the tangent when the polar coordinate equation of the curve is given.

Example 13 Prove the angle between the polar radius of the logarithmic spiral $r = e^{m\theta}$ ($m > 0$) and the tangent line is a constant (Figure 2-13).

Proof $r' = me^{m\theta}$, substitute r and r' into Equation (7), we have

$$\tan\psi = \frac{r(\theta)}{r'(\theta)} = \frac{e^{m\theta}}{me^{m\theta}} = \frac{1}{m}.$$

Therefore, the angle ψ between the polar radius of the logarithmic spiral and the tangent line is a constant.

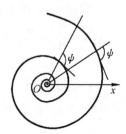

Figure 2-13

Exercises 2-4

1. Find the derivatives $\dfrac{dy}{dx}$ of the implicit functions determined by the following equations:

   (1) $e^y = \sin(x+y)$;
   (2) $\sqrt{x} + \sqrt{y} = \sqrt{a}$;
   (3) $y = xe^y + \cos(xy)$;
   (4) $\arctan \dfrac{y}{x} = \ln\sqrt{x^2 + y^2}$.

2. Find the derivatives of the following functions by the logarithmic derivation method:

   (1) $y = \sqrt{\dfrac{3x-2}{(5-2x)(x-1)}}$;
   (2) $y = \dfrac{(2x+3)\sqrt[4]{x-6}}{\sqrt[3]{x+1}}$;
   (3) $y = (x + \sqrt{1+x^2})^x$;
   (4) $y = \sqrt{x\sin x \sqrt{1-e^x}}$.

3. Find the derivatives of the functions defined by the following parametric equations:

   (1) $\begin{cases} x = \sqrt{1+\theta}, \\ y = \sqrt{1-\theta}; \end{cases}$
   (2) $\begin{cases} x = e^{-t}, \\ y = e^t; \end{cases}$
   (3) $\begin{cases} x = \dfrac{1}{1+t}, \\ y = \dfrac{t}{1+t}. \end{cases}$

4. Let $\begin{cases} x = f(t) - \pi, \\ y = f(e^{3t} - 1), \end{cases}$ where f is derivable, and $f'(0) \neq 0$. Find $\left.\dfrac{dy}{dx}\right|_{t=0}$.

5. Find the equations of tangent line and normal line of the following curves at the specific points:

   (1) $\begin{cases} x = \ln \sin t, \\ y = \cos t, \end{cases} t = \dfrac{\pi}{2}$;
   (2) $\begin{cases} x = \dfrac{3t}{1+t^2}, \\ y = \dfrac{3t^2}{1+t^2}, \end{cases} t = \dfrac{\pi}{2}$.

6. Find the derivatives $\dfrac{d\rho}{d\theta}$ of the functions expressed by the following polar equations:

   (1) $\begin{cases} \theta = \omega t, \\ \rho = \rho_0 + vt; \end{cases}$
   (2) $\begin{cases} \theta = \tan \alpha - \alpha, \\ \rho = R\sec \alpha. \end{cases}$

7. Find the second order derivatives $\dfrac{d^2 y}{dx^2}$ of the following implicit functions:

   (1) $y = 1 + xe^y$;
   (2) $y = \cos(x+y)$;
   (3) $y = \tan(x+y)$;
   (4) $e^{x+y} = xy$.

8. Find the second order derivatives $\dfrac{d^2 y}{dx^2}$ of the following functions defined by the following parametric equations:

   (1) $\begin{cases} x = t, \\ y = 4t^2 + 1; \end{cases}$
   (2) $\begin{cases} x = \sqrt{1+\theta^2}, \\ y = \sqrt{1-\theta^2}; \end{cases}$

(3) $\begin{cases} x=\ln(1+t^2), \\ y=\arctan t; \end{cases}$ (4) $\begin{cases} x=f'(t), \\ y=tf'(t)-f(t), \end{cases}$ where $f''(t)$ exists and is not 0.

9. Find the third order derivatives $\dfrac{d^3 y}{dx^3}$ of the functions defined by the following parametric equations:

(1) $\begin{cases} x=t^2, \\ y=t^3-t; \end{cases}$ (2) $\begin{cases} x=1-e^\theta, \\ y=\theta-e^{-\theta}. \end{cases}$

10. Prove the function defined by the parametric equations $x=e^t\sin t$, $y=e^t\cos t$ meets the following equation

$$y''(x+y)^2=2(xy'-y).$$

2.5 Differential of the function

In the theoretical study and practical application, we often encounter such a problem: when the independent variable x has the increment Δx, find the increment $\Delta y=f(x+\Delta x)-f(x)$ of the function $y=f(x)$. However, for some complicated functions it is not easy to find difference $f(x+\Delta x)-f(x)$. If we try to express Δy as a linear function of Δx, thus the complex problem is mapped into a simple one. In this section, it will be shown that the essential idea behind the differential is to approximate a linear one in a small interval. This is a fundamental ideal in calculus and also in many science and engineering applications.

2.5.1 Concept of the differential

Cited Example 1 A square sheet of metal is affected by the changes of temperature. When the length of side changes from x_0 to $x_0+\Delta x$, find the increment of area under the influence of temperature.

Denote the length of the side by x and area by S, then $S=x^2$. The change of the area can be viewed as the increment ΔS of the function $S=x^2$ corresponding to the increment Δx of independent variable x. Obviously,

$$\Delta S=(x_0+\Delta x)^2-x_0^2=2x_0\Delta x+(\Delta x)^2.$$

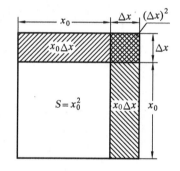

Figure 2-14

It can be seen from the above equation that, ΔS is divided into two parts: one term $2x_0\Delta x$ is a linear function of Δx, which is the sum of areas of two rectangles drawn oblique lines as shown in Figure 2-14. The second term $(\Delta x)^2$ in Figure 2-14 is the area of the small square with crossed slash. When $\Delta x\to 0$, it is an infinitesimal of higher order than Δx. Therefore, the increment of the area ΔS can be approximately replaced by the first term $2x_0\Delta x$.

Cited Example 2 Assume that for the function $y=x^3$, the increment of independent variable at x_0 is Δx. Find the increment of the function Δy.

Because, $\Delta y=(x_0+\Delta x)^3-x_0^3=3x_0^2\Delta x+3x_0(\Delta x)^2+(\Delta x)^3,$

as $\Delta x\to 0$, $3x_0(\Delta x)^2+(\Delta x)^3$ is the higher order infinitesimal of Δx. Therefore,

$\Delta y \approx 3x_0^2 \Delta x.$

> **Definition** Let the function $y=f(x)$ be defined in the neighborhood of x_0. If the increment
> $$\Delta y = f(x_0+\Delta x)-f(x_0)$$
> of the function $y=f(x)$ can be expressed as
> $$\Delta y = A\Delta x + o(\Delta x), \qquad (1)$$
> where, A is a constant independent on Δx, then the function $y=f(x)$ is called differentiable at x_0 and $A\Delta x$ is called differential of the function $y=f(x)$ at x_0, denoted by $\mathrm{d}y$, that is,
> $$\mathrm{d}y = A\Delta x \quad \text{or} \quad \mathrm{d}f(x)=A\Delta x. \qquad (2)$$

2.5.2 The necessary and sufficient condition for differential

From the definition of differential, we can see that the differential of the function f at x_0, $A\Delta x$, is the linear and main part of its increment Δy on the interval $[x_0, x_0+\Delta x]$ (or $[x_0-\Delta x, x_0]$). Now we want to know what is the condition for a function to have a differential and what is the value of the constant A?

> **Theorem 1** The necessary and sufficient condition for the function $y=f(x)$ to be differentiable at x_0 is that $f(x)$ is derivable at x_0, and $\mathrm{d}y=f'(x_0)\Delta x$.

Proof Necessity. Because $f(x)$ is differentiable at x_0, then we have
$$\Delta y = f(x_0+\Delta x)-f(x_0) = A \cdot \Delta x + o(\Delta x).$$
Divided by Δx on both sides, and let $\Delta x \to 0$, we obtain
$$\lim_{\Delta x \to 0} \frac{\Delta y}{\Delta x} = A.$$
Therefore $f(x)$ is derivable at x_0, and $A=f'(x_0)$. Thus,
$$\mathrm{d}y = f'(x_0)\Delta x.$$

Sufficiency. Because $f(x)$ is derivative at x_0. According to Equation (12) in Section 2.1.5, we know that
$$\Delta y = f'(x_0)\Delta x + \alpha \Delta x,$$
where $\lim_{\Delta x \to 0} \alpha = 0$, thus $\lim_{\Delta x \to 0} \frac{\alpha \Delta x}{\Delta x}=0$, namely, $\alpha \Delta x = o(\Delta x)$. According to the definition of differential, $y=f(x)$ is differentiable at x_0, and
$$\mathrm{d}y = f'(x_0)\Delta x.$$

This theorem indicates that, for a function of one variable, differentiability is equivalent to derivability, and we may use the two terms interchangeably in the calculus of one variable. The method of finding the differential and the method of derivation will both be called differentiation.

If f is differentiable at every point on the interval I, then f is said to be differentiable

on I, and denoted as
$$dy = f'(x)\Delta x. \tag{3}$$
Specially, when $y = f(x) = x$, $f'(x) = 1$, then we have
$$dy = dx = 1 \cdot \Delta x, \Delta x = dx.$$
Thus the differential of the function $f(x)$ can be written as
$$dy = f'(x)dx, \tag{4}$$
that is,
$$\frac{dy}{dx} = f'(x).$$

That is to say, the derivative of the function $y = f(x)$ can be regarded as a quotient of the differential of the function and the differential of the independent variable.

Note Although differential and derivative have a close relationship with each other, they have obvious differences: derivative is the change rate of the function at a point. And differential is the main part of the function increment caused by the increment of the independent variable; the derivative value is only related to x, while the differential is not only related to x, but also the increment Δx, and $|\Delta x|$ may not be very small.

2.5.3 Geometric meaning of differential

In Figure 2-15, the graph of the function $y = f(x)$ is a curve and the derivative $f'(x_0)$ represents the slope of the tangent line to the curve at the point $M_0(x_0, f(x_0))$, that is $\tan\theta$.

Therefore, we have
$$NM = f(x_0 + \Delta x) - f(x_0) = \Delta y,$$
$$NT = \tan\theta \cdot \Delta x = f'(x_0)\Delta x = dy,$$
$$TM = \Delta y - dy = \alpha\Delta x = o(\Delta x).$$

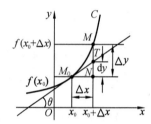

Figure 2-15

That is to say, the differential of a function $y = f(x)$ at x_0 represents the increment of the ordinate of the tangent line to the curve $y = f(x)$ at the corresponding point M_0.

When $|\Delta x| \ll 1$, if replace approximately the increment Δy by the differential $dy = f'(x_0)\Delta x$, we have
$$f(x_0 + \Delta x) \approx f(x_0) + f'(x_0)\Delta x. \tag{5}$$

Geometrically, this means replacing approximately the curve $\widehat{M_0 M}$ by the tangent line $M_0 T$ in the neighborhood of the point M_0. This is called local linearization and is a very important and useful idea in science and engineering.

Denote $x_0 + \Delta x = x$, that is $\Delta x = x - x_0$, then we have
$$f(x) \approx f(x_0) + f'(x_0)(x - x_0). \tag{6}$$

Specially, when $x_0 = 0, \Delta x = x - x_0 = x$. According to Equation (6), we obtain
$$f(x) \approx f(0) + f'(0)x. \tag{7}$$

Example 1 Assume the function $y = x^3$. Find:

(1) the differential of the function;

(2) the differential of the function at $x = 2$;

(3) the increment Δy, differential dy, and $\Delta y - dy$ at $x = 2$ when $\Delta x = 0.01$.

Solution (1) $dy = y'dx = 3x^2 dx$;

(2) $dy\Big|_{x=2} = 3x^2\Big|_{x=2} dx = 12dx$;

(3) $dy\Big|_{\substack{x=2 \\ \Delta x=0.01}} = 3x^2 dx\Big|_{\substack{x=2 \\ \Delta x=0.01}} = 12 \times 0.01 = 0.12$,

$$\Delta y = (2+0.01)^3 - 2^3 = 0.120\ 601,$$

$$\Delta y - dy = 0.000\ 601.$$

It is obvious that using dy to approximately replace Δy can make the absolute error less than 10^{-3}.

2.5.4 Differential rules of elementary functions

Because $dy = f'(x)dx$, we know that the differential of the function can be obtained by multiplying dx by the derivative of the function. Similarly, the rules for rational operations on differentials can also be obtained from the rules of rational operations on derivatives.

> **Theorem 2** If $u = u(x)$, $v = v(x)$ are differentiable.
> (1) $d(u \pm v) = du \pm dv$;
> (2) $d(uv) = udv + vdu$, especially when $v = C$ (constant), we have $d(Cu) = Cdu$;
> (3) $d\left(\dfrac{u}{v}\right) = \dfrac{vdu - udv}{v^2}$ $(v \neq 0)$.

According to every derivative rule, there is a corresponding differential rule obtained from the former by multiplying dx. We illustrate the major rules in the following table.

Derivative rules	Differential rules
$(C)' = 0$	$dC = 0$
$(x^\mu)' = \mu x^{\mu-1}$	$d(x^\mu) = \mu x^{\mu-1} dx$
$(a^x)' = a^x \ln a$	$d(a^x) = a^x \ln a dx$
$(e^x)' = e^x$	$d(e^x) = e^x dx$
$(\log_a x)' = \dfrac{1}{x \ln a}$	$d(\log_a x) = \dfrac{1}{x \ln a} dx$
$(\ln x)' = \dfrac{1}{x}$	$d(\ln x) = \dfrac{1}{x} dx$
$(\sin x)' = \cos x$	$d(\sin x) = \cos x dx$
$(\tan x)' = \sec^2 x$	$d(\tan x) = \sec^2 x dx$
$(\cot x)' = -\csc^2 x$	$d(\cot x) = -\csc^2 x dx$
$(\sec x)' = \sec x \tan x$	$d(\sec x) = \sec x \tan x dx$
$(\csc x)' = -\csc x \cot x$	$d(\csc x) = -\csc x \cot x dx$
$(\arcsin x)' = \dfrac{1}{\sqrt{1-x^2}}$	$d(\arcsin x) = \dfrac{1}{\sqrt{1-x^2}} dx$

Continued

Derivative rules	Differential rules
$(\arccos x)' = -\dfrac{1}{\sqrt{1-x^2}}$	$d(\arccos x) = -\dfrac{1}{\sqrt{1-x^2}}dx$
$(\arctan x)' = \dfrac{1}{1+x^2}$	$d(\arctan x) = \dfrac{1}{1+x^2}dx$
$(\operatorname{arccot} x)' = -\dfrac{1}{1+x^2}$	$d(\operatorname{arccot} x) = -\dfrac{1}{1+x^2}dx$
$(\operatorname{sh} x)' = \operatorname{ch} x$	$d(\operatorname{sh} x) = \operatorname{ch} x dx$
$(\operatorname{ch} x)' = \operatorname{sh} x$	$d(\operatorname{ch} x) = \operatorname{sh} x dx$
$(\operatorname{th} x)' = \dfrac{1}{\operatorname{ch}^2 x}$	$d(\operatorname{th} x) = \dfrac{1}{\operatorname{ch}^2 x}dx$
$(\operatorname{arsh} x)' = \dfrac{1}{\sqrt{1+x^2}}$	$d(\operatorname{arsh} x) = \dfrac{1}{\sqrt{1+x^2}}dx$
$(\operatorname{arch} x)' = \dfrac{1}{\sqrt{x^2-1}}$	$d(\operatorname{arch} x) = \dfrac{1}{\sqrt{x^2-1}}dx$
$(\operatorname{arth} x)' = \dfrac{1}{1-x^2}$	$d(\operatorname{arth} x) = \dfrac{1}{1-x^2}dx$

Example 2 Find dy if $y = e^x \sin x$.

Solution Let $u = e^x$, $v = \sin x$, then
$$dy = d(u \cdot v) = udv + vdu = e^x d(\sin x) + \sin x d(e^x)$$
$$= e^x \cos x dx + e^x \sin x dx = e^x (\sin x + \cos x) dx,$$
or
$$y' = (e^x \sin x)' = e^x \sin x + e^x \cos x = e^x (\sin x + \cos x).$$
Therefore,
$$dy = y' dx = e^x (\sin x + \cos x) dx.$$

Next we discuss the differential rule of composite function.

Let $y = f(u)$, $u = \varphi(x)$ are both derivable. Then the derivative of the composite function $y = f[\varphi(x)]$ is
$$\frac{dy}{dx} = f'[\varphi(x)]\varphi'(x),$$
hence its differential is
$$dy = f'[\varphi(x)]\varphi'(x)dx.$$
We notice that
$$f'[\varphi(x)] = f'(u), du = \varphi'(x)dx,$$
therefore,
$$dy = f'(u)du.$$

That is to say, whether u is an independent variable or a function of another variable, the differential of the function has the same form as $dy = f'(u)du$. This property is called the invariance of the differential form.

The invariance of the differential form is of great significance in both theory and calculation.

Example 3 Assume $y = \ln \cos x^2$. Find dy.

Solution

$$dy = \frac{1}{\cos x^2} d(\cos x^2) = \frac{1}{\cos x^2}(-\sin x^2) d(x^2)$$

$$= -\tan x^2 \cdot 2x dx = -2x \tan x^2 dx.$$

Example 4 Suppose that $y = y(x)$ is defined by $\sin(xy) + \ln(x+y^2) = 1$, find y' by using the invariance of the differential form.

Solution

$$d[\sin(xy)] + d[\ln(x+y^2)] = 0,$$

$$\cos(xy) d(xy) + \frac{1}{x+y^2} d(x+y^2) = 0,$$

$$\cos(xy)(x dy + y dx) + \frac{1}{x+y^2}(dx + 2y dy) = 0,$$

$$\left[x\cos(xy) + \frac{2y}{x+y^2}\right] dy + \left[y\cos(xy) + \frac{1}{x+y^2}\right] dx = 0,$$

$$dy = -\frac{y\cos(xy) + \frac{1}{x+y^2}}{x\cos(xy) + \frac{2y}{x+y^2}} dx.$$

Therefore, $\quad y' = -\dfrac{y\cos(xy) + \dfrac{1}{x+y^2}}{x\cos(xy) + \dfrac{2y}{x+y^2}} = -\dfrac{(x+y^2)y\cos(xy) + 1}{(x+y^2)x\cos(xy) + 2y}.$

2.5.5 Application of the differential

1) Linear approximate computation

We have seen that a curve lies very close to its tangent line near the point of tangency. In fact, by zooming in toward a point on the graph of a differentiable function, we noticed that the graph looks more and more like its tangent line. This observation is the basis for a method of finding approximate values of functions. From the geometric meaning of the differential, we have

$$f(x) \approx f(x_0) + f'(x_0)(x - x_0).$$

This is the basis of the solutions to all the examples that follows.

Example 5 There is a ball with radius 1 m. When the radius increases 1 cm, find the increment of the volume of the ball.

Solution The volume of the ball is $V = \dfrac{4}{3}\pi r^3$. By the approximate equality, we obtain

$$\Delta V \approx dV = 4\pi r^2 \Delta r.$$

Since $r = 1$ m, $\Delta r = 0.01$ m, substitute them into the above equation, we shall have

$$\Delta V \approx 4\pi \times 1^2 \times 0.01 \approx 0.13 \text{ m}^3.$$

Example 6 Find an approximate value of $\sqrt{325}$.

Solution Because $\sqrt{325} = \sqrt{324+1} = 18\sqrt{1 + \dfrac{1}{324}}.$

Let $f(x)=\sqrt{1+x}$, then we have $f(0)=1$, $x=\dfrac{1}{324}$.

$$f(x)\approx f(0)+f'(0)x=1+\dfrac{1}{2\sqrt{1+x}}\bigg|_{x=0}\times x=1+\dfrac{1}{2}x,$$

$$\sqrt{1+\dfrac{1}{324}}\approx 1+\dfrac{1}{2}\times\dfrac{1}{324}=\dfrac{649}{648}.$$

Therefore, $\sqrt{325}\approx 18\times\dfrac{649}{648}\approx 18.03.$

We list some useful first degree approximations as follows

$$\sin x\approx x;\quad \arcsin x\approx x;$$
$$\tan x\approx x;\quad \arctan x\approx x;$$
$$e^x\approx 1+x;\quad \ln(1+x)\approx x;$$
$$\sqrt[n]{1+x}\approx 1+\dfrac{1}{n}x.$$

Example 7 Find an approximate value of $\sin 1°$.

Solution $1°=\dfrac{\pi}{180}$ rad. According to $\sin x\approx x$, we have

$$\sin 1°=\sin\dfrac{\pi}{180}\approx\dfrac{\pi}{180}\approx 0.017\ 5.$$

Example 8 (Mass-energy conversion) The statement of Newton's second law of motion $F=\dfrac{d}{dt}(mv)=m\dfrac{dv}{dt}=ma$ assumes that the mass is a constant (namely, fixed), but strictly speaking, this is not correct. Because the mass of the object increases with the growth of its speed. In the formula modified by Einstein, the mass is $m=\dfrac{m_0}{\sqrt{1-\dfrac{v^2}{c^2}}}$, in which the "static mass m_0" means the mass of the object when it is not moving no quality, and c is the speed of light, about 3×10^8 m/s. By approximation

$$\dfrac{1}{\sqrt{1-x^2}}\approx 1+\dfrac{1}{2}x^2.$$

Estimate the increment of the mass Δm after considering the speed v.

Solution When v is quite small compared to c, $\dfrac{v^2}{c^2}$ approaches zero. Therefore, we can make use of $\dfrac{1}{\sqrt{1-v^2/c^2}}\approx 1+\dfrac{1}{2}\left(\dfrac{v^2}{c^2}\right)$, safely.

Then we have

$$m=\dfrac{m_0}{\sqrt{1-\dfrac{v^2}{c^2}}}\approx m_0\left[1+\dfrac{1}{2}\left(\dfrac{v^2}{c^2}\right)\right]=m_0+\dfrac{1}{2}m_0v^2\left(\dfrac{1}{c^2}\right),$$

namely, $$m\approx m_0+\dfrac{1}{2}m_0v^2\left(\dfrac{1}{c^2}\right).$$

The equation above represents the increment of the mass after considering the speed.

In Newton's physics, $\frac{1}{2}m_0 v^2$ is the kinetic energy of the object (E_k). If we rewrite $m \approx m_0 + \frac{1}{2}m_0 v^2 \left(\frac{1}{c^2}\right)$ into the form of $(m-m_0)c^2 \approx \frac{1}{2}m_0 v^2$, then we shall have

$$(m-m_0)c^2 \approx \frac{1}{2}m_0 v^2 = \Delta E_k,$$

or
$$(\Delta m)c^2 \approx \Delta E_k.$$

In other words, the change of the kinetic energy when the speed changes from 0 to v is ΔE_k, which is approximately equal to $(\Delta m)c^2$. Because $c = 3 \times 10^8$ m/s, $\Delta E_k \approx 9 \times 10^{16} \Delta m$ J.

Thus, small quality changes can create large energy changes.

2) Error estimation

Differential can be used to solve the error estimation problem in measurement problems. In measurement problems, the size of some amount can be measured directly, and these data is called the direct measurement data, while some amount is calculated according to the formula, the data of which is called the indirect measurement data. For example, the ball diameter D can be measured by vernier caliper directly (it is the direct measurement data), and the volume of it is calculated by the formula $V = \frac{\pi}{6}D^3$ (it is the indirect measurement data).

Because of the precision of the instrument, measurement conditions and measurement methods can be affected by many factors. The direct measurement data will have errors, and indirect measurement data calculated according to it will also have errors, which is called the indirect error.

Assume the direct measurement data is a and the precise value is A. Then we call $|A-a|$ the absolute error of a, and the ratio $\frac{|A-a|}{|a|}$ is called the relative error of a.

In fact, because the exact value is unknown, it is often impossible to know $|A-a|$. The upper bound of $|A-a|$ can only be estimated by the case, and is called the absolute error limit or absolute error of a, recorded as δ_A, namely,

$$|A-a| \leqslant \delta_A.$$

$\frac{\delta_A}{|a|}$ is called the relative error limit or relative error of a.

Assume x is got from direct measurement. y is calculated from the formula $y = f(x)$. Now we calculate the error of y caused by the error of x. If the absolute error of x is δ_x, that is

$$|\Delta x| \leqslant \delta_x,$$

then when $y' \neq 0$,

$$|\Delta y| \approx |dy| = |f'(x)||\Delta x| \leqslant |f'(x)|\delta_x.$$

Namely, the absolute error of y is,

$$\delta_y = |f'(x)|\delta_x,$$

thus the relative error of y is

$$\frac{\delta_y}{|y|} = \frac{|f'(x)|}{|f(x)|}\delta_x.$$

Example 9 Assume the diameter of the circle plate is $x = 5.2$ cm and the absolute error of it is $\delta_x = 0.05$ cm. Try to estimate the absolute and relative errors of the area of the circle plate.

Solution According to the area formula $y = \frac{\pi}{4}x^2$, we know that the absolute error of the area is

$$\delta_y = |y'| \cdot \delta_x = \frac{\pi}{2}x\delta_x = \frac{\pi}{2} \times 5.2 \times 0.05 \approx 0.41 \text{ cm}^2,$$

and the relative error is

$$\frac{\delta_y}{|y|} = \frac{\frac{\pi}{2}x\delta_x}{\frac{\pi}{4}x^2} = 2\frac{\delta_x}{|x|} = 2 \times \frac{0.05}{5.2} \approx 0.0192 \approx 2\%.$$

Example 10 (Open blocked arteries) In the late 1830s, the French physiologist Jean Poiseuille found the rule of the relationship between expending the radius of the partly blocked artery and recovering the normal flow of the blood. His research formula $V = kr^4$ shows that the volume V that flows through the capillary at a fixed pressure in the time unit is equal to a constant multiplying the fourth power of the radius r. Try to find how much the effect is on V when the radius r increases by 10%?

Solution The relationship of the differentials of r and V and is expressed by the function

$$dV = \frac{dV}{dr}dr = 4kr^3 \, dr.$$

The relative change of V is

$$\frac{dV}{V} = \frac{4kr^3 \, dr}{kr^4} = 4\frac{dr}{r}.$$

The relative change of V is four times of that of r, so 10% increase of r will cause 40% increase of the flow.

Exercises 2-5

1. Given $y = x^2 + 1$, find Δy and dy at $x = 3$ and $\Delta x = 0.1, 0.01, -0.2$ respectively. Calculate the values of Δy and dy.

2. Find the differentials of the following functions:

(1) $y = x\ln x$;

(2) $y = \frac{a}{x} + \arctan\frac{a}{x}$;

(3) $y = e^{-x}\cos(3-x)$;

(4) $y = \arctan\frac{1-x^2}{1+x^2}$;

(5) $y = e^{\sin^2 x}$;

(6) $y = 5^{\ln \tan x}$;

(7) $y = x\arccos\sqrt{1-x^2}$;

(8) $y = f\left(\arctan\frac{1}{x}\right)$, where f is derivable.

3. Assume $y=y(x)$ is defined by the following equations.

(1) $xy+y^2-\cos y=1$, find dy;

(2) $x^3+y^3-\sin 3x+6y=0$, find $dy\big|_{x=0}$.

4. Assume u,v are both differentiable functions of x. Find dy.

(1) $y=uv^{-2}$; (2) $y=\arctan\dfrac{u}{v}$;

(3) $y=\ln[\sin(u^2+v^2)]$; (4) $y=f(uv)$, f is derivable.

5. Find the approximate values of the following functional values:

(1) $\ln 1.002$; (2) $\sqrt[3]{997}$;

(3) $\sin 29°$.

6. When $|x|$ is small enough, prove the following approximation formulas:

(1) $\sin x \approx x$; (2) $\arcsin x \approx x$;

(3) $\tan x \approx x$; (4) $\arctan x \approx x$;

(5) $\ln(1+x) \approx x$; (6) $e^x \approx 1+x$;

(7) $(1+x)^a \approx 1+ax$.

7. Find the approximate values of the following functional values:

(1) $\sin 0.02$; (2) $\arcsin 0.01$;

(3) $e^{0.05}$; (4) $\arctan 0.03$.

8. Assume the center angle of the fan is $\alpha=60°$. The radius is $r=100$ cm. If r does not change, α decreases by $0.5°$. What is the increment of the area of the fan? If α does not change, r increases 1 cm, what is the increment of the area of the fan?

9. According to the electrical knowledge, the relationship between electric power P, voltage U, and resistance R is $P=\dfrac{U^2}{R}$. There is an electric furnace whose rated power is $P=1$ kW, receiving the supply voltage of $U=220$ V. When the power fluctuation is less than 5%, how much the electrical power deviates from the rated power?

10. The calculation of the volume V of the ball is accurate to 1%. What is the relative error of the radius R of the ball obtained from the volume?

2.6 The application of derivative in economic analysis

2.6.1 Marginal analysis

1) The marginal concept

In economic problems, we often use the concept of change rate, and includes average change rate and instantaneous change rate. Average change rate is the ratio of the increment of the function and the increment of independent variable, for example, the change rate of annual output, cost, profit, etc.. And the instantaneous change rate is the derivative of the function with respect to the independent variable, which is the limit of the average change rate when the increment of the independent variable approaches zero.

Assume the function $y=f(x)$ is derivable at x_0, then the average change rate of the function at $(x_0, x_0+\Delta x)$ is

$$\frac{\Delta y}{\Delta x} = \frac{f(x_0+\Delta x)-f(x_0)}{\Delta x}.$$

It represents the average change rate of the function $y=f(x)$ at $(x_0, x_0+\Delta x)(\Delta x>0)$. And the instantaneous change rate of the function at $x=x_0$ is

$$f'(x_0) = \lim_{\Delta x \to 0} \frac{f(x_0+\Delta x)-f(x_0)}{\Delta x}.$$

In economics, it is called the marginal function value of the function $y=f(x)$ at $x=x_0$. It represents the change speed of the function at $x=x_0$.

At the point $x=x_0$, when x increases one unit from x_0, the corresponding increment Δy of y is accurately $\Delta y \big|_{\Delta x=1}^{x=x_0}$. When the change unit of x is small enough, or the ratio of the increment of x to x_0 is small enough, according to the application of differential, we get

$$\Delta y \bigg|_{\Delta x=1}^{x=x_0} \approx dy \bigg|_{\Delta x=1}^{x=x_0} = f'(x)\Delta x \bigg|_{\Delta x=1}^{x=x_0} = f'(x_0).$$

When $\Delta x = -1$, it means that x decreases from x_0 one unit.

It indicates that when x changes one unit at $x=0$, the function $y=f(x)$ approximately changes $f'(x_0)$ units.

Definition 1 Assume the function $f(x)$ is derivable, the derivative $f'(x)$ is called the marginal function of $f(x)$. $f'(x_0)$ is called the value of the marginal function.

The meaning of the marginal function value $f'(x_0)$: when x changes one unit at $x=x_0$, $f(x)$ changes $f'(x_0)$ units.

Example 1 Assume $y=2x^2$, find and interpret the marginal function value of y at $x=5$.

Solution $y'=4x$, $y'\big|_{x=5}=20$. The marginal function value of y at $x=5$ is 20. It represents that x increases (or decreases) one unit at $x=5$, and y (approximately) increases (or decreases) 20 units.

2) The common marginal functions in economics

(1) Marginal cost

The derivative of the total cost function $C(Q)$

$$C'(Q) = \lim_{\Delta Q \to 0} \frac{C(Q+\Delta Q)-C(Q)}{\Delta Q}$$

is called the marginal cost, where Q is the number of units of a product produced in some time interval.

Sometimes it is desirable to carry out marginal analysis relative to average cost (cost per unit).

The derivative of average cost $\overline{C}(Q)$ is

$$\overline{C}'(Q) = \left[\frac{C(Q)}{Q}\right]' = \frac{QC'(Q)-C(Q)}{Q^2},$$

which is called the marginal average cost.

Generally, the total cost $C(Q)$ is the sum of the fixed cost C_0 and the variable cost $C_1(Q)$, that is
$$C(Q)=C_0+C_1(Q).$$
Therefore, the marginal cost is
$$C'(Q)=[C_0+C_1(Q)]'=C_1'(Q).$$
Obviously, the marginal cost is irrelevant to the fixed cost.

According to the meaning of the marginal cost and concept of derivative, we have the following conclusions:

① The marginal cost is just related to the variable cost, unrelated the fixed cost.

② The price of some product is P,

if $C'(x)<P$, then the company can continue to increase the production;

if $C'(x)>P$, then the company should not keep increasing the production, but improve the quality of products, improve the ex-factory price or lower the cost of production.

Example 2 The total cost of producing Q units of some product is given by the cost funtion
$$C(Q)=200+5Q+\frac{1}{20}Q^2.$$
Calculate the total cost, average cost and marginal cost when $Q=20$, and interpret the results of marginal cost.

Solution According to the total cost function $C(Q)=200+5Q+\frac{1}{20}Q^2$, we have
$$C(20)=320,\ \bar{C}(20)=\frac{C(20)}{20}=16,$$
$$C'(20)=C'(Q)\bigg|_{Q=20}=\left(200+5Q+\frac{1}{20}Q^2\right)'\bigg|_{Q=20}=\left(5+\frac{1}{10}Q\right)\bigg|_{Q=20}=7.$$
The economic meaning of marginal cost $C'(20)=7$ is: at a production level of 20 units, the total production costs are increasing at the rate of 7 per unit of the product.

Example 3 The total cost of producing Q units of some product is $C(Q)=1\ 100+\frac{Q^2}{1\ 200}$.

① Find the total cost and average cost of producing 900 units;

② Find the average change rate of the total cost when producing 900 to 1 000 units;

③ Use marginal cost to approximate the cost of producing the 901st product.

Solution ① The total cost of producing 900 units is
$$C(900)=1\ 100+\frac{900^2}{1\ 200}=1\ 775.$$
The average cost is
$$\bar{C}(900)=\frac{C(900)}{900}\approx 1.97.$$

② The change rate of the total cost when producing 900 to 1000 units is
$$\frac{\Delta C(Q)}{\Delta Q}=\frac{C(1\,000)-C(900)}{1\,000-900}=\frac{1\,933-1\,775}{100}=1.58.$$

③ The marginal cost function
$$C'(Q)=\frac{2Q}{1\,200}=\frac{Q}{600},$$
evaluated at $Q=900$, approximates the cost of producing the 901st unit: $C'(900)=1.5$.

At a production level of 900 units of product, the total production costs are increasing at the rate of 1.5 per product.

(2) Marginal profit

The derivative of the total profit function $R(Q)$
$$R'(Q)=\lim_{\Delta Q\to 0}\frac{\Delta R}{\Delta Q}=\lim_{\Delta Q\to 0}\frac{R(Q+\Delta Q)-R(Q)}{\Delta Q},$$
is called the marginal profit. Marginal profit is the instantaneous rate of change of profit relative to production at a given production level.

Assume the price is P, and P is a function of Q. Then, $R(Q)=PQ=QP(Q)$. Thus the marginal profit is
$$R'(Q)=P(Q)+QP'(Q).$$

Example 4 Assume the demand function of some product is $P=20-\frac{Q}{5}$, where P is the price and Q is the number of units of a product.

① Find the total profit, average profit and marginal profit when Q is 15.

② Find the average change rate of the profit when Q increases from 15 to 20.

Solution ① The total profit function is
$$R(Q)=QP(Q)=20Q-\frac{Q^2}{5}.$$

The average profit is
$$\bar{R}(Q)=\frac{R(Q)}{Q}=20-\frac{Q}{5}.$$

The marginal profit is
$$R'(Q)=20-\frac{2}{5}Q.$$

When $Q=15$, the total profit is $R(15)=255$, the average profit is $\bar{R}(15)=17$, and the marginal profit is $R'(15)=14$.

② When Q changes from 15 to 20, the average change rate is
$$\frac{\Delta R}{\Delta Q}=\frac{R(20)-R(15)}{20-15}=\frac{320-255}{5}=13.$$

(3) The marginal demand

If $Q=f(P)$ is a demand function, then the derivative of the demand amount Q with respect to the price P,
$$\frac{dQ}{dP}=f'(P)=\lim_{\Delta P\to 0}\frac{f(P+\Delta P)-f(P)}{\Delta P}$$

is called the marginal demand function.

The inverse function $P = f^{-1}(Q)$ of $Q = f(P)$ is called the price function. The derivative of price with respect to demand $\dfrac{dP}{dQ}$ is called the marginal price function. According to derivative rule of inverse function, we have

$$\frac{dP}{dQ} = \frac{1}{\dfrac{dQ}{dP}} \quad \text{or} \quad [f^{-1}(Q)]' = \frac{1}{f'(P)}.$$

Example 5 The demand function of some product is $Q = Q(P) = 75 - P^2$. Find the marginal demand when $P = 5$, and explain its economic meaning.

Solution $Q'(P) = \dfrac{dQ}{dP} = -2P$. When $P = 5$, the marginal demand is

$$Q'(5) = Q'(P)\bigg|_{P=5} = -10.$$

Its economic meaning is: when $P = 5$, if the price increases (or decreases) by one unit, the demand will decrease (or increase) by ten units.

Example 6 The demand function of some product is $Q = 6\ 000 - \dfrac{P^2}{8}$. Find the marginal demand when $P = 12$, and explain its economic meaning.

Solution The marginal demand function is

$$Q'(P) = -\frac{P}{4}.$$

When $P = 12$, the marginal demand is $Q'(12) = -3$. It represents that when the price is 12, the price increases (or decreases) by one unit, the demand will decrease (or increase) by three units.

(4) Marginal profit

The derivative of the total profit function $L = L(Q)$,

$$L'(Q) = \lim_{\Delta Q \to 0} \frac{L(Q + \Delta Q) - L(Q)}{\Delta Q},$$

is called the marginal profit. It is the instantaneous rate of change of profit relative to production at a given level.

Generally, the total profit function $L(Q)$ is equal to the difference of total revenue function $R(Q)$ and the total cost function $C(Q)$, that is, $L(Q) = R(Q) - C(Q)$. Then the marginal profit is

$$L'(Q) = R'(Q) - C'(Q).$$

Marginal profit is the difference between marginal revenue and marginal cost.

Example 7 A company's market research department recommends the manufacture and marketing of a new product. After suitable test marketing, the research department presents the following price-demand equation: $Q = 200 - 2P$. In the price-demand equation, the demand Q (unit: t) is given as a function of price P (unit: Yuan). The financial department provides the cost function

$$C(Q) = 500 + 20Q.$$

where 500 Yuan is the estimate of fixed cost and 20 Yuan is the estimate of variable costs per ton. Find the marginal profit when Q is respectively 50 t, 80 t and 100 t, and explain the economic meaning.

Solution Since the total profit function is

$$L(Q) = R(Q) - C(Q) = QP(Q) - C(Q) = 80Q - \frac{1}{2}Q^2 - 500,$$

then the marginal profit is

$$L'(Q) = 80 - Q.$$

Therefore,

$$L'(50) = L'(Q)\bigg|_{Q=50} = (80 - Q)\bigg|_{Q=50} = 30;$$

$$L'(80) = L'(Q)\bigg|_{Q=80} = (80 - Q)\bigg|_{Q=80} = 0;$$

$$L'(100) = L'(Q)\bigg|_{Q=100} = (80 - Q)\bigg|_{Q=100} = -20.$$

The economic meaning of the above results is: $L'(50) = 30$ means that at a demand level of 50 t, the total profit are increasing at the rate of 30 Yuan per ton. $L'(80) = 0$ means that at a demand level of 80 t, the total profit are increasing at the rate of 0 Yuan per ton. $L'(100) = -20$ means that at a demand level of 100 t, the total profit are decreasing at the rate of 20 Yuan per ton.

We can learn from this example that, if $L'(Q) > 0$, increase one more ton on the basis of the fact that the demand is Q, the total profit will increase; if $L'(Q) < 0$, increase one more ton on the basis of the fact that the demand is Q, the total profit will decrease. Therefore, for a company, it isn't true that the larger the demand is, the greater the profit is. When does an increase in production lead to the largest increase in profit? The further discussion of this question will be continued in Chapter 4.

2.6.2 Elasticity analysis

1) The concept of elasticity

Definition 2 Assume the function $y = f(x)$ is derivable at the point $x = x_0$. The ratio $\frac{\Delta y / y_0}{\Delta x / x_0}$ of the relative change amount $\frac{\Delta y}{y_0} = \frac{f(x_0 + \Delta x) - f(x_0)}{f(x_0)}$ of the function and the relative change amount of the independent variable $\frac{\Delta x}{x_0}$ is called the average relative change rate of $f(x)$ between two points from $x = x_0$ to $x = x_0 + \Delta x$. It is also called the elasticity between the two points. When $\Delta x \to 0$, the limit of $\frac{\Delta y / y_0}{\Delta x / x_0}$ is called the relative change rate of $f(x)$ at $x = x_0$, also called the relative derivative or the elastic, recorded as $\frac{Ey}{Ex}\bigg|_{x=x_0}$ or $\frac{E}{Ex} f(x_0)$. Namely,

$$\left.\frac{Ey}{Ex}\right|_{x=x_0} = \lim_{\Delta x \to 0} \frac{\frac{\Delta y}{y_0}}{\frac{\Delta x}{x_0}} = \lim_{\Delta x \to 0} \frac{\Delta y}{\Delta x} \cdot \frac{x_0}{y_0} = f'(x_0) \frac{x_0}{f(x_0)}.$$

For a general x, if $y = f(x)$ is derivable, then we have

$$\frac{Ey}{Ex} = \lim_{\Delta x \to 0} \frac{\frac{\Delta y}{y}}{\frac{\Delta x}{x}} = \lim_{\Delta x \to 0} \frac{\Delta y}{\Delta x} \cdot \frac{x}{y} = y' \frac{x}{y}.$$

It is the function of x, called the elasticity function of $y = f(x)$.

The elasticity $\frac{E}{Ex} f(x)$ of the function $f(x)$ at the point x indicates the impact of change scope $\frac{\Delta x}{x}$ of x on that of $f(x)$, also the intensity or sensitivity of $f(x)$ responding to the change of x.

$\frac{E}{Ex} f(x_0)$ represents that at the point $x = x_0$, when x changes 1%, $f(x)$ approximately changes $\frac{E}{Ex} f(x_0) \%$. In explaining the concrete meaning of elasticity in application problems, we often ignore the word "approximately".

Example 8 Find the elasticity function $\frac{Ey}{Ex}$ of $y = x^{\alpha}$ (α is a constant), and explain its meaning.

Solution
$$y' = \alpha x^{\alpha-1},$$
$$\frac{Ey}{Ex} = y' \frac{x}{y} = \alpha x^{\alpha-1} \frac{x}{x^{\alpha}} = \alpha.$$

Its meaning is, whatever x is, when x changes 1%, the function $y = x^{\alpha}$ always changes $\alpha\%$. According to it, we know that the elasticity function of power function is a constant. So it is called the invariant elasticity function.

2) The elasticity function of common functions (a, b, c, λ are constants)

(1) The elasticity function of the constant function $f(x) = c$ is $\frac{Ec}{Ex} = 0$;

(2) The elasticity function of the linear function $f(x) = ax + b$ is $\frac{E(ax+b)}{Ex} = \frac{ax}{ax+b}$;

(3) The elasticity function of the power function $f(x) = ax^{\lambda}$ is $\frac{E(ax^{\lambda})}{Ex} = \lambda$;

(4) The elasticity function of the exponential function $f(x) = ba^{\lambda x}$ is $\frac{E(ba^{\lambda x})}{Ex} = \lambda x \ln a$;

(5) The elasticity function of the logarithmic function $f(x) = b\ln ax$ is $\frac{E(b\ln ax)}{Ex} = \frac{1}{\ln ax}$;

(6) The elasticity function of the sine function $f(x) = \sin x$ is $\frac{E(\sin x)}{Ex} = x\cot x$;

(7) The elasticity function of the cosine function $f(x)=\cos x$ is $\dfrac{E(\cos x)}{Ex}=-x\tan x$.

3) The arithmetic of elasticity

(1) $\dfrac{E[f_1(x)\pm f_2(x)]}{Ex}=\dfrac{f_1(x)\dfrac{Ef_1(x)}{Ex}\pm f_2(x)\dfrac{Ef_2(x)}{Ex}}{f_1(x)\pm f_2(x)}$;

(2) $\dfrac{E[f_1(x)f_2(x)]}{Ex}=\dfrac{Ef_1(x)}{Ex}+\dfrac{Ef_2(x)}{Ex}$;

(3) $\dfrac{E\left[\dfrac{f_1(x)}{f_2(x)}\right]}{Ex}=\dfrac{Ef_1(x)}{Ex}-\dfrac{Ef_2(x)}{Ex}$ $(f_2(x)\neq 0)$.

4) Common elasticity functions in economic analysis

(1) Elasticity of demand to price

The elasticity is a widely used concept in economic management. Often using it as a tool to analyze the economic law and economic problems. When the function in definition is the demand function $Q=f(P)$, the elasticity at this time is the elasticity of demand towards price.

Assume the demand function of a commodity is $Q=f(P)$, and is derivable at P_0. In general, $Q=f(P)$ is monotonously decreasing, ΔP and ΔQ have opposite signs. So $\dfrac{\Delta Q/Q_0}{\Delta P/P_0}$ and $\dfrac{P_0}{f(P_0)}f'(P_0)$ are both not positive numbers. In order to express the elasticity with positive numbers, we call

$$\bar{\eta}[P_0,P_0+\Delta P]=-\dfrac{P_0}{Q_0}\dfrac{\Delta Q}{\Delta P}$$

the elasticity of the commodity between two points P_0 and $P_0+\Delta P$. And we call

$$\eta\bigg|_{P=P_0}=\eta(P_0)=-\dfrac{P_0}{f(P_0)}f'(P_0)$$

the demand elasticity of the commodity at the point P_0, and the elasticity at the point P is

$$\eta=-\dfrac{P}{Q}\dfrac{dQ}{dP}=-\dfrac{P}{f(P)}f'(P),$$

which is called the demand elasticity function (simply called the demand elasticity).

Note Sometimes, the elasticity of demand towards price is also denoted as

$$E_d=\dfrac{EQ}{EP}=-\dfrac{P}{Q}\dfrac{dQ}{dP}.$$

Example 9 Assume the demand function of some product is $Q=75-P^2$. Find

① $\bar{\eta}[5,8]$;

② $\eta(P)$;

③ $\eta(3)$, $\eta(5)$ and $\eta(8)$, and explain the economic meanings.

Solution ① Known $P_0=5$, then
$$Q_0=75-P_0^2=50.$$

When $P=8$, $Q=75-P^2=11$, so $\Delta P=P-P_0=3$, $\Delta Q=Q-Q_0=-39$,

$$\overline{\eta}[5,8] = -\frac{P_0}{Q_0}\frac{\Delta Q}{\Delta P} = -\frac{5}{50} \cdot \frac{-39}{3} = 1.3.$$

It represents that the price P of the commodity increases from 5 to 8. In that domain, every time P increases 1% from 5, the demand amount Q averagely decreases 1.3% from 50.

② $\eta(P) = -\dfrac{P}{f(P)}f'(P) = -\dfrac{P}{75-P^2}(-2P) = \dfrac{2P^2}{75-P^2}$.

③ $\eta(3) = \dfrac{3}{11}$, it represents that when $P=3$, the price increases (or decreases) 1%, and the demand amount decreases (or increases) $\dfrac{3}{11}$%;

$\eta(5) = 1$, it represents that when $P=5$, the price increases (or decreases) 1%, and the demand amount decreases (or increases) 1%;

$\eta(8) = \dfrac{128}{11}$, it represents that when $P=8$, the price increases (or decreases) 1%, and the demand amount decreases (or increases) $\dfrac{128}{11}$%.

When the elastic function of a function is a constant, it is called an invariant elasticity function.

In economic analysis, it is generally believed that demand elasticity of a commodity has a direct impact on total income. According to the magnitude of demand elasticity, it can be divided into the following three cases:

Case I If the demand elasticity of a commodity is $\eta(P_0) > 1$, then we say the demand for the commodity is elastic to price, namely, the change of price will cause a big change in demand. If the price increases by 10%, the demand will decrease by more than 10%. So the total revenue decreases. Conversely, if the price decreases by 10%, the demand amount will increase by more than 10%. So the total revenue will increase. That is to say, for the elastic commodity, reducing price will make the total revenue increase, but raising price will decrease the total income.

Case II If the demand elasticity is $\eta(P_0) = 1$, then the commodity has elasticity under the price level P_0. The percentage of the increase of its price is the same as that of the decrease of demand. And the total income is unchanged whether the price raises or reduces.

Case III If the demand elasticity is $\eta(P_0) < 1$, the commodity demand under the price level P_0 which is inelastic to price. The changes of price can only cause small changes in demand. If the price increases by 10%, the demand decreases by less than 10%. So the total income increases. On the contrary, the total income decreases. For the commodities which lack flexibility, raising price will make the total income increase, and reducing price make total income decrease.

(2) Elasticity of supply to price

Assume the supply S of the commodity is the function $S = f(P)$ of the price P. Then

the definition of elasticity of the supply towards the price is
$$\frac{ES}{EP} = \frac{P}{S}\frac{dS}{dP},$$
also simply called the supply elasticity.

Example 10 Assume the supply function of some commodity is $S = -20 + 5P$. Calculate the supply elasticity function and the elasticity of supply when $P = 6$, and explain the economic meaning.

Solution The supply elasticity function is
$$\frac{ES}{EP} = \frac{P}{S}\frac{dS}{dP} = \frac{5P}{-20+5P},$$

Thus, $$\left.\frac{ES}{EP}\right|_{P=6} = \left.\frac{5P}{-20+5P}\right|_{P=6} = 3.$$

It means, when $P = 6$, the price increases (decreases) 1%, the supply amount will increase (decrease) 3%.

(3) Elasticity of the income to price

Assume the demand function of some commodity is $Q = f(P)$. Then the function of income towards price is
$$R(P) = PQ = Pf(P)$$
and the elasticity definition of the price is
$$\frac{ER}{EP} = \frac{P}{R}\frac{dR}{dP},$$
also simply called the income elasticity.

Example 11 Known the demand function of some commodity is $Q = 50 - 2P$. Calculate the income elasticity function of the commodity and the income elasticity when $P = 10$, $P = 12.5$ and $P = 15$. And explain the economic meaning.

Solution The income function is $R(P) = PQ = 50P - 2P^2$. Therefore, according to the definition, we can have the income elasticity function, which is
$$\frac{ER}{EP} = \frac{P}{R}\frac{dR}{dP} = \frac{50-4P}{50-2P}.$$

Therefore, $$\left.\frac{ER}{EP}\right|_{P=10} = \left.\frac{50-4P}{50-2P}\right|_{P=10} = \frac{1}{3};$$

$$\left.\frac{ER}{EP}\right|_{P=12.5} = \left.\frac{50-4P}{50-2P}\right|_{P=12.5} = 0;$$

$$\left.\frac{ER}{EP}\right|_{P=15} = \left.\frac{50-4P}{50-2P}\right|_{P=15} = -\frac{1}{2}.$$

The above indicates: when $P = 10$, the price increases (decreases) 1%, and the total income will increase (decrease) $\frac{1}{3}\%$. When $P = 12.5$, the price increases (decreases) 1%, the total income will not change. When $P = 15$, the price increases (decreases) 1%, and the total income will decrease (increase) 0.5%.

Example 12 Assume the demand function of some commodity is a derived function

$Q=Q(P)$, and the marginal function is $R=R(P)=PQ(P)$. Try to prove
$$\frac{EQ}{EP}+\frac{ER}{EP}=1.$$

Prove Because
$$\frac{ER}{EP}=\frac{P}{R}\frac{dR}{dP}=\frac{1}{Q(P)}[PQ(P)]'=\frac{1}{Q(P)}[Q(P)+PQ'(P)]$$
$$=1-\left[-\frac{P}{Q(P)}Q'(P)\right],$$

then, we have
$$\frac{ER}{EP}=1-\frac{EQ}{EP},$$

That is
$$\frac{EQ}{EP}+\frac{ER}{EP}=1.$$

Exercises 2-6

1. Assume the fixed cost of the product produced by some factory every month is 1 000(Yuan), and the variable cost of producing Q units products is $0.01Q^2 + 10Q$ (Yuan). If the price of every unit product is 30 Yuan. Try to find: the output when the marginal cost, the profit function, and the marginal profit is zero.

2. The total cost function and the total income function of producing some product are respectively $C(Q)=100+2Q+0.02Q^2$ (Yuan) and $R(Q)=7Q+0.01Q^2$ (Yuan). Calculate the marginal profit function and the marginal profit when the output is respectively 200 kg, 250 kg and 300 kg on that day. And explain its economic meaning.

3. Assume the cost C is the function $C(Q)=400+3Q+\frac{1}{2}Q^2$ of output Q, and demand amount is the function $P=100Q^{-\frac{1}{2}}$ of P. Find:

(1) the marginal cost, the marginal income, the marginal profit, and explain its economic meaning;

(2) the elasticity of income towards price.

4. Assume the demand function of some product is $Q=10\ 000e^{-0.02P}$. Find:

(1) the marginal demand function and the marginal demand when the price is $P=100$;

(2) the demand elasticity;

(3) the demand elasticity value when the price is $P=100$, and interpret the economic meaning.

5. Point out that in the following demand and supply functions, what is the value of the price P when the demand is high elastic or low elastic.

(1) $Q=100(2-\sqrt{P})$; (2) $P=\sqrt{a-bQ}(a,b>0)$.

6. Find the elasticity of the following functions (in which, k, a are constants).

(1) $y=kx^a$; (2) $y=e^{kx}$;

(3) $y=4-\sqrt{x}$; (4) $y=10\sqrt{9-x}$.

Summary

1. Main contents

Derivative and differential are two of the most important basic concepts in differential calculus. This chapter mainly introduces the related concepts and geometry meaning of derivative and differential and their calculation methods.

(1) The concept of derivative. The derivative of the function $y=f(x)$ at the point x_0 is the ratio limit of the increments of the function Δy and the independent variable Δx when $\Delta x \to 0$. That is to say,

$$f'(x_0)=\lim_{\Delta x \to 0}\frac{\Delta y}{\Delta x}=\lim_{\Delta x \to 0}\frac{f(x_0+\Delta x)-f(x_0)}{\Delta x}.$$

It describes the speed of change of the function—change rate when the independent variable x has an increment Δx.

If the function $f(x)$ are derivable at the point x_0, then $f(x)$ is continuous at the point x_0.

(2) The concept of the one sided function gives the sufficient and necessary condition of the fact that the function $y=f(x)$ is derivable at the point x_0 that the right and left derivatives of $f(x)$ at the point x_0 exist and are the same.

(3) According to the definition of derivative, calculate the derivative of some simple functions, and establish the four algorithms of derivative and the derivation rule of function composition and inverse function.

Assume $u(x)$, $v(x)$ are derivable at x. And then,

$$(u \pm v)'=u' \pm v'; (u \cdot v)'=u'v+uv'; \frac{u}{v}=\frac{u'v-uv'}{v^2} \ (v \neq 0).$$

Assume the function $x=\varphi(y)$ is monotonously continuous in some domain I. If $x=\varphi(y)$ is derivable at the point y, and $\varphi'(y) \neq 0$, then its inverse function $y=\varphi^{-1}(x)=f(x)$ is also derivable at the corresponding point $x \in I'$, and $f'(x)=\dfrac{1}{\varphi'(y)}$.

If $u=\varphi(x)$ is derivable at the point x, and $y=f(u)$ is derivable when $u=\varphi(x)$, then $y=f[\varphi(x)]$ is derivable at the point x, and its derivative is $\dfrac{dy}{dx}=f'(u) \cdot \varphi'(x)$, namely,

$$\frac{dy}{dx}=\frac{dy}{du} \cdot \frac{du}{dx}.$$

According to the definition and the above methods, we can have the derivative formulas of basic elementary functions and some common elementary functions.

(4) The concept of implicit function and its derivation method, logarithmic derivative method, the derivation method determined by parameter equations.

Assume the equation $F(x,y)=0$ determines the implicit function $y=f(x)$. Take derivation of both sides of the equation $F(x,y)=0$. Then we have the equation which includes y', and calculate y'. Therefore, we have the derivative of that implicit function.

For the power function $y=[\varphi(x)]^{\psi(x)}$ and the more complex irrational function, we can take logs and then calculate the derivative, which can make the derivation process more simple. This method is called the logarithmic derivative method.

If the parameter function $\begin{cases} x=\varphi(t), \\ y=\psi(t) \end{cases}$ determines that y is the function of x. $x=\varphi(t)$, $y=\psi(t)$ are all derivatives. And $\varphi'(t) \neq 0$. Then

$$\frac{dy}{dx} = \frac{dy}{dt} \cdot \frac{1}{\frac{dx}{dt}} = \frac{\varphi'(t)}{\psi'(t)}.$$

(5) The concept of related change rate and its calculation method.

(6) The concept of high-order derivative and its calculation method.

The derivative $y=f(x)$ of the derivative of $f'(x)$ is called the second order derivative of $f(x)$. Namely,

$$y'' = \frac{dy'}{dx} = \lim_{\Delta x \to 0} \frac{f'(x+\Delta x) - f'(x)}{\Delta x}.$$

The derivative of the second order derivative of $f(x)$ is called the third order derivative of $f(x)$ ······ The derivative of the $n-1$ order derivative of $f(x)$ is called the n order derivative of $f(x)$.

(7) The concept of differential, the relation between derivable and differentiable, differential form invariance, differential rule and some common differential formulas.

If $\Delta y = f(x_0 + \Delta x) - f(x_0) = A\Delta x + o(\Delta x)$, in which A is a constant which has nothing to do with Δx, then we say that $y=f(x)$ is differentiable at the point x_0. The sufficient and necessary condition of the fact that the function $y=f(x)$ is differentiable at the point x_0 is that $f(x)$ is derivable at the point x_0, and $dy=f'(x_0)dx$.

The differential describes the approximate value of the increment of the function $y=f(x)$ when the independent variable has an increment Δx.

(8) Approximate calculation and error estimation by using differential.

2. Basic requirements

(1) Understand the concept of derivative and differential, the geometric meaning of derivative, and the relation between differentiability and continuity of function.

(2) Skillfully use the basic formula of derivative. Grasp the four operations of derivative, compound function derivation rule, inverse function derivation rule. Understand the four operations of differential and the invariance of differential form.

(3) Learn to calculate the derivative of the implicit function and the function determined by parameter equation.

(4) Understand the concept of higher order derivative and can find the high order derivatives of some functions.

(5) Can find the increment of functions and the approximate value of function by differential approximation formula.

Quiz

1. Find the derivatives of the following functions:

   (1) $y=e^x(x^2-x+e^{-x})$;

   (2) $y=\sqrt{x+\sqrt[3]{x+\sqrt[3]{x}}}$;

   (3) $y=\ln\dfrac{1+\sqrt{\sin x}}{1-\sqrt{\sin x}}+2\arctan\sqrt{\sin x}$;

   (4) $y=\sqrt{t}\arccos\sqrt{t}$;

   (5) $y=x^{\sin x}$;

   (6) $y=\left(1+\dfrac{1}{x}\right)^x$.

2. Find the derivatives $\dfrac{dy}{dx}$ of the implicit functions defined by the following equations:

   (1) $y^5+xy^4+2y-x^3=1$;

   (2) $\sin(xy)-\ln\dfrac{x+1}{y}=1$.

3. Find the tangent equation and the normal line equation of the curve $2x^2-3y^2=1$ at the point $y=1$.

4. Find the derivatives $\dfrac{dy}{dx}$ of the functions defined by the following parametric equations:

   (1) $\begin{cases}x=\sqrt{t^2+1},\\ y=\dfrac{t-1}{\sqrt{t^2+1}};\end{cases}$

   (2) $\begin{cases}x=a(1+\cos\theta)\sin\theta,\\ y=a(1+\cos\theta)\cos\theta.\end{cases}$

5. Find the second order derivative $\dfrac{d^2y}{dx^2}\bigg|_{x=0}$ of the function $y=y(x)$ defined by the equation $e^x-e^y=xy$.

6. Find the second order derivative $\dfrac{d^2y}{dx^2}$ of the function $y=y(x)$ defined by the parametric equation $\begin{cases}x=a\cos^3\theta,\\ y=a\sin^3\theta.\end{cases}$

7. Assume $y=\dfrac{2-x}{(1-2x)(1+x)}$. Find $y^{(n)}$.

8. Suppose that the equation $x-y=\ln(x+y)$ is defined by the function $y=y(x)$ or $x=x(y)$. Find dy, dx and $\dfrac{dy}{dx},\dfrac{dx}{dy}$ by using the invariance of differential form.

9. The radius of a balloon contracts 2% because of air leakage. Try to find what percentage of the surface area is reduced?

10. Find the marginal function and the elasticity function of the following functions:

    (1) $y=x^2e^{-x}$;

    (2) $y=x^ae^{-b(x+c)}$ (a,b,c are constants).

11. The maximum daily capacity of a chemical plant is 1 000 tons, and total daily cost C of products (unit: Yuan) is a function of daily production x (unit: t)

    $$C=C(x)=1\,000+7x+50\sqrt{x},\ x\in[0,1\,000].$$

    (1) Find the marginal cost when the daily production is 100;

(2) Find the average unit cost when the daily production is 100.

12. The total cost C of producing products per week (unit: thousand Yuan) is a function of weekly production Q (unit: hundred items)
$$C=C(Q)=100+12Q+Q^2.$$
If the price of every 100 items is 40 000 Yuan, find the profit function and the weekly production when the marginal profit is zero.

Exercises

1. Find the derivatives of the following functions:

(1) $y=\dfrac{\sin x+\cos x}{3^x}$;

(2) $y=\ln\sqrt{\dfrac{1-\sin x}{1+\cos x}}$;

(3) $y=a^{a^x}$;

(4) $y=x^{x^x}$;

(5) $y=\log_x e+x^{\frac{1}{x}}$;

(6) $y=\begin{cases} xe^{-x^2}-1, & x\leqslant 0, \\ \sin x, & x>0. \end{cases}$

2. Assume $\varphi(x)$ is continuous at $x=a$. Is $f(x)=|x-a|\varphi(x)$ derivable at $x=a$?

3. Determine the constants a and b, such that the function
$$f(x)=\begin{cases} \arctan(ax), & x>0, \\ x^2+2x+b, & x\leqslant 0 \end{cases}$$
is continuous and derivable at $x=0$.

4. Find a and b, such that the function $f(x)=\lim\limits_{n\to\infty}\dfrac{x^2 e^{n(x-1)}+ax+b}{e^{n(x-1)}+1}$ is derivable. Find $f'(x)$.

5. Try to prove the sum of the intercepts of the two axes which is cut by tangent line at any point on the parabola $x^{\frac{1}{2}}+y^{\frac{1}{2}}=a^{\frac{1}{2}}$ is a.

6. Let $r=a(1+\cos\theta)$. Find $\dfrac{dy}{dx}$.

7. Prove that the function $y=y(x)$ defined by the parametric equations $\begin{cases} x=e^t\sin t, \\ y=e^t\cos t \end{cases}$ satisfies the following equation:
$$(x+y)^2\dfrac{d^2 y}{dx^2}=2\left(x\dfrac{dy}{dx}-y\right).$$

8. Assume $y=f(x+y)$, where $f(x)$ is second-order differentiable. Find $\dfrac{d^2 y}{dx^2}$.

9. Assume $y=(\arcsin x)^2$. Try to prove the following expression
$$(1-x^2)y^{(n+1)}-(2n-1)xy^{(n)}-(n-1)^2 y^{(n-1)}=0$$
is correct. And find $y'(0), y''(0), \cdots, y^{(n)}(0)$.

10. Try to use x to represent the differential dy of the function $y=\arctan\left(\dfrac{5}{3}\tan\dfrac{x}{2}\right)$.

11. Assume the curve equation is

$$\begin{cases} x=t^2-1, \\ y=t^3-t. \end{cases}$$

(1) Find the equation of its horizontal tangent;

(2) Prove that the two tangents are orthogonal to each other at the intersection of the curve.

12. A person is walking across a bridge which is 20 m high away from the water surface at a speed of 2 m/s. At some moment, directly below the person, there is a boat driving in the vertical direction of the bridge at the speed of $\frac{4}{3}$ m/s. Find the separation rate of the boat and the person at the end of the fifth second.

13. Known the steel expansion coefficient is 0.000 011. The total length of an accurate all-steel pendulum is 24.83 cm at 20 ℃. When the room temperature rises to 35 ℃ in summer, how much slower will the clock be every day? When the temperature reaches -10 ℃ in winter, how much faster will the clock be every day?

14. The supply function of some commodity is $S=f(P)=8+2P$. Calculate the elasticity of supply when $P=4$.

15. Assume the total income R of the commodity is the function of the sales volume Q:

$$R(Q)=104Q-0.4Q^2.$$

Find: (1) the marginal profit of the total profit when the sales volume is Q;

(2) the marginal profit of the total profit Q when the sales volume is $Q=50$;

(3) the marginal profit of the total profit when the sales volume is $Q=100$.

16. Assume the weekly demand amount Q (unit: kg) of chocolates is the function of the price P (unit: Yuan).

$$Q=f(P)=\frac{1\,000}{(2P+1)^2}.$$

Find: (1) the marginal demand amount of chocolates when $P=9.5$.

(2) The demand elasticity of chocolates when $P=10$.

Chapter 3 The mean value theorem

We know the derivative of a function reflects a local property of the function. In order to investigate the global properties of a function in an interval, we will learn some mean value theorems in this section. The mean value theorems of differential calculus include Rolle's Theorem, Lagrange's Theorem, Cauchy's Theorem and Taylor's Theorem. These theorems are the theoretical basis for many applications of the derivatives.

3.1 Mean value theorems the differential calculus

Suppose that $y = f(x)$ ($x \in [a, b]$) is a smooth continuous curve over $[a,b]$. If the functional values at the two end points a and b are equal, that is, $f(a)=f(b)$, then there is at least one point on the curve such that the tangent of the curve at the point is parallel to the x-axis. As shown in Figure 3-1, the tangent lines at P_1, P_2, P_3, P_4, P_5 are parallel to the secant line $\overline{AB}$. For convenience of explanation we begin by introducing the following lemma.

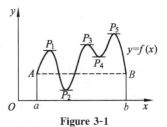

Figure 3-1

3.1.1 Fermat Lemma

Fermat Lemma Suppose that $y = f(x)$ is differentiable at x_0. If $f(x_0)$ is a local maximum (or minimal) on neighborhood $U(x_0)$ of x_0, then $f'(x_0)=0$.

Figure 3-2 shows the graph of a function f with a local maximum at P and a local minimum at Q. It appears that at the maximum and minimum points the tangent lines are horizontal and therefore each has slope 0. Fermat lemma shows that this is always true for differentiable functions.

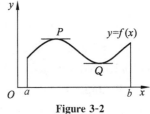

Figure 3-2

Proof Suppose that $f(x) \leqslant f(x_0)$, $\forall x \in U(x_0)$. It is known that a function $f(x)$ is derivable at x_0 if and only if the right derivative of $f(x)$ at x_0 equals the left derivative of $f(x)$ at x_0, that is, $f'_+(x_0)=f'_-(x_0)$. According to the

119

definition of derivative, we have

$$f'_+(x_0) = \lim_{x \to x_0^+} \frac{f(x)-f(x_0)}{x-x_0}, \quad \forall x \in (x_0, x_0+\delta) \subset U(x_0) \quad (\delta > 0).$$

Since $f(x) \leqslant f(x_0)$, so $\frac{f(x)-f(x_0)}{x-x_0}$, thus $f'_+(x_0) \leqslant 0$.

When $\forall x \in (x_0-\delta, x_0) \subset U(x_0)$, $\frac{f(x)-f(x_0)}{x-x_0} \geqslant 0$.

Thus, $f'_-(x_0) = \lim_{x \to x_0^-} \frac{f(x)-f(x_0)}{x-x_0} \geqslant 0$.

Because $f(x)$ is derivable at x_0, $f'_+(x_0) = f'_-(x_0) = f'(x_0)$. Therefore, $f'(x_0) = 0$.
Similarly, we can prove the case $f(x) \geqslant f(x_0)$.

3.1.2 Rolle's Theorem

In the following, we will introduce Rolle's Theorem. Let us first observe a geometric phenomenon. Suppose that $y = f(x)(x \in [a,b])$ is a smooth continuous curve over $[a,b]$ (Figure 3-3). If the functional values at the two points $x=a$, $x=b$ are equal, that is, $f(a)=f(b)$, then there is at least one point P on the curve such that the tangent of the curve at point P is parallel to the x-axis. This phenomenon is described in the following theorem.

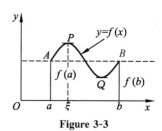

Figure 3-3

Theorem 1 (Rolle's Theorem) Suppose the function $y = f(x)$ is continuous on the closed interval $[a,b]$ and derivable in the open interval (a,b), and $f(a) = f(b)$, then there exists at least one point ξ at least in (a,b), such that $f'(\xi) = 0$.

Proof Since $y = f(x)$ is continuous on the closed interval $[a,b]$, so it have the maximum M and minimum m on $[a,b]$.

If $M = m$, $f(x)$ is a constant on $[a,b]$, then we have $f'(\xi) = 0$.

If $M > m$, then at least one of M and m is not equal to $f(a) = f(b)$. Suppose $M \neq f(a) = f(b)$, then the maximal value M is on the interval (a,b). That is to say, there exists at least one ξ in the interval (a,b), such that $f(\xi) = M$. Therefore, $\forall x \in [a,b]$, $f(x) \leqslant f(\xi)$. And because $f(x)$ is derivable in (a,b), from Fermat Lemma, we obtain

$$f'(\xi) = 0.$$

Note If any one of the three conditions is not satisfied, the conclusion of Rolle's Theorem is not necessarily true. And it is easy to see that the three conditions are sufficient conditions.

Example 1 To prove the accuracy of Rolle's Theorem by using the function $f(x) = x^2 - 5x + 4$ on the interval $[2,3]$.

Solution $f(x) = x^2 - 5x + 4$ is a polynomial function. It is obvious that the function is continuous and derivable in $(-\infty, +\infty)$.

Because $f(2)=f(3)=-2$, so function $f(x)=x^2-5x+4$ can meet all the conditions of Rolle's Theorem on the interval $[2,3]$. Therefore, there exists $\xi \in (2,3)$, such that $f'(\xi)=0$. Actually, from $f'(x)=2x-5=0$, we get

$$\xi=\frac{5}{2}\in(2,3).$$

Then we have $f'(\xi)=0$.

Example 2 Let $f(x)=(x+1)(x-1)(x-2)(x-3)$, try to find how many roots of $f'(x)=0$ without finding $f'(x)$ and intervals the roots in.

Solution Since $f(x)$ is continuous on $[-1,1]$ and derivable in $(-1,1)$, and $f(-1)=f(1)$, according to Rolle's Theorem, there exists at least one point $\xi_1 \in (-1,1)$, such that

$$f'(\xi_1)=0.$$

Similary, there exists at least one point $\xi_2 \in (1,2)$ and $\xi_3 \in (2,3)$, such that

$$f'(\xi_2)=0, \ f'(\xi_3)=0.$$

Because $f(x)$ is a quadruplicate function of x, $f'(x)$ is a cubic algebraic equation of x. It has at least three real roots. It has been proved that it certainly has three real roots and they respectively locate in the intervals $(-1,1)$, $(1,2)$ and $(2,3)$.

Example 3 Prove equation $x^5-5x+1=0$ has only one positive real root, which is less than 1.

Proof Suppose $f(x)=x^5-5x+1$, $f(x)$ is continuous on $[0,1]$. $f(0)=1$, $f(1)=-3$. From Zero Point Theorem, there exists at least a point $x_0 \in (0,1)$, such that $f(x_0)=0$. Then x_0 is the positive real root of the equation that is less than 1.

Then prove x_0 is the only one positive real root for the equation. Using the proof by contradiction, suppose there exists $x_1 \in (0,1)$, $x_1 \neq x_0$, such that $f(x_1)=0$. It is easy to find that the function $f(x)$ satisfies the conditions of Rolle's Theorem in the interval between x_0, x_1. So there exists at least one point ξ (locating between x_0, x_1), such that $f'(\xi)=0$. However, $f'(x)=5(x^4-1)<0$, $x \in (0,1)$. This is a contradiction. Thus, x_0 is the only one positive real root less than 1.

3.1.3 Lagrange's Theorem

Since the third condition $f(a)=f(b)$ in Rolle's Theorem is so special that it limits the application of Rolle's Theorem. If we only consider the first two conditions, that is, if $f(a) \neq f(b)$, is it possible to find a point on the curve $y=f(x)$, at which the tangent line is parallel to the secant $\overline{AB}$(Figure 3-3)?

Because the slope of the secant $\overline{AB}$ is equal to $\dfrac{f(b)-f(a)}{b-a}$, but $f'(x)$ represents the slope of the tangent at a point of the curve $y=f(x)$. So the question above can be described as: is it possible to find a point ξ in the open interval (a,b), such that

$$f'(\xi)=\frac{f(b)-f(a)}{b-a}$$

or
$$f(b)-f(a)=f'(\xi)(b-a).$$

If we only consider the first two conditions in Rolle's Theorem, we obtain the following important theorem, which was firstly proposed by French mathematician, Joseph Lagrange.

Theorem 2 (Lagrange's Theorem) If the function $y=f(x)$ is continuous on the closed interval $[a,b]$ and is differentiable in the open interval (a, b), then there exists at least one point ξ in (a, b), such that
$$f(b)-f(a)=f'(\xi)(b-a) \quad \text{or} \quad f'(\xi)=\frac{f(b)-f(a)}{b-a}.$$

Proof The equation of the curve is
$$y=f(x),\ x\in[a,b],$$
and the equation of the secant $\overline{AB}$ (Figure 3-4) is
$$\bar{y}-f(a)=\frac{f(b)-f(a)}{b-a}(x-a).$$

Write the function $F(x)$ as
$$F(x)=y-\bar{y}=f(x)-\left[f(a)+\frac{f(b)-f(a)}{b-a}(x-a)\right],$$

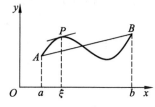

Figure 3-4

which is the difference between $f(x)$ and the secant line $\overline{AB}$. It is obvious that $F(x)$ is continuous on $[a,b]$ and is differentiable on (a,b), and $F(a)=F(b)=0$, according to Rolle's Theorem there exists at least one point $\xi\in(a,b)$, such that $F'(\xi)=0$, that is,
$$f'(\xi)=\frac{f(b)-f(a)}{b-a}.$$

When $f(a)=f(b)$, Lagrange's Theorem is transformed into Rolle's Theorem. Therefore Rolle's Theorem is a special case of Lagrange's Theorem. In other words, the Lagrange's Theorem is a generalization of Rolle's Theorem.

Formula
$$f(b)-f(a)=f'(\xi)(b-a) \quad (a<\xi<b)$$
is called the Lagrange formula. If $b<a$, $f(x)$ is continuous on $[b,a]$ and is differentiable in (b,a), then
$$f(a)-f(b)=f'(\xi)(a-b) \quad (b<\xi<a).$$
Therefore, no matter whether $a<b$ or $b<a$, the Lagrange formula can be written as
$$f(b)-f(a)=f'(\xi)(b-a) \quad (\xi \text{ is between } a \text{ and } b).$$
Since ξ is between a and b, so it can be written as $\xi=a+\theta(b-a)(0<\theta<1)$, namely
$$f(b)-f(a)=f'[a+\theta(b-a)](b-a) \quad (0<\theta<1).$$
If we choose $a=x$, $b=x+\Delta x$, then the Lagrange formula may be written as
$$\Delta y=f(x+\Delta x)-f(x)=f'(x+\theta\Delta x)\Delta x \quad (0<\theta<1).$$
This formula is called the finite increment formula.

Corollary 1 If $f'(x)=0$ for all x in an interval (a,b), then f is constant in (a, b).

Proof Let x_1 and x_2 be any two numbers in (a, b) with $x_1 < x_2$. Since f is differentiable in (a, b), it must be differentiable on (x_1, x_2) and continuous on $[x_1, x_2]$. By applying the Lagrange's Theorem to f on the interval $[x_1, x_2]$, there exists at least one point ξ,
$$f(x_2)-f(x_1)=f'(\xi)(x_2-x_1) \quad (x_1<\xi<x_2).$$
Since $f'(x)=0$, for all x, we have $f'(\xi)=0$, and so
$$f(x_2)-f(x_1)=0 \Rightarrow f(x_2)=f(x_1).$$
Therefore, f has the same value at any two points x_1, x_2 on (a, b). This means that f is constant on (a, b).

The interval (a, b) in Corollary 1 can be replaced by $[a, b]$, $(a, +\infty)$, $[a, b)$, $(-\infty, +\infty)$ and any other interval.

From the Chapter 2, we know that "the derivative of a constant function is zero". Corollary 1 shows that its inverse proposition is also true.

Corollary 2 If $f'(x)=g'(x)$ for all x in an interval (a, b), then $f(x)-g(x)$ is a constant in (a, b), that is, $f(x)=g(x)+C$, where C is a constant.

Proof Because
$$[f(x)-g(x)]'=f'(x)-g'(x)=0,$$
from Corollary 1, we know that $f(x)-g(x)$ is a constant in (a, b), denoted as C, that is
$$f(x)-g(x)=C, \quad x\in(a,b),$$
then
$$f(x)=g(x)+C, \quad x\in(a,b).$$

Example 4 Prove the following inequality: $|\arctan a-\arctan b|\leqslant|a-b|$.

Proof Let $f(x)=\arctan x$, then $f(x)$ is continuous and differential on the interval $(-\infty, \infty)$. According to Lagrange's Theorem, we have
$$\arctan a-\arctan b=(\arctan x)'\Big|_{x=\xi}(a-b) \quad (\xi \text{ is between } a \text{ and } b),$$
that is
$$\arctan a-\arctan b=\frac{a-b}{1+\xi^2}.$$
Because $\frac{1}{1+\xi^2}\leqslant 1$, so
$$|\arctan a-\arctan b|=\frac{|a-b|}{1+\xi^2}\leqslant|a-b|.$$

Example 5 Prove the equality:
$$\arcsin x+\arccos x=\frac{\pi}{2} \quad (-1\leqslant x\leqslant 1).$$

Proof Suppose $f(x)=\arcsin x+\arccos x$, then
$$f'(x)=\frac{1}{\sqrt{1-x^2}}-\frac{1}{\sqrt{1-x^2}}=0,$$
thus $f(x)\equiv C$ (constant) on $[-1, 1]$.

Because $f(0)=f(-1)=f(1)=\frac{\pi}{2}$, so $f(x)=\frac{\pi}{2}$, that is

$$\arcsin x + \arccos x = \frac{\pi}{2} \ (-1 \leqslant x \leqslant 1).$$

Example 6 There is a monotonic decreasing derivative function $f'(x)$ of $f(x)$ on a closed interval $[0, c]$, and $f(0)=0$. Prove the inequality
$$f(a)+f(b)>f(a+b)$$
holds for any a, b satisfying $0<a<b<a+b$.

Proof By Lagrange's Theorem, there exists at least one point ξ_1 on $[0, a]$, such that
$$f(a)=f(a)-f(0)=f'(\xi_1)\cdot a \quad (0<\xi_1<a),$$
and there exists at least one point ξ_2 on $[b, a+b]$, such that
$$f(a+b)-f(b)=f'(\xi_2)\cdot a \quad (b<\xi_2<a+b).$$

Because $f'(x)$ is monotonic decreasing on $[0, c]$, and for
$$0<\xi_1<a<b<\xi_2<a+b<c,$$
we obtain
$$f'(\xi_2)<f'(\xi_1),$$
therefore
$$f(a+b)-f(b)-f(a)=[f'(\xi_2)-f'(\xi_1)]\cdot a<0,$$
that is
$$f(a)+f(b)>f(a+b).$$

3.1.4 Cauchy's Theorem

If we express the curve $\overset{\frown}{AB}$ (Figure 3-5) by using the following parametric equations
$$\begin{cases} X=g(x), \\ Y=f(x), \end{cases} a \leqslant x \leqslant b.$$

Then the slope of the tangent line at the point (X, Y) on the curve is

$$\frac{dY}{dX}=\frac{f'(x)}{g'(x)}.$$

The slope of secant $\overline{AB}$ is

$$\frac{f(b)-f(a)}{g(b)-g(a)}.$$

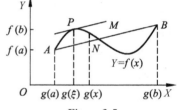

Figure 3-5

Suppose point P corresponds to the parameter $x=\xi$, by Lagrange's formula, the geometric fact may be written as
$$\frac{f(b)-f(a)}{g(b)-g(a)}=\frac{f'(\xi)}{g'(\xi)}.$$

This is the Cauchy Theorem which will be introduced as follows.

Theorem 3 (Cauchy's Theorem) Suppose that $f, g:[a,b]\to \mathbf{R}$, satisfy the following conditions:
(1) they are continuous on the interval $[a, b]$;
(2) they are differentiable in the interval (a, b);
(3) $g'(x)\neq 0, \forall x \in (a,b)$.

Then there exists at least one point $\xi \in (a,b)$, such that
$$f'(\xi)[g(b)-g(a)]=g'(\xi)[f(b)-f(a)].$$

It is not difficult to see that Cauchy's Theorem is a generalization of the Lagrange's Theorem when $g(x)=x$.

Proof If we want to prove there exists $\xi \in (a, b)$, such that
$$f'(\xi)[g(b)-g(a)] = g'(\xi)[f(b)-f(a)],$$
we only need to prove
$$\{f(x)[g(b)-g(a)] - g(x)[f(b)-f(a)]\}'|_{x=\xi} = 0.$$
Let $\varphi(x) = f(x)[g(b)-g(a)] - g(x)[f(b)-f(a)]$, it is obvious that $\varphi(x)$ is continuous on $[a, b]$ and derivable in (a, b), and $\varphi(b)=\varphi(a)$. From Rolle's Theorem, there exists at least one point $\xi \in (a, b)$, such that
$$\varphi'(\xi) = 0.$$
While $\varphi'(x) = f'(x)[g(b)-g(a)] - g'(x)[f(b)-f(a)]$, so there exists $\xi \in (a, b)$, such that
$$f'(\xi)[g(b)-g(a)] = g'(\xi)[f(b)-f(a)].$$
Because $g'(x) \neq 0$, $x \in (a,b)$, then according to Rolle's Theorem, $g(b) \neq g(a)$, therefore the conclusion of Cauchy's Theorem can be written as
$$\frac{f(b)-f(a)}{g(b)-g(a)} = \frac{f'(\xi)}{g'(\xi)}, \quad \xi \text{ is between } a \text{ and } b.$$

Example 7 Suppose that $\varphi(x)$ is continuous on $[0,1]$, and differentiable in $(0,1)$, prove that there exists ξ in $(0,1)$, such that
$$\varphi'(\xi) = 2\xi[\varphi(1) - \varphi(0)].$$

Proof To prove there exists $\xi \in (0, 1)$ such that $\varphi'(\xi) = 2\xi[\varphi(1) - \varphi(0)]$, we only need to prove there exists $\xi \in (0,1)$ such that
$$\varphi'(\xi)(1^2 - 0^2) = 2\xi[\varphi(1) - \varphi(0)],$$
$$\frac{\varphi'(\xi)}{2\xi} = \frac{\varphi(1) - \varphi(0)}{1^2 - 0^2}.$$
Therefore let $f(x) = \varphi(x)$, $g(x) = x^2$, by using Cauchy's Theorem we obtain
$$\varphi'(\xi) = 2\xi[\varphi(1) - \varphi(0)], \quad \xi \in (0,1).$$

Exercises 3-1

1. Verify Roller's Theorem for function $f(x) = x^3 + 4x^2 - 7x - 10$ on $[-1, 2]$.

2. (1) Verify Lagrange's Theorem for function $f(x) = \ln x$ on $[1, e]$;

   (2) Verify Lagrange's Theorem for function $f(x) = \arctan x$ on $[0, 1]$.

3. Use Rolle's Theorem to prove: equation $x^3 - 3x + c = 0$ cannot have two different real roots on closed interval $[0,1]$, c is a arbitrary real number.

4. Suppose that $f(x)$ has second order derivative on interval $[1, 2]$, and $f(2) = f(1) = 0$, $F(x) = (x-1)^2 f(x)$. Prove that there exists at least one point $\xi \in (1, 2)$ such that $F''(\xi) = 0$.

5. Suppose that for all real numbers x, y, inequality $|f(y) - f(x)| \leqslant M|y-x|^2$ (M is a positive constant) holds. Prove that: when $-\infty < x < +\infty$, $f(x) \equiv $ constant.

6. If the roots of real coefficient polynomial
$$p_n(x) = a_0 x^n + a_1 x^{n-1} + \cdots + a_{n-1} x + a_n \quad (a_0 \neq 0)$$
are all real roots, prove all-order derivatives $p'_n(x)$, $p''_n(x)$, $\cdots$, $p_n^{(n-1)}(x)$ only have real roots.

7. Prove the following inequalities:

(1) $nb^{n-1}(a-b) < a^n - b^n < na^{n-1}(a-b)$ $(b > a > 0, n > 1)$;

(2) $e^x > 1 + x$ $(x \neq 0)$;

(3) $\dfrac{a-b}{a} \leqslant \ln \dfrac{a}{b} \leqslant \dfrac{a-b}{b}$ $(0 < b \leqslant a)$;

(4) $(a+b)\ln \dfrac{a+b}{2} < a\ln a + b\ln b$ $(0 < a < b)$.

8. Prove:

(1) $\arctan x = \arcsin \dfrac{x}{\sqrt{1+x^2}}$;

(2) $3\arccos x - \arccos(3x - 4x^3) = \pi$ $\left(|x| \leqslant \dfrac{1}{2}\right)$.

9. Suppose that $f(x)$, $g(x)$ are differentiable, $f(0) = g(0)$, and when $x > 0$, $f'(x) > g'(x)$. Prove that $f(x) > g(x)$, $\forall x > 0$.

10. Suppose that $f(x)$ is continuous on $[a,b]$, and second-order derivable in (a, b), the linear segment connecting $A(a, f(a))$ and $B(b, f(b))$ and the curve segment $\overset{\frown}{AB}$ interseit at $C(c, f(c))$, $c \in (a, b)$. Prove there exists at least one point $\xi \in (a, b)$, such that $f''(\xi) = 0$.

11. Suppose that $x_1 x_2 > 0$, to prove there exists ξ between x_1 and x_2, so that $x_1 e^{x_2} - x_2 e^{x_1} = (1 - \xi) e^{\xi} (x_1 - x_2)$.

12. Prove that the equation
$$x^n + a_1 x^{n-1} + a_2 x^{n-2} + \cdots + a_{n-1} x - a_n = 0 \quad (a_i \geqslant 0, i = 1, 2, \cdots, n-1, a_n > 0)$$
has a unique positive root.

3.2 Taylor's Theorem

Using a simple function to approximate a complicated function is one of the basic and very important ideas in mathematics. Taylor's Theorem works, which is a basic theorem to show how to approximate a given differentiable function by means of polynomials which has important applications in theoretical research and approximate calculations.

We have known that, if function $f(x)$ is differentiable x_0, then
$$f(x) = f(x_0) + f'(x_0)(x - x_0) + o(x - x_0) = p_1(x_0) + o(x - x_0),$$
that is to say, if we use a linear polynomial $p_1(x) = f(x_0) + f'(x_0)(x - x_0)$ to approximate $f(x)$ in the neighborhood of x_0, satisfying $p_1(x_0) = f(x_0)$, $p'_1(x_0) = f'(x_0)$. The approximate error is an infinitesimal of higher order with respect to $(x - x_0)$. However, in practice, the precision is not enough by using the linear polynomial to approximate the function. We need to find a quadratic polynomial or the polynomial with higher degree to

approximate.

Suppose that $f(x)$ is n times differentiable at x_0, can we find a suitable polynomial of degree $n(n>1)$,

$$p_n(x)=a_0+a_1(x-x_0)+a_2(x-x_0)^2+\cdots+a_n(x-x_0)^n \tag{1}$$

to approximate the function $f(x)$, such that the difference between $p_n(x)$ and $f(x)$ is an infinitesimal of higher order will respect to $(x-x_0)^n$? Firstly we suppose $f(x)=p_n(x)$, that is,

$$f(x)=p_n(x)=a_0+a_1(x-x_0)+a_2(x-x_0)^2+\cdots+a_n(x-x_0)^n.$$

Then we can obtain

$$p_n(x_0)=f(x_0),\ p_n'(x_0)=f'(x_0),\ p_n''(x_0)=f''(x_0),\ \cdots,p_n^{(n)}(x_0)=f^{(n)}(x_0).$$

According to these equations, the coefficients of $p_n(x)$ can be obtained:

$$a_0=p_n(x_0)=f(x_0),$$

$$a_1=p_n'(x_0)=\frac{f'(x_0)}{1!},$$

$$a_2=\frac{p_n''(x_0)}{2!}=\frac{f''(x_0)}{2!},$$

$$\cdots\cdots\cdots$$

$$a_n=\frac{p_n^{(n)}(x_0)}{n!}=\frac{f^{(n)}(x_0)}{n!}.$$

Therefore, the coefficients of $p_n(x)$ various orders derivatives at x_0, that is

$$p_n(x)=f(x_0)+\frac{f'(x_0)}{1!}(x-x_0)+\frac{f''(x_0)}{2!}(x-x_0)^2+\cdots+\frac{f^{(n)}(x_0)}{n!}(x-x_0)^n.$$

Suppose that the function $f(x)$ is differentiable of order n at the point x_0. Then the polynomial

$$p_n(x)=f(x_0)+\frac{f'(x_0)}{1!}(x-x_0)+\frac{f''(x_0)}{2!}(x-x_0)^2+\cdots+\frac{f^{(n)}(x_0)}{n!}(x-x_0)^n \tag{2}$$

is called the Taylor polynomial of degree n of the function $f(x)$ at the point x_0. The coefficients $\frac{f^{(k)}(x_0)}{k!}(k=1,2,\cdots,n)$ are called Taylor coefficients of the function f at the point x_0.

Theorem (Taylor's Theorem with Lagrange remainder) Suppose that the function $f(x)$ is differentiable of order $n+1$ in an interval (a,b) and $x_0\in(a,b)$. Then for any $x\in(a,b)$, there exists at least one point ξ, which lies between the points x and x_0, such that

$$f(x)=f(x_0)+f'(x_0)(x-x_0)+\frac{f''(x_0)}{2!}(x-x_0)^2+\cdots+$$

$$\frac{f^{(n)}(x_0)}{n!}(x-x_0)^n+\frac{f^{(n+1)}(\xi)}{(n+1)!}(x-x_0)^{n+1}. \tag{3}$$

Proof Let
$$F(t)=f(x)-\left[f(t)+f'(t)(x-t)+\cdots+\frac{f^{(n)}(t)}{n!}(x-t)^n\right],$$
$$G(t)=(x-t)^{n+1}.$$

Then we only need to prove
$$F(x_0)=\frac{f^{(n+1)}(\xi)}{(n+1)!}G(x_0) \quad \text{or} \quad \frac{F(x_0)}{G(x_0)}=\frac{f^{(n+1)}(\xi)}{(n+1)!}.$$

Suppose $x_0<x$, then $F(t)$ and $G(t)$ are continuous on $[x_0, x]$ and differentiable in (x_0, x), and
$$F'(t)=-\frac{f^{(n+1)}(t)}{n!}(x-t)^n,$$
$$G'(t)=-(n+1)(x-t)^n\neq 0.$$

Since $F(x)=G(x)=0$, by using Cauchy's Theorem, we obtain
$$\frac{F(x_0)}{G(x_0)}=\frac{F(x_0)-F(x)}{G(x_0)-G(x)}=\frac{f'(\xi)}{G'(\xi)}=\frac{f^{(n+1)}(\xi)}{(n+1)!},$$
where $\xi\in(x_0, x)\subset(a,b)$.

The expression (3) is called Taylor formula of degree n for the function $f(x)$ in a neighborhood of x_0. The remainder
$$R_n(x)=f(x)-p_n(x)=\frac{f^{(n+1)}(\xi)}{(n+1)!}(x-x_0)^{n+1},$$
where $\xi=x_0+\theta(x-x_0)$ $(0<\theta<1)$, is called Lagrange remainder. So the Expression (3) is also called Taylor Formula with Lagrange remainder.

When $n=0$, Formula (3) is the expression of Lagrange's Theorem. Therefore Taylor's Theorem is an extension of Lagrange's Theorem.

Because
$$\lim_{x\to x_0}\frac{R_n(x)}{(x-x_0)^n}=\lim_{x\to x_0}\frac{f^{(n+1)}(\xi)}{(n+1)!}(x-x_0)=0,$$
thus when $x\to x_0$, $R_n(x)$ is the higher order infinitesimal with respect to $(x-x_0)^n$. Thus when we do not need the accurate expression of the reminder term, Taylor formula can be written as
$$f(x)=f(x_0)+f'(x_0)(x-x_0)+\frac{f''(x_0)}{2!}(x-x_0)^2+\cdots+$$
$$\frac{f^{(n)}(x_0)}{n!}(x-x_0)^n+o[(x-x_0)^n], \tag{4}$$
where the remainder term $R_n(x)=o[(x-x_0)^n]$ is called as the Peano remainder of the function $f(x)$. Formula (4) is called the Taylor formula with Peano remainder of the function f at the point x_0.

For the special case $x_0=0$, the Taylor formula wiht Lagrange remainder becomes
$$f(x)=f(0)+f'(0)x+\frac{f''(0)}{2!}x^2+\cdots+\frac{f^{(n)}(0)}{n!}x^n+\frac{f^{(n+1)}(\theta x)}{(n+1)!}x^{n+1} \quad (0<\theta<1), \tag{5}$$
which is called Maclaurin formula with Lagrange remainder.

Example 1 Find the degree n Maclaurin formula with Lagrange remainder of $f(x)=e^x$.

Solution Because
$$f'(x)=f''(x)=\cdots=f^{(n)}(x)=e^x,$$
thus
$$f(0)=f'(0)=f''(0)=\cdots=f^{(n)}(0)=1.$$

Therefore, the Maclaurin formula with Lagrange remainder is
$$e^x=1+x+\frac{x^2}{2!}+\cdots+\frac{x^n}{n!}+\frac{e^{\theta x}}{(n+1)!}x^{n+1} \quad (0<\theta<1).$$

Maclaurin formula with Peano remainder is
$$e^x=1+x+\frac{x^2}{2!}+\cdots+\frac{x^n}{n!}+o(x^n).$$

Therefore, e^x is approximated by the following n-order approximate polynomial
$$e^x\approx 1+x+\frac{x^2}{2!}+\cdots+\frac{x^n}{n!},$$
the error can be denoted as
$$|R_n(x)|=\left|\frac{e^{\theta x}}{(n+1)!}x^{n+1}\right|<\frac{e^{|x|}}{(n+1)!}|x|^{n+1} \quad (0<\theta<1).$$

Especially when $x=1$, the approximate formula of irrational number e is
$$e\approx 1+1+\frac{1}{2!}+\cdots+\frac{1}{n!},$$
the error is
$$|R_n|=\frac{e}{(n+1)!}<\frac{3}{(n+1)!}.$$

When $n=10$, then $e\approx 2.718\,282$, the error is less than 10^{-6}.

Example 2 Find the degree n Maclaurin formula with Lagrange remainder of $f(x)=\sin x$.

Solution Because
$$f^{(k)}(x)=(\sin x)^{(k)}=\sin\left(x+\frac{k\pi}{2}\right) \quad (k\in \mathbf{N}^+),$$
then
$$f^{(k)}(0)=\begin{cases}(-1)^{m-1}, & k=2m-1,\\ 0, & k=2m\end{cases} \quad (m\in \mathbf{N}^+).$$

Thus, the degree $n(n=2m)$ Maclaurin formula of $f(x)=\sin x$ is
$$\sin x=x-\frac{x^3}{3!}+\frac{x^5}{5!}+\cdots+(-1)^{m-1}\frac{x^{2m-1}}{(2m-1)!}+R_{2m}(x),$$
where the Lagrange remainder is
$$R_{2m}(x)=\frac{\sin\left[\theta x+(2m+1)\frac{\pi}{2}\right]}{(2m+1)!}x^{2m+1} \quad (0<\theta<1),$$
and the Peano remainder is
$$R_{2m}(x)=o(x^{2m}) \quad (x\to 0).$$

Especially when $m=1$, the approximate formula

$$\sin x \approx x,$$

and the error is

$$|R_2(x)| = \left|\frac{\sin\left(\theta x + \frac{3\pi}{2}\right)}{3!}x^3\right| \leqslant \frac{|x|^3}{6} \quad (0<\theta<1).$$

When $m=2$ and $m=3$, we get the degree 3 and degree 5 polynomials of $\sin x$,

$$\sin x \approx x - \frac{1}{3!}x^3 \quad \text{and} \quad \sin x \approx x - \frac{1}{3!}x^3 + \frac{1}{5!}x^5,$$

the errors are respectively less than $\frac{1}{5!}|x|^5$ and $\frac{1}{7!}|x|^7$.

Exercises 3-2

1. Find the degree 2 Taylor formula of the function $y=\tan x$ at $x_0=0$.

2. Find the degree 3 Taylor formula of the function $y=\sqrt{x}$ at $x_0=4$.

3. Find the degree $2n$ Maclaurin formula of the function $y=\sin^2 x$.

4. Find the degree $2n$ Maclaurin formula of the function $y=\dfrac{e^x+e^{-x}}{2}$.

5. Find the following limits by using the Maclaurin formula with Peano remainder.

   (1) $\lim\limits_{x\to 0}\dfrac{3x-\sin 3x}{(1-\cos x)\ln(1+2x)}$;

   (2) $\lim\limits_{x\to 0}\dfrac{x-\sin x}{x^2(e^x-1)}$;

   (3) $\lim\limits_{x\to 0}\dfrac{\sqrt[4]{1+x^2}-\sqrt[4]{1-x^2}}{x^2}$;

   (4) $\lim\limits_{x\to 0}\dfrac{e^{x^3}-1-x^3}{\sin^6 2x}$.

6. Find a quadratic trinomial $p(x)$, which can be closest to 2^x at the neighborhood of $x=0$ (find the Taylor formula of 2^x).

Summary

1. Main contents

Basic Differential calculus theorem is the basic theory of differential calculus. It reveals the inner link between the function and derivative. It is the effective tool for studying the function derivative with derivative. And it is the important bridge to connect the partial nature of derivative and the whole nature of the function in the domain.

(1) The conditions of Rolle's Theorem, Lagrange's Theorem and Cauchy's Theorem are all sufficient and not necessary. If lacking one of the conditions of the theorems, the conclusions may not be true.

(2) Rolle's Theorem, Lagrange's Theorem and Cauchy Theorem have the same geometric background: curve $\overparen{AB}$ is continuous and derivabe (except two endpoints), there exists at least a point whose tangent is parallel to the secant $\overline{AB}$.

(3) Taylor formula embodies a method of thought by using a polynomial to approach a function. The relation of Rolle's Theorem, Lagrange's Theorem, Cauchy's Theorem and Taylor's Theorem is as follows:

| Rolle's Theorem $f'(\xi)=0$ | $\xrightarrow[f(a)=f(b)]{\text{generalize}}$ | Lagrange's Theorem $\dfrac{f(b)-f(a)}{b-a}=f'(\xi)$ | $\xrightarrow[g(x)=x]{\text{generalize}}$ | Cauchy's Theorem $\dfrac{f(b)-f(a)}{g(b)-g(a)}=\dfrac{f'(\xi)}{g'(\xi)}$ |

generalize $\Big\Updownarrow n=0$

Taylor's Theorem
$$f(x)=f(x_0)+f'(x)(x-x_0)+\dfrac{f''(x_0)}{2!}(x-x_0)^2+\cdots+\dfrac{f^{(n)}(x_0)}{n!}(x-x_0)^n+R_n(x)$$

2. Basic requirements

(1) Understand Rolle's Theorem and Lagrange's Theorem.

(2) Understand Cauchy's Theorem and Taylor's Theorem.

(3) Apply Taylor's Theorem to solve some simple practical problems.

Quiz

1. Prove that the function $y=x^3$ satisfies Lagrange's Theorem on the closed interval $[0, 1]$, and find ξ.

2. Prove the inequality by using Lagrange's Theorem.
$$\dfrac{x}{1+x}<\ln(1+x)<x \quad (x>0).$$

3. Prove
$$\arctan x+\arctan \dfrac{1-x}{1+x}=\begin{cases}\dfrac{\pi}{4}, & x>-1,\\ -\dfrac{3\pi}{4}, & x<-1.\end{cases}$$

4. Write down the Maclaurin formula of order 3 of the function $y=\arcsin x$.

5. Suppose the function $f(x)$ is continuous on the interval $[0,1]$, and is derivable in $(0, 1)$, $f(1)=0$. Prove that there exists a point $\xi \in (0, 1)$, such that $f(\xi)=-\xi f'(\xi)$.

6. Suppose that $f(x)$ is continuous on $[a, b]$, and is derivable in (a, b), $0<a<b$. Prove that there exists a point $\xi \in (a, b)$, such that
$$f(b)-f(a)=\xi\left(\ln \dfrac{b}{a}\right)f'(\xi) \quad (a<\xi<b).$$

7. The following limits by using Maclaurin formula with Peano remainder:

(1) $\lim\limits_{x\to 0}\dfrac{\cos x-e^{-\frac{x^2}{2}}}{x^4}$;

(2) $\lim\limits_{x\to 0}\dfrac{\cos x^2-x^2\cos x-1}{\sin x^2}$.

8. Prove the inequality by using Taylor formula: $\sqrt{1+x}>1+\dfrac{x}{2}-\dfrac{x^2}{8}$ $(x>0)$.

Exercises

1. Prove that it is impossible for the polynomial $f(x)=x^3-3x+a$ to have two zero

points on $[0, 1]$.

2. Assume the function $f(x)=(x-1)(x-2)(x-3)$, prove that there exists ξ in $(1,3)$, such that $f''(\xi)=0$.

3. Suppose $f(x)$ is derivable on the interval $[0, 1]$, and $0<f(x)<1$. When $x \in (0, 1)$, $f'(x)\neq 1$. Prove that the equation $f(x)=x$ has a unique root in $(0,1)$.

4. Suppose that $f(x)$ is derivable on the interval $[a, b](a<b)$, and $f(a)=f(b)$. Prove that there exists $\xi \in (a, b)$, such that $f(a)-f(\xi)=\xi f'(\xi)$.

5. Suppose that $f(x)$ is continuous on $[0, 1]$, and is differentiable in $(0, 1)$, $|f'(x)|<1$ and $f(0)=f(1)$. Prove that $\forall x_1, x_2 \in [0, 1]$, $|f(x_1)-f(x_2)|<\dfrac{1}{2}$.

6. Suppose that $f(x)$ is twice differentiable, $f(0)=f(1)=0$. When $x \in (0,1)$, $|f''(x)|\leqslant A$. Prove that $|f'(x)|\leqslant \dfrac{A}{2}$, $\forall x \in [0,1]$.

Chapter 4 Applications of derivatives

In this chapter, based on the fundamental theorems of differential calculus, now we are in a better position to pursue the applications of differentiation in greater depth. Here we learn how derivatives affect the shape of a graph of a function, including the monotonicity of functions and convexity of curves, extreme values, in particular, how they help us locate maximum and minimum values of functions. Based on the change of the function in the interval, we learn the method of drawing the graph of a function. And we learn a measurement for the degree of "bend" of a curve, and find an approximate solution to the equation.

4.1 Indeterminate form limit

4.1.1 The indeterminate form of type $\dfrac{0}{0}$

In general, if we have a limit of the form $\lim\limits_{x \to x_0}\dfrac{f(x)}{g(x)}$, where both $f(x) \to 0(\pm\infty)$ and $g(x) \to 0(\pm\infty)$ as $x \to x_0$, then this limit may or may not exist and is called an intermediate form of type $\dfrac{0}{0}\left(\dfrac{\infty}{\infty}\right)$. Before we have learned the type $\dfrac{0}{0}$, for example

$$\lim_{x \to 0}\dfrac{\sin x}{x}\overset{\frac{0}{0}}{=}1,$$

and

$$\lim_{x \to \infty}\dfrac{x - \sin x}{x^3}\overset{\frac{0}{0}}{=\!=\!=}\lim_{x \to 0}\dfrac{x - \left(x - \dfrac{1}{3!}x^3\right) + o(x^3)}{x^3} = \dfrac{1}{3!}.$$

How to find the limits of the following indeterminate forms will be discussed in this section:

$$\lim_{x \to 1}\dfrac{1-x^3}{e^{x-1} - \cos(x-1)}\left(\text{type } \dfrac{0}{0}\right), \lim_{x \to \infty}\dfrac{\dfrac{\pi}{2} - \arctan x}{\dfrac{1}{x}}\left(\text{type } \dfrac{0}{0}\right), \lim_{x \to \infty}\dfrac{x^2}{e^x}\left(\text{type } \dfrac{\infty}{\infty}\right),$$

$$\lim_{x \to 0}x^x(\text{type } 0^0), \lim_{x \to 0^+}\left[\dfrac{1}{\ln(1+x)} - \dfrac{1}{x}\right](\text{type } \infty - \infty), \lim_{x \to 0^+}x^\mu \ln x(\mu > 0)(\text{type } 0 \cdot \infty).$$

In sum, there are three types of "the indeterminate forms":

(1) $\lim\limits_{x \to x_0} \dfrac{f(x)}{g(x)}$ is type $\dfrac{0}{0}$ or $\dfrac{\infty}{\infty}$.

(2) $\lim\limits_{x \to x_0} f(x) \cdot g(x)$ is type $0 \cdot \infty$ and $\lim\limits_{x \to x_0}[f(x)-g(x)]$ is type $\infty - \infty$.

(3) $\lim\limits_{x \to x_0} f(x)^{g(x)}$ is type 0^0 or type 1^∞ or type ∞^0.

The above x_0 can also be ∞.

For the second and the third types of indeterminate form, they are always transformed to the first type. For the first type $\left(\dfrac{0}{0} \text{ or } \dfrac{\infty}{\infty}\right)$, its limit can be found by using the method derived by Cauchy's Theorem. In the following we will introduce a very simple and powerful method to find this kind of limits.

Theorem 1 (L' Hospital's Rule) If $f(x)$ and $g(x)$ satisfy the following conditions:

(1) $\lim\limits_{x \to x_0} f(x)=0$, $\lim\limits_{x \to x_0} g(x)=0$;

(2) they are derivable in $\mathring{U}(x_0, \delta)$ of x_0, and $g'(x) \neq 0$;

(3) $\lim\limits_{x \to x_0} \dfrac{f'(x)}{g'(x)} = l$ (l is a finite number or ∞),

therefore
$$\lim_{x \to x_0} \dfrac{f(x)}{g(x)} = \lim_{x \to x_0} \dfrac{f'(x)}{g'(x)} = l.$$

Proof Assumption (1) shows that $\lim\limits_{x \to x_0} f(x) = \lim\limits_{x \to x_0} g(x) = 0$, if $f(x_0) = g(x_0) = 0$, $f(x)$ and $g(x)$ will be continuous at the point x_0. Supposing x is an arbitrary point in the neighborhood of x_0, then on the closed interval whose end points are x and x_0, $f(x)$ and $g(x)$ satisfy the conditions of Cauchy's Theorem, therefore there exists one point ξ between x_0 and x, such that

$$\dfrac{f(x)}{g(x)} = \dfrac{f(x)-f(x_0)}{g(x)-g(x_0)} = \dfrac{f'(\xi)}{g'(\xi)}.$$

Because ξ is between x_0 and x, thus as x approaches x_0, ξ approaches x_0. From condition (3) we obtain:

$$\lim_{x \to x_0} \dfrac{f(x)}{g(x)} = \lim_{\xi \to x_0} \dfrac{f'(\xi)}{g'(\xi)} = \lim_{x \to x_0} \dfrac{f'(x)}{g'(x)} = l.$$

Theorem 2 Suppose that $f(x)$ and $g(x)$ satisfy the following conditions:

(1) $\lim\limits_{x \to \infty} f(x)=0$, $\lim\limits_{x \to \infty} g(x)=0$;

(2) there exists a positive number X, when $|x|>X$, $f(x)$ and $g(x)$ are both derivable and $g'(x) \neq 0$;

(3) $\lim\limits_{x \to \infty} \dfrac{f'(x)}{g'(x)} = l$ (l is a finite number or ∞),

therefore
$$\lim_{x \to \infty} \dfrac{f(x)}{g(x)} = \lim_{x \to \infty} \dfrac{f'(x)}{g'(x)} = l.$$

Proof Let $x = \dfrac{1}{y}$. Because as $x \to \infty$, $y \to 0$, then we have

$$\lim_{x \to \infty} \frac{f(x)}{g(x)} = \lim_{y \to 0} \frac{f\left(\dfrac{1}{y}\right)}{g\left(\dfrac{1}{y}\right)}.$$

The L'Hospital's Rule can be used on the right side of the above formula, thus we obtain

$$\lim_{x \to \infty} \frac{f(x)}{g(x)} = \lim_{y \to 0} \frac{f\left(\dfrac{1}{y}\right)}{g\left(\dfrac{1}{y}\right)} = \lim_{y \to 0} \frac{f'\left(\dfrac{1}{y}\right)\left(-\dfrac{1}{y^2}\right)}{g'\left(\dfrac{1}{y}\right)\left(-\dfrac{1}{y^2}\right)} = \lim_{y \to 0} \frac{f'\left(\dfrac{1}{y}\right)}{g'\left(\dfrac{1}{y}\right)} = \lim_{x \to \infty} \frac{f'(x)}{g'(x)} = l.$$

From L'Hospital's Rule, we get that the limit of a quotient of functions is equal to the limit of the quotient of their derivatives, provided that the given conditions are satisfied. It is especially important to verify the conditions regarding the limits of $f(x)$ and $g(x)$ before using L'Hospital's Rule.

Example 1 Find $\lim\limits_{x \to 0} \dfrac{e^{2x} - 1}{\ln(1+x)}$.

Solution This is the type $\dfrac{0}{0}$ which satisfies the conditions of Theorem 1, by using L'Hospital's Rule, we obtain

$$\lim_{x \to 0} \frac{e^{2x} - 1}{\ln(1+x)} = \lim_{x \to 0} \frac{2e^{2x}}{\dfrac{1}{1+x}} = 2.$$

Example 2 Find $\lim\limits_{x \to 1} \dfrac{x^3 - 3x + 2}{x^3 - x^2 - x + 1}$.

Solution

$$\lim_{x \to 1} \frac{x^3 - 3x + 2}{x^3 - x^2 - x + 1} = \lim_{x \to 1} \frac{3x^2 - 3}{3x^2 - 2x - 1} = \lim_{x \to 1} \frac{6x}{6x - 2} = \frac{3}{2}.$$

Example 3 Find $\lim\limits_{x \to 0} \dfrac{\tan x - x}{x - \sin x}$.

Solution

$$\lim_{x \to 0} \frac{\tan x - x}{x - \sin x} = \lim_{x \to 0} \frac{\sec^2 x - 1}{1 - \cos x} = \lim_{x \to 0} \frac{1 - \cos^2 x}{\cos^2 x (1 - \cos x)} = \lim_{x \to 0} \frac{1 + \cos x}{\cos^2 x} = 2.$$

Example 4 Find $\lim\limits_{x \to +\infty} \dfrac{\dfrac{\pi}{2} - \arctan x}{\dfrac{1}{x}}$.

Solution

$$\lim_{x \to +\infty} \frac{\dfrac{\pi}{2} - \arctan x}{\dfrac{1}{x}} = \lim_{x \to +\infty} \frac{-\dfrac{1}{1+x^2}}{-\dfrac{1}{x^2}} = \lim_{x \to +\infty} \frac{x^2}{1+x^2} = 1.$$

Example 5 Find $\lim\limits_{x \to 0} \dfrac{\sin x}{x^2}$.

Solution

$$\lim_{x\to 0}\frac{\sin x}{x^2}=\lim_{x\to 0}\frac{\cos x}{2x}=\infty.$$

4.1.2 The indeterminate form of type $\frac{\infty}{\infty}$

If when $x\to x_0$ (or $x\to\infty$), $f(x)$ and $g(x)$ all approach ∞, then $\frac{f(x)}{g(x)}$ is called an indeterminate form of type $\frac{\infty}{\infty}$ as $x\to x_0$ ($x\to\infty$).

For this type of indeterminate form, the L'Hospital's Rule is still valid.

> **Theorem 3** If $f(x)$ and $g(x)$ satisfy the following conditions:
> (1) $\lim\limits_{x\to x_0}f(x)=\infty$, $\lim\limits_{x\to x_0}g(x)=\infty$;
>
> (2) they are derivable in $\overset{\circ}{U}(x_0,\delta)$ of x_0, and $g'(x)\neq 0$;
> (3) $\lim\limits_{x\to x_0}\frac{f'(x)}{g'(x)}=l$ (l is a finite number or ∞),
>
> then
> $$\lim_{x\to x_0}\frac{f(x)}{g(x)}=\lim_{x\to x_0}\frac{f'(x)}{g'(x)}=l.$$

Example 6 Find $\lim\limits_{x\to+\infty}\frac{x^\mu}{e^x}$ ($\mu>0$).

Solution This is type $\frac{\infty}{\infty}$. To positive constant μ, there always exists a natural number n, such that $n-1<\mu\leqslant n$, and $\mu-n\leqslant 0$. By using L'Hospital's Rule for n times, we obtain

$$\lim_{x\to+\infty}\frac{x^\mu}{e^x}=\lim_{x\to+\infty}\frac{\mu(\mu-1)\cdot\cdots\cdot(\mu-n+1)x^{\mu-n}}{e^x}=0.$$

This example indicates that, when x approaches $+\infty$, no matter how big the positive constant μ is, x^μ's approaching infinity is much slower than e^x.

Example 7 Find $\lim\limits_{x\to+\infty}\frac{\ln x}{x^\mu}$ ($\mu>0$).

Solution This is type $\frac{\infty}{\infty}$. By using L'Hospital's Rule, we get

$$\lim_{x\to+\infty}\frac{\ln x}{x^\mu}=\lim_{x\to+\infty}\frac{\frac{1}{x}}{\mu x^{\mu-1}}=\lim_{x\to+\infty}\frac{1}{\mu x^\mu}=0.$$

The result shows that no matter how small the positive constant μ is, when x approaches $+\infty$, x^μ's approaching infinity is much faster than $\ln x$. From Examples 6 and 7, by comparing the three functions, when x approaches $+\infty$, $\ln x$ is increasing the slowest and e^x is increasing the fastest.

4.1.3 Other indeterminate forms

Besides the types $\dfrac{0}{0}$ and $\dfrac{\infty}{\infty}$, there are five other types of indeterminate form, namely

$$0 \cdot \infty, \ \infty - \infty, \ 0^0, \ 1^\infty, \ \infty^0.$$

These types of indeterminate form can be changed into the form of $\dfrac{0}{0}$ or $\dfrac{\infty}{\infty}$.

If $\lim\limits_{x \to x_0} f(x) = 0$, $\lim\limits_{x \to x_0} g(x) = \infty$, $f(x)g(x)$ is an indeterminate form of type $0 \cdot \infty$ as x approaches x_0. Because

$$f(x)g(x) = \dfrac{f(x)}{\dfrac{1}{g(x)}} \quad \text{or} \quad \dfrac{g(x)}{\dfrac{1}{f(x)}},$$

type $0 \cdot \infty$ can be transformed into type $\dfrac{0}{0}$ or $\dfrac{\infty}{\infty}$.

If $\lim\limits_{x \to x_0} f(x) = +\infty$, $\lim\limits_{x \to x_0} g(x) = +\infty$, $f(x) - g(x)$ is an indeterminate form of type $\infty - \infty$ as x approaches x_0. The difference can be written as

$$f(x) - g(x) = \dfrac{1}{\dfrac{1}{f(x)}} - \dfrac{1}{\dfrac{1}{g(x)}} = \dfrac{\dfrac{1}{g(x)} - \dfrac{1}{f(x)}}{\dfrac{1}{f(x)} \cdot \dfrac{1}{g(x)}},$$

then it is transformed into the type $\dfrac{0}{0}$.

If $y = f(x)^{g(x)}$ is the type 0^0, 1^∞ or ∞^0, it can be changed into type $0 \cdot \infty$ by taking logarithm

$$\ln y = g(x) \ln f(x).$$

Suppose that the limit is k, $+\infty$ or $-\infty$, then the limit of the indeterminate form is e^k, $+\infty$ or 0.

Example 8 Find $\lim\limits_{x \to \frac{\pi}{2}} (\sec x - \tan x)$.

Solution This is the type $\infty - \infty$. Because

$$\sec x - \tan x = \dfrac{1 - \sin x}{\cos x},$$

as $x \to \dfrac{\pi}{2}$, the right side of the above expression is type $\dfrac{0}{0}$. By using L'Hospital's Rule, we obtain

$$\lim_{x \to \frac{\pi}{2}} (\sec x - \tan x) = \lim_{x \to \frac{\pi}{2}} \dfrac{1 - \sin x}{\cos x} = \lim_{x \to \frac{\pi}{2}} \dfrac{\cos x}{\sin x} = 0.$$

Example 9 Find $\lim\limits_{x \to 1} x^{\frac{1}{1-x}}$.

Solution This is the type 1^∞. Let $y = x^{\frac{1}{1-x}}$ and take logarithms on both sides, we have

$$\ln y = \dfrac{\ln x}{1 - x}.$$

As x approaches 1, the right side is the type $\dfrac{0}{0}$. Because

$$\lim_{x\to 1}\frac{\ln x}{1-x}=\lim_{x\to 1}\frac{\frac{1}{x}}{-1}=-1,$$

therefore
$$\lim_{x\to 1}x^{\frac{1}{1-x}}=\lim_{x\to 1}e^{\frac{\ln x}{1-x}}=e^{-1}.$$

Example 10 Find $\lim\limits_{x\to 0^+}x^x$.

Solution This is the type 0^0. Because $x^x=e^{x\ln x}$, thus
$$\lim_{x\to 0^+}x^x=\lim_{x\to 0^+}e^{x\ln x}=e^{\lim\limits_{x\to 0^+}x\ln x}=e^{\lim\limits_{x\to 0^+}\frac{\ln x}{\frac{1}{x}}}$$
$$=e^{\lim\limits_{x\to 0^+}(-x)}=e^0=1.$$

Example 11 Find $\lim\limits_{x\to 0^+}\left(\frac{1}{\tan x}\right)^{\sin x}$.

Solution This is the type ∞^0.
$$\lim_{x\to 0^+}\left(\frac{1}{\tan x}\right)^{\sin x}=\lim_{x\to 0^+}e^{-\sin x\ln\tan x}=e^{-\lim\limits_{x\to 0^+}\frac{\ln\tan x}{\frac{1}{\sin x}}},$$

$$\lim_{x\to 0^+}\frac{\ln\tan x}{\frac{1}{\sin x}}=\lim_{x\to 0^+}\frac{\frac{\sec^2 x}{\tan x}}{-\frac{\cos x}{\sin^2 x}}=\lim_{x\to 0^+}\left(-\frac{\sin x}{\cos x}\right)=0.$$

Therefore
$$\lim_{x\to 0^+}\left(\frac{1}{\tan x}\right)^{\sin x}=e^0=1.$$

The L'Hospital's Rule is an effective method to find the limit of indeterminate forms. When we use the L'Hospital's Rule, we should combine it with the important limits obtained before. When the rule is used continuously, the expression should be simplified and take advantage of the equivalence infinitesimal quantity. Thus, the operation would be simplified.

Example 12 Find the limit of $\lim\limits_{x\to 0}\dfrac{\tan x-x}{x^2\sin x}$.

Solution
$$\lim_{x\to 0}\frac{\tan x-x}{x^2\sin x}=\lim_{x\to 0}\frac{\tan x-x}{x^3}\ (\text{as } x\to 0,\ \sin x\sim x)$$
$$=\lim_{x\to 0}\frac{\sec^2 x-1}{3x^2}=\lim_{x\to 0}\frac{2\sec^2 x\tan x}{6x}$$
$$=\frac{1}{3}\lim_{x\to 0}\sec^2 x\cdot\lim_{x\to 0}\frac{\tan x}{x}=\frac{1}{3}.$$

Example 13 Find $\lim\limits_{x\to\infty}\dfrac{x+\sin x}{x}$.

Solution This is type $\dfrac{\infty}{\infty}$. By using L'Hospital's Rule, we get
$$\lim_{x\to\infty}\frac{x+\sin x}{x}=\lim_{x\to\infty}(1+\cos x).$$

The limit of the right side does not exist, neither does the left. However it is wrong. Because this function does not meet the third condition of Theorem 3, the L'Hospital's Rule cannot be used here. In fact,

$$\lim_{x\to\infty}\frac{x+\sin x}{x}=\lim_{x\to\infty}\left(1+\frac{\sin x}{x}\right)=1.$$

Only when the indeterminate form is type $\frac{0}{0}$ or $\frac{\infty}{\infty}$, can L'Hospital's Rule can be used directly.

Other indeterminate forms like type $0\cdot\infty$, $\infty-\infty$, 0^0, ∞^0 and 1^∞ should be transformed into type $\frac{0}{0}$ or $\frac{\infty}{\infty}$ before the usage of L'Hospital's Rule.

Exercises 4-1

1. Find the following limits:

(1) $\lim\limits_{x\to 0}\dfrac{e^x-e^{-x}}{\sin x}$;

(2) $\lim\limits_{x\to 0}\dfrac{\tan x-x}{x-\sin x}$;

(3) $\lim\limits_{x\to 0}\dfrac{e^{\alpha x}-\cos \alpha x}{e^{\beta x}-\cos \beta x}$;

(4) $\lim\limits_{x\to \frac{\pi}{4}}\dfrac{\tan x-1}{\sin 4x}$;

(5) $\lim\limits_{x\to 0}\dfrac{(e^{x^2}-1)\sin x^2}{x^2(1-\cos x)}$;

(6) $\lim\limits_{x\to 0}\dfrac{1-\cos x^2}{x^2\sin x^2}$;

(7) $\lim\limits_{x\to 0^+}\dfrac{\ln \tan 7x}{\ln \tan 2x}$;

(8) $\lim\limits_{x\to \frac{\pi}{2}}\dfrac{\tan x}{\tan 3x}$;

(9) $\lim\limits_{x\to\infty}x(e^{-\frac{1}{x}}-1)$;

(10) $\lim\limits_{x\to 1}\left(\dfrac{x}{x-1}-\dfrac{1}{\ln x}\right)$;

(11) $\lim\limits_{x\to \frac{\pi}{2}^-}(\cos x)^{\frac{\pi}{2}-x}$;

(12) $\lim\limits_{m\to\infty}\left(\cos\dfrac{x}{m}\right)^m$;

(13) $\lim\limits_{x\to \frac{\pi}{2}^-}(\tan x)^{2x-\pi}$;

(14) $\lim\limits_{x\to 0}\left(\dfrac{\cot x}{x}-\dfrac{1}{x^2}\right)$;

(15) $\lim\limits_{x\to 0}\left(\dfrac{2}{\pi}\arccos x\right)^{\frac{1}{x}}$;

(16) $\lim\limits_{x\to 0}\left(\cot x-\dfrac{1}{x}\right)$;

(17) $\lim\limits_{x\to 1^-}\ln x\ln(1-x)$;

(18) $\lim\limits_{x\to 0}\left[\dfrac{(1+x)^{\frac{1}{x}}}{e}\right]^{\frac{1}{x}}$;

(19) $\lim\limits_{x\to\infty}\left[x-x^2\ln\left(1+\dfrac{1}{x}\right)\right]$;

(20) $\lim\limits_{x\to 1^-}\dfrac{\ln(1-x)+\tan\dfrac{\pi x}{2}}{\cot \pi x}$.

2. Do the following limits exist? Can the L'Hospital's Rule be used? Why? If the limit exists, please find it.

(1) $\lim\limits_{x\to 0}x\sin\dfrac{k}{x}$;

(2) $\lim\limits_{x\to 0}\dfrac{x^2\cos\dfrac{1}{x}}{\tan x}$;

(3) $\lim\limits_{x\to\infty}\dfrac{x-2\cos x}{x+2\cos x}$;

(4) $\lim\limits_{x\to 0}\dfrac{e^x+e^{-x}}{e^x-e^{-x}}$.

3. When $x\to+\infty$, compare the orders of the following infinitely large quantities.

(1) $(\ln x)^k$ and x^a ($k>0, a>0$);

(2) x^a ($a>0$) and $e^{\sqrt{x}}$;

(3) e^x and x^x;

(4) $\ln \sin \dfrac{a}{x}$ and $\ln \sin \dfrac{b}{x}$ $(a>0, b>0)$.

4.2　Monotonicity and local extreme value

On the basis of Lagrange's Theorem and Taylor formula we are ready to study the properties of a function on an interval by means of derivatives. We will discuss the monotonicity of functions and local extreme values.

4.2.1　Monotonicity of functions

Suppose $f(x)$ is continuous on $[a, b]$ and derivable on (a, b). In order to know how the derivative of $f(x)$ can tell us where a function is increasing or decreasing, consider the graph in Figure 4-1, the tangent line of $f(x)$ has positive slope and so $f'(x) \geqslant 0$. As shown in Figure 4-2, the tangent line of $f(x)$ has negative slope and $f'(x) \leqslant 0$ ($f'(x)=0$ only exist on few points). Through these two figures, we find that the monotonicity of derivable function is related to the sign of its first derivative. It appears that $f(x)$ increases when $f'(x)$ is positive and decreases when $f'(x)$ is negative. To prove this, we will introduce the following theorem based on Lagrange's Theorem.

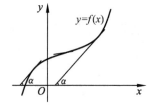
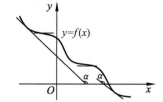

Figure 4-1　　　　　　　　**Figure 4-2**

Theorem 1　Suppose $f(x)$ is continuous on $[a, b]$ and derivable on (a, b),
(1) If $f'(x)>0$ for all $x \in (a, b)$, then $f(x)$ is monotone increasing on $[a, b]$;
(2) If $f'(x)<0$ for all $x \in (a, b)$, then $f(x)$ is monotone decreasing on $[a, b]$.

Proof　(1) Let x_1, x_2 be any two numbers on $[a, b]$ (suppose $x_1 < x_2$), by Lagrange's Theorem, we obtain
$$f(x_2) - f(x_1) = f'(\xi)(x_2 - x_1) \quad (x_1 < \xi < x_2).$$
Because $f'(\xi) > 0$, $x_2 - x_1 > 0$, thus $f(x_2) - f(x_1) > 0$ ($f(x_2) > f(x_1)$). Therefore, $f(x)$ is monotone increasing on $[a, b]$.

Conclusion (2) can be proved in the same way.

According to Theorem 1, for determining the monotonicity of a given function $f(x)$, we should first find the roots of the equation $f'(x_0)=0$, and the points where the function $f(x)$ is not derivable, and then partition the domain by these roots and points. Because $f'(x)$ has definite sign in each of the subintervals, the monotonicity of $f(x)$ can be easily determined by the theorem. Therefore, the monotonicity of functions can be discussed as follows:

(1) Determine the domain of the continuous function $f(x)$.

(2) Find the root of $f'(x)=0$ and the point at which $f'(x)$ doesn't exist, and then partition the domain into several subintervals by these roots and points.

(3) Determine the monotonicity of $f(x)$ by the sign of $f'(x)$ on each interval.

It should be noted that the closed interval $[a,b]$ in Theorem 1 can be applied to open interval (a, b) and infinite interval.

Generally, if the roots for $f'(x)=0$ is finite and $f'(x)>0(<0)$ holds for the remaining points, Theorem 1 is still true.

Example 1 Discuss the monotonicity of $y=\arctan x - x$.

Solution
$$y'=\frac{1}{1+x^2}-1=\frac{-x^2}{1+x^2} \leqslant 0.$$

When $x \in (-\infty, +\infty)$, $y' \leqslant 0$ and only when $x=0$, the equality holds. So $y=\arctan x - x$ decreases on $(-\infty, +\infty)$.

Example 2 Determine the monotonicity intervals of $f(x)=2x^3-9x^2+12x-3$.

Solution This function is continuous on $(-\infty, +\infty)$, and it has continuous derivate
$$f'(x)=6x^2-18x+12=6(x-1)(x-2).$$
Let $f'(x)=0$, $x=1,2$. So number of the points at which $f'(x)=0$ is finite (two), and $f'(x)>0$ on $(-\infty,1)$; $f'(x)<0$ on $(1,2)$; $f'(x)>0$ on $(2,+\infty)$. So the function increase monotonously on $(-\infty, 1)$ and $(2, +\infty)$, decrease on $(1,2)$. The above discussion is presented in table as follows:

x	$(-\infty, 1)$	1	$(1,2)$	2	$(2,+\infty)$
$f'(x)$	+	0	−	0	+
$f(x)$	increase monotonously ↗	2	decrease monotonously ↘	1	increase monotonoously ↗

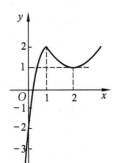

Figure 4-3

The graph of $y=f(x)$ is shown as Figure 4-3.

Example 3 Prove the following inequalities.

(1) When $x>1$, $\ln x > \dfrac{2(x-1)}{x+1}$;

(2) When $x \in (0,1)$, $(1+x)\ln^2(1+x) < x^2$.

Proof (1) Let $f(x)=\ln x - \dfrac{2(x-1)}{x+1}$, when $x>1$,
$$f'(x)=\frac{1}{x}-\frac{4}{(x+1)^2}=\frac{(x-1)^2}{x(x+1)^2}>0.$$
When $x>1, f'(x)>0$, $f(x)$ increases monotonously. So $f(x)>f(1)=0$,
$$\ln x - \frac{2(x-1)}{x+1} > 0.$$

(2) Let $f(x)=x^2-(1+x)\ln^2(1+x)$, when $x\in(0,1)$,
$$f'(x)=2x-\ln^2(1+x)-2\ln(1+x),$$
$$f''(x)=2-\frac{2\ln(1+x)}{1+x}-\frac{2}{1+x}=\frac{2}{1+x}[x-\ln(1+x)]>0.$$

When $x>0$, $\ln(1+x)<x$, so when $x>0$, $f'(x)$ increases monotonously. Because $f'(0)=0$, $f'(x)>0$. So when $x>0$, $f(x)$ increases monotonously. Because $f(0)=0$, $f(x)>0$,
$$(1+x)\ln^2(1+x)<x^2.$$

4.2.2 The extreme value of function

If a function has both monotonically increasing part and monotonically decreasing part of its defined interval, the curve will appear as the top of a mountain or the bottom of a valley at some point (Figure 4-4). The curve is the top of a mountain when $x=x_2$, x_4, but is the bottom of a valley when $x=x_1$, x_3, x_5. The value of function at these points are larger or smaller than adjacent points. In Example 2, the value of

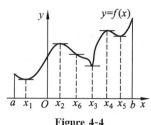

Figure 4-4

adjacent points is smaller than $x=1$, the value of adjacent points is larger than $x=2$. So we call the value $f(1)$ is local maximum and call the value $f(2)$ is local minimum. A strict definition of the extreme value of function is given below.

> **Definition** Suppose $f(x)$ is defined in $U(x_0)$. If when $x\in \mathring{U}(x_0)$, $f(x)<f(x_0)$, we call x_0 maximal point of $f(x)$ and $f(x_0)$ is a maxima of $f(x)$; if when $x\in \mathring{U}(x_0)$, $f(x)>f(x_0)$, we call x_0 minimal point of $f(x)$ and $f(x_0)$ is a minima of $f(x)$.
>
> Local maximum and local minimum values are collectively called extreme value. The point at which the function obtains extreme values are called extreme points.

Note The extreme value of a function is the character within a local range which is only considered in the neighborhood of some point. As shown in Figure 4-7, x_1, x_3, x_5 are minimal points and x_2, x_4 are maximal points. In the same function, minima is maybe larger than maxima (Figure 4-7, $f(x_2)<f(x_5)$). It is difficult to find the extreme value of $f(x)$ by definition. However, as shown in Figure 4-7, extreme point is a turning point which from monotonic increasing to monotonic decreasing or from monotonic decreasing to monotonic increasing. It means that $f'(x)$ changes from positive to negative or from negative to positive when it passes $f'(x_0)$, so $f'(x_0)=0$. The necessary and sufficient condition of obtaining extreme value will be discussed below.

> **Theorem 2 (Necessary condition)** Suppose $f(x)$ is derivable and obtain extreme value at the point x_0, $f'(x)=0$.

Theorem 2 is Fermat's Theorem in last chapter.

Generally, the point at which the first derivative is zero (real root of $f'(x)=0$) is the stationary point of function. From Theorem 2 we can know that: extreme point of derivable function must be its stationary point. But stationary point is not necessarily its extreme point. For example, $f(x)=x^3$, $f'(x)=3x^2$. Because $f'(x)>0$, $f(x)=x^3$ increases monotonically on $(-\infty, +\infty)$. When $x=0$, $f'(x)=0$. $x=0$ is the stationary point but not extreme point of this function. Therefore, Theorem 2 is necessary but not sufficient condition of obtaining extreme value.

Note At the point where derivative does not exist, the function may has extreme value. For example, $f(x)=|x|$. From chapter 3 we call know that: the derivative does not exist when $x=0$, but it is the minimal point of the function. The function obtain minima $f(0)=0$ at this point. So extreme point must be stationary point of the function or the point at which derivative does not exist. These two points are collectivelly called suspected point of extreme point.

Suspected point may be extreme point or may not. How to judge whether suspected point is extreme point? Two sufficient conditions of judging extreme point of function will be given below.

Theorem 3 (Sufficient condition Ⅰ) Suppose at the point x_0, $f(x)$ is continuous and derivable within some deleted neighborhood $\mathring{U}(x_0)$. When $\mathring{U}(x_0)$, then:

(1) If when $x<x_0$, $f'(x)>0$. But when $x>x_0$, $f'(x)<0$. At the point x_0, $f(x)$ obtains the local minimum $f(x_0)$.

(2) If when $x<x_0$, $f'(x)<0$. But when $x>x_0$, $f'(x)>0$. At the point x_0, $f(x)$ obtains the local minimum $f(x_0)$.

(3) If the sign of $f'(x)$ doesn't change, x_0 is not extreme point.

Proof (1) Suppose x_1 is any point in the neighborhood given by theorem. From Lagrange's Theorem we can know that
$$f(x_1)-f(x_0)=f'(\xi)(x_1-x_0),$$
ξ is between x_1 and x_0.

If $x_1<x_0$, $x_1<\xi<x_0$, ξ is on the left half of given neighborhood. We know $f'(\xi)>0$, so $f(x_1)-f(x_0)<0$ which means $f(x_1)<f(x_0)$.

If $x_1>x_0$, $x_0<\xi<x_1$, ξ is on the right half of given neighborhood. We know $f'(\xi)<0$, so $f(x_1)-f(x_0)<0$ which means $f(x_1)<f(x_0)$.

When $x \in \mathring{U}(x_0)$, $f(x_1)<f(x_0)$, so x_0 is the maximal point.

(2) In the same way, (2) can be proved.

(3) On the given neighborhood, whether $x<x_0$ or $x>x_0$, the sign of $f'(x)$ is same. That means $f(x)$ always increases or decreases continuously, so x_0 is not extreme point.

Theorem 3 can simply described as: around the suspected point x_0, if the sign of derivative of $f(x)$ changes, x_0 is extreme point, otherwise its not extreme point.

Example 4 Find the extreme value of $f(x)=x^3-3x^2-9x+5$.

Solution Firstly, find the stationary point and the point at which derivative does not exist. Because
$$f'(x)=3x^2-6x-9=3(x+1)(x-3),$$
let $f'(x)=0$, we can get stationary points $x=-1$, $x=3$ (there is no point at which derivative does not exist).

Secondly, divide the domain into several intervals by stationary point and the point at which derivative does not exist. Here the domain of $f(x)$ is $(-\infty,+\infty)$. Let's divide the domain into three intervals:
$$(-\infty,-1),\ (-1,3),\ (3,+\infty).$$

Thirdly, judge the sign of $f'(x)$ and whether stationary point or point at which derivative does not exist is extreme point. Generally, the sign of $f'(x)$ is listed.

x	$(-\infty,-1)$	-1	$(-1,3)$	3	$(3,+\infty)$
$f'(x)$	+	0	−	0	+
$f(x)$	↗	10 local maximum	↘	−22 local minimum	↗

The sign of $f'(x)$ on each interval and the monotonic increasing and decreasing of function are indicated clearly in the table. According to Theorem 3, we know that $x=-1$ is the maximal point, $f(-1)=10$ is maxima, $x=3$ is minimal point and $f(3)=-22$ is minimal point.

When the second derivative of $f(x)$ at the stationary point exists and is not zero, whether $f(x)$ has extreme value at stationary point or not can be judged by the following theorem.

Theorem 4 (Sufficient condition Ⅱ) Suppose $f(x)$ has the second derivative at the point x_0, and $f'(x_0)=0$, $f''(x_0)\neq 0$,

(1) when $f''(x_0)<0$, $f(x)$ obtains local maximum at the point x_0;

(2) when $f''(x_0)>0$, $f(x)$ obtains local minimum at the point x_0.

Proof (1) Because $f''(x_0)<0$, according to definition of the second derivative,
$$f''(x_0)=\lim_{x\to x_0}\frac{f'(x)-f'(x_0)}{x-x_0}<0.$$
According to the characteristic of the limit of function, on some deleted neighborhood of x_0,
$$\frac{f'(x)-f'(x_0)}{x-x_0}<0.$$
Because $f'(x_0)=0$, $\dfrac{f'(x)}{x-x_0}<0$. Therefore, to x_0 which is different from x on this

neighborhood, the sign of $f'(x_0)$ and $x-x_0$ are opposite. So if $x-x_0<0$ which means $x<x_0$, $f'(x)>0$; if $x-x_0>0$ which means $x>x_0$, $f'(x)<0$. According to Theorem 3, $f(x)$ obtains local maximum at the point x_0.

In the same way, (2) can be proved.

Example 5 Find the extreme value of $f(x)=e^x\cos x$.

Solution $f'(x)=e^x\cos x-e^x\sin x$.

Let $f'(x)=0$, we can obtain the stationary point $x=k\pi+\dfrac{\pi}{4}$, k is an integer. Because $f''(x)=-2e^x\sin x$, when $x=2k\pi+\dfrac{\pi}{4}$, $f''(x)<0$, $x=2k\pi+\dfrac{\pi}{4}$ is the maximal point and the local maximum of function is

$$f\left(2k\pi+\dfrac{\pi}{4}\right)=\dfrac{\sqrt{2}}{2}e^{2k\pi+\frac{\pi}{4}};$$

when $x=(2k+1)\pi+\dfrac{\pi}{4}$, $f''(x)>0$, so $x=(2k+1)\pi+\dfrac{\pi}{4}$ is the minimal point and the local minimum of the function is

$$f\left[(2k+1)\pi+\dfrac{\pi}{4}\right]=-\dfrac{\sqrt{2}}{2}e^{(2k+1)\pi+\frac{\pi}{4}}\ (k=0,\pm1,\pm2,\cdots).$$

Summing up the process of solving above problems, we can know that the steps of finding extreme value are as follows:

(1) find the derivative of function;

(2) find the zero point (stationary point) and the point at which derivate does not exist;

(3) apply Theorem 3 or 4 to judge whether stationary point or the point at which derivate does not exist is the extreme point, and then find the extreme value.

4.2.3 Maximum value and minimum value of function

The maximum and minimum value is the property of function in global range of the domain. In some practical problems, we often encounter various problems of finding maximum and minimum value of some function. The methods of finding maximum and minimum value and their application will be discussed in this chapter.

Suppose the function $y=f(x)$ is continuous on $[a,b]$, according to the property of continuous function on closed interval, $f(x)$ must have maximum and minimum values on $[a,b]$. If the local maximum (local minimum) value is obtained within (a,b), the maximum (minimum) value must be the local maximum (local minimum) of the function. But the maximum (minimum) value may be

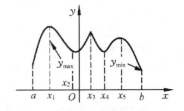

Figure 4-5

obtained at the end of interval. Thus the maximum (minimum) value of $f(x)$ on $[a,b]$ can be found by comparing all extreme values of $f(x)$ on $[a,b]$, and the value of $f(x)$ at two endpoints $f(a),f(b)$. As shown in Figure 4-5, $f(x)$ is continuous on $[a,b]$ and is not derivable at the point x_3, extreme points are x_1, x_2, x_3, x_4, x_5, endpoints are a, b. Comparing the value of function at each point, we can know that the minimum is obtained when $x=b$, and the maximum is obtained when $x=x_1$.

Usually, to continuous function on $[a, b]$, extreme point on (a, b) does not need to be taken into consideration. We only need to find the function value of all suspected points and of endpoints, and then compare them, the maximum and minimum value can be obtained.

When there is only one extreme value of $y=f(x)$ in its span, if it is local maximum (local minimum), it is maximum (minimum) of the function.

Example 6 Find maximum and minimum of $f(x)=x^4-8x^2-2$ on $[-1, 4]$.

Solution Let
$$f'(x)=4x^3-16x=4x(x^2-4)=4x(x-2)(x+2)=0,$$
we can find stationary points $x=0, -2, 2$. Because $x=-2$ is not within the domain, it is rejected. Comparing the value of function at stationary points and endpoints of interval. Because $f(0)=-2, f(2)=-18, f(-1)=-9, f(4)=126$.

Thus, at the point $x=2$, the minimum is -18; at the point $x=4$, the maximum is 126.

Example 7 Prove the inequality $x^\alpha - \alpha x \leqslant 1-\alpha$ $(x>0, 0<\alpha<1)$.

Solution Let $f(x)=x^\alpha-\alpha x-(1-\alpha)$,
$$f'(x)=\alpha x^{\alpha-1}-\alpha=\alpha(x^{\alpha-1}-1).$$

Let $f'(x)=0$, the only stationary point $x=1$ is obtained. When $0<x<1$, $f'(x)>0$; When $x>1$, $f'(x)<0$, $f(1)$ is the maximum of $f(x)$ on $(0, +\infty)$, which means $f(x) \leqslant f(1)=0$. So $x^\alpha-\alpha x-(1-\alpha)\leqslant 0$, which means $x^\alpha-\alpha x \leqslant 1-\alpha$ $(x>0, 0<\alpha<1)$.

Finding the maximum and minimum in practical problems is the problem which appears frequently in practice. To find the maximum and minimum in practical problems, we need to transform practical problems into mathematical problems. According to the realities, we should establish its mathematical model and then determine the range of actual variable. Because functions in practical problems are usually derivable, the maximum and minimum in practical problem are often obtained at stationary points and endpoints of intervals. If the range of function is open interval, the maximum and minimum are obtained at stationary points.

Example 8 Enclose a rectangular field with fence by take the straight bank of river as one side. There is some fence which is 36 m, what the area of the largest field is?

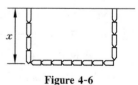

Figure 4-6

Solution Suppose the length of vertical side with the river bank in rectangular field is x m, the length of another side is $(36-2x)$ m. As shown in Figure 4-6, suppose the area of rectangular field is S m^2,
$$S=x(36-2x)=36x-2x^2, \quad 0<x<\frac{36}{2}=18.$$

$$S'(x)=36-4x.$$

Let $S'(x)=0$, then $x=9$.

$x=9$ is the only stationary point. From actual meaning of the problem, the maximum of the area exists. By applying sufficient condition of extreme value, we can know that it is the local maximum within the open interval $(0, 18)$. Because $S''(x)=-4$, $S''(9)<0$, $x=9$ is the maximal point.

Therefore $S(9)$ is the maximum on $(0,18)$. When one side of the field is 9 and another side is 18, the area of field is the largest and the area is $9\times 18=162$ m^2.

Example 9 If we want to build a bottomless cylindrical gas tank, how to choose its diameter d and height h (Figure 4-7) can we use the least material?

Solution The least material means the surface area is the smallest. Suppose the surface area is S, because

$$V=\pi\left(\frac{d}{2}\right)^2 h,$$

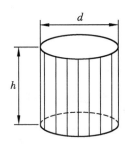

Figure 4-7

$h=\dfrac{4V}{\pi d^2}$. Because the gas tank is bottomless, the surface area of gas tank

$$S=\pi dh+\pi\left(\frac{d}{2}\right)^2,$$

$$S=\frac{4V}{d}+\frac{\pi}{4}d^2, \quad 0<d<+\infty,$$

$$S'(d)=-\frac{4V}{d^2}+\frac{\pi}{2}d.$$

Let $S'(d)=0$, we obtain $d=2\sqrt[3]{\dfrac{V}{\pi}}$.

This is the only stationary point of the function of surface area. From the meaning of practical problem, minimum must exist. So the value of function at atstationary point is the minimum. That means when

$$d=2\sqrt[3]{\frac{V}{\pi}}, \quad h=\frac{4V}{\pi\left(2\sqrt[3]{\dfrac{V}{\pi}}\right)^2}=\sqrt[3]{\frac{V}{\pi}} \quad (d=2h),$$

the material we use is the least.

Example 10 (Solution of oil well problem) In the preliminary analysis at the beginning of this section, we draw a map and roughly figured out the cost. Here we suppose the length of underwater pipeline is x, the length of overland pipeline is y (Figure 4-8).

The point opposite the right angle of oil well is the key to representing the relation between x and y. From $x^2=12^2+(20-y)^2$, we get

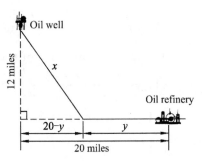

Figure 4-8

$$x=\sqrt{144+(20-y)^2}.$$

In this model, only positive roots make sense.

The cost of pipeline

$$c = 50\ 000x + 30\ 000y$$
$$= 50\ 000\sqrt{144+(20-y)^2} + 30\ 000y \text{ dollars}.$$

Our aim is to find minimum of $c(y)$ on $0 \leqslant y \leqslant 20$. The first derivative of c is

$$c' = 50\ 000 \cdot \frac{1}{2} \cdot \frac{2(20-y)(-1)}{\sqrt{144+(20-y)^2}} + 30\ 000$$

$$= -50\ 000\,\frac{20-y}{\sqrt{144+(20-y)^2}} + 30\ 000.$$

Let $c' = 0$, we can obtain that

$$y = 11 \quad \text{or} \quad y = 29.$$

Only $y=11$ is within the interval. The value at critical point $y=11$ and at the endpoint is

$$c(11) = 1\ 080\ 000, \quad c(0) = 1\ 166\ 190, \quad c(20) = 1\ 200\ 000.$$

The smallest cost is 1 080 000 dollars. We can get that only by using underwater pipelines to 11 miles away from the refinery.

Exercises 4-2

1. Find monotonic interval of the following functions:

   (1) $y = x^3 - 3x^2 - 9x + 14$; (2) $y = \dfrac{x}{1+x^2}$;

   (3) $y = \sqrt{2x - x^2}$; (4) $y = x - 2\sin x,\ 0 \leqslant x \leqslant 2\pi$;

   (5) $y = x - \ln(1+x)$; (6) $y = x - e^x$.

2. Prove: if $f(x)$ and $g(x)$ are continuous at the point a and are derivable on (a, b). And when $f(a) = g(a)$, $x \in (a, b)$, $f'(x) > g'(x)$. So when $x \in (a, b)$, $f(x) > g(x)$.

3. Prove the following inequalities:

   (1) when $x > 1$, $2\sqrt{x} > 3 - \dfrac{1}{x}$; (2) when $x > 0$, $\ln(1+x) > \dfrac{\arctan x}{1+x}$;

   (3) when $0 < x < \dfrac{\pi}{2}$, $\tan x > x + \dfrac{x^3}{3}$; (4) when $x > 0$, $1 - x + \dfrac{x^2}{2} > e^{-x} > 1 - x$;

   (5) when $x > 0$, $x - \dfrac{x^2}{2} < \ln(1+x) < x$.

4. Find extreme value of the following functions:

   (1) $y = \dfrac{(x-2)(x-3)}{x^2}$; (2) $y = \sqrt[3]{(x^2-a^2)^2}$ (a is positive constant);

   (3) $y = x + \sqrt{1-x}$; (4) $y = \cos x + \sin x \left(-\dfrac{\pi}{2} \leqslant x \leqslant \dfrac{\pi}{2}\right)$;

   (5) $y = x^2 e^{-x^2}$; (6) $y = x - \ln(1+x^2)$;

   (7) $y = x^{\frac{1}{x}}$ ($x > 0$).

5. Prove that the equation $\sin x = x$ has only one real root.

6. Prove that the equation $e^x - x - 1 = 0$ has only one real root.

7. Discuss how many real roots does the equation $\ln x = ax$ have (a is positive constant).

8. Prove that when $b > e\ln a$, the equation $a^x = bx (a>1)$ has two real roots; when $0 \leqslant b \leqslant e\ln a$, it has no real roots; when $b<0$, it has only one real root.

9. We have known that $f(x)$ is continuous on $[a, +\infty]$. When $x>a$, $f'(x)>k$ (k is positive constant), and $f(a)<0$. Prove $f(x)$ has only one stationary point on $\left(a, a-\dfrac{f(a)}{k}\right)$.

10. We have known that $f(0)=0$, $f'(x)$ increases monotonically. Prove that when $x>0$, $g(x)=\dfrac{f(x)}{x}$ increases monotonically.

11. When the constant a is what, $x=\dfrac{\pi}{3}$ is the extreme point of $f(x)=a\sin x+\dfrac{1}{3}\sin 2x$. Is it the maxima or minimum. And find it.

12. We have known that $f(x)=(x-x_0)^n \varphi(x)$ (n is natural number), $\varphi(x)$ is continuous at the point x_0.
  (1) Prove $f(x)$ is derivable at the point x_0.
  (2) If $\varphi(x_0) \neq 0$, whether the point x_0 is the extreme point of $f(x)$ or not? Why?

13. We have known that $f(x)=3x^2+\dfrac{A}{x^3}$, A is positive constant. When A is at least what, to $x>0$, $f(x) \geqslant 20$?

14. We have known that there is a the nth order non-zero derivative of $f(x)$ at the point x_0, and $f'(x_0)=f''(x_0)=\cdots=f^{(n-1)}(x_0)=0$. Prove: if n is an odd number, the point x_0 is not the extreme point; if n is an even number, when $f^{(n)}(x_0)<0$, the point x_0 is the maximal point; when $f^{(n)}(x_0)>0$, the point x_0 is the minimal point.

15. We have known that $f(x)$ is nonnegative, the constant $c>0$, prove $F(x)=cf^2(x)$ and $f(x)$ have the same extreme point.

16. Find the maximum and minimum of the following functions on specified interval (if exist):
  (1) $y=x+\cos x$, $[0, 2\pi]$;
  (2) $y=|x^2-3x+2|$, $[-10, 10]$;
  (3) $y=2\tan x-\tan^2 x$, $[0, \dfrac{\pi}{2})$;
  (4) $y=xe^{-x^2}$, $(-\infty, +\infty)$.

17. Find the maximum of $x_n = \sqrt[n]{n}$ ($n=1, 2, \cdots$).

18. Prove the following inequlity:
  (1) when $-2 \leqslant x \leqslant 2$, $|3x-x^3| \leqslant 2$;
  (2) when $0 \leqslant x \leqslant \dfrac{\pi}{2}$, $1 \leqslant \sin x + \cos x \leqslant \sqrt{2}$;
  (3) $x^4-2x^3+2x+1 \geqslant \dfrac{5}{16}$;
  (4) when $x \leqslant 1$, $4ax^3+3(b-4a)x^2+6(2a-b)x-2(a-b)<0$, a, b are constants, and

$0 < b < 2a$.

19. To make a covered rectangular box, the volume is 72 cm³, the ratio of the sides of bottom is 1 : 2. When length, width and height of the box are what, the surface area of the box will be the smallest.

20. A round iron sheet of radius R to be a conical container by taking off a fan. When the central angle of the cut-off fan is what, the volume of conical container will be the largest?

21. The length of AB is 100 km. It's 20 km from the factory C to A. AC is perpendicular to AB (Figure 4-9). Now build a road from someplace D to the factory C, the material should be transported from B to the factory C. We have known that the ratio of rail freight to road freight is 3 : 5, how to choose the place D to make cost the least?

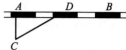

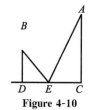

Figure 4-9

22. The factory A and B is on the same side of the river bank. The vertical distance between A, B, and the shore $AC = 1\,000$ m, $BD = 400$ m and $CD = 800$ m (Figure 4-10). Now we need to build a water tower at someplace E on the line CD to provide water for the factory A and B. How to choose the place of E to make the length of water pipe the shortest?

Figure 4-10

23. The lever fulcrum is at one end, and the point of force is at the other end (Figure 4-11). The weight of the hanging object is 490 kg which is 1 m from fulcrum, the weight of lever is uniformly distributed, the linear density is 5 kg/m. Find the minimum force F that balances the lever and the length of lever.

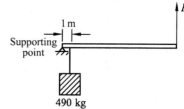

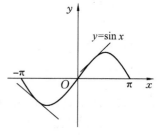

Figure 4-11

4.3 The convexity, concavity and inflection point of function

In addition of rise and fall, function also has the characteristic of convexity and concavity. For instance, as shown in Figure 4-12, observing the shape of the graph of $y = \sin x$, $x \in (-\pi, \pi)$. It curves upward on $(-\pi, 0)$, we call that concavity. It curves down on $(0, \pi)$, we call that convexity. The dividing point of these two curves is $O(0, 0)$. When curve is concave, from the figure, the whole curve is above the tangent of any point. When curve is convex, from the figure, the whole curve is below the tangent of any point.

Figure 4-12

Definition 1 Suppose the function $f(x)$ is derivable on the interval I. If the curve of this function on I is above the tangent of each point, we call $y = f(x)$ is concave on I

(concave arc for short). If the curve of this function on I is below the tangent of each point, we call $y=f(x)$ is convex on I (convex arc for short).

As shown in Figure 4-13, we can see that the strings connecting two arbitrary points on concave curve are always above the arc between two points, but the convex curve is just opposite.

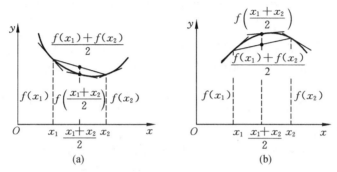

Figure 4-13

Thus, it is easy to obtain the following equivalent definitions:

Definition 1' The sufficient and necessary condition of the concavity of the curve $y=f(x)$ on I is that to arbitrary points x_1, x_2, on I, then
$$\frac{f(x_1)+f(x_2)}{2}>f\left(\frac{x_1+x_2}{2}\right).$$

The sufficient and necessary condition of the convexity of the curve $y=f(x)$ on I is that to arbitrary points x_1, x_2 on I, then
$$\frac{f(x_1)+f(x_2)}{2}<f\left(\frac{x_1+x_2}{2}\right).$$

From Figure 4-16, the slope of the tangent on the concave curve (from x_1 to x_2) increases with the increase of x and the slope of tangent is $f'(x)$. We can obtain that $f'(x)$ increases monotonically. So $f''(x)$ on this interval; the slope of the tangent on the convex curve (from x_1 to x_2) decreases with the increase of x. We can obtain that $f'(x)$ decreases monotonically. So $f''(x)$ on this interval. After researches, in turn, this conclusion is also true.

Therefore, there are theorems of judging the concavity and convexity of curves by applying the second derivative.

Theorem 1 Suppose the function $y=f(x)$ is continuous on $[a, b]$ and has the second derivative on (a,b).

(1) If $f''(x)>0$ on (a, b), $y=f(x)$ is concave on $[a, b]$;

(2) If $f''(x)<0$ on (a, b), $y=f(x)$ is convex on $[a, b]$.

Proof (1) Suppose x_1 and x_2 are any two points on $[a,b]$, and $x_1 < x_2$, denote that $\frac{x_1+x_2}{2} = x_0$ and $x_2 - x_0 = x_0 - x_1 = h$, so $x_1 = x_0 - h$ and $x_2 = x_0 + h$. According to Lagrange's Theorem, we can obtain that

$$f(x_0+h) - f(x_0) = f'(x_0+\theta_1 h)h,$$
$$f(x_0) - f(x_0-h) = f'(x_0-\theta_2 h)h.$$

Here $0 < \theta_1 < 1$, $0 < \theta_2 < 1$. According to substraction of two equalities, we can obtain that

$$f(x_0+h) + f(x_0-h) - 2f(x_0) = [f'(x_0+\theta_1 h) - f'(x_0-\theta_2 h)]h.$$

To $f'(x)$, by applying Lagrange's Theorem on $[x_0 - \theta_2 h, x_0 + \theta_1 h]$, we can obtain that

$$[f'(x_0+\theta_1 h) - f'(x_0-\theta_2 h)]h = f''(\xi)(\theta_1+\theta_2)h^2,$$

Here $x_0 - \theta_2 h < \xi < x_0 + \theta_1 h$.

According to the assumption (1) in Theorem 1, $f''(\xi) > 0$, so

$$f(x_0+h) + f(x_0-h) - 2f(x_0) > 0,$$

which is

$$\frac{f(x_0+h) + f(x_0-h)}{2} > f(x_0),$$

thus

$$\frac{f(x_1) + f(x_2)}{2} > f\left(\frac{x_1+x_2}{2}\right),$$

From Theorem 1, we can know that the graph of $y = f(x)$ on $[a, b]$ is concave.

(2) can be proved in the same way.

In order to find the concave and convex interval of $f(x)$, actually to find the same sign interval(the interval of which the sign will not change) of y''. Like determining the same sign interval of y' when discuss monotonicity of function, find the point at which $y'' = 0$ and y'' doesn't exist first, and then list the sign of y'' on each interval.

Example 1 Find concavity and convexity of the curve $y = \ln x$.

Solution Because $y' = \frac{1}{x}$, $y'' = -\frac{1}{x^2}$, $y'' < 0$ on the domain of function $(0, +\infty)$. From Theorem 2, we can obtain that $y = \ln x$ is convex on $(0, +\infty)$.

Example 2 Judge the concavity and convexity of the curve $y = x^{\frac{1}{3}}$.

Solution When $x \neq 0$, $y'' = -\frac{2}{9}x^{-\frac{5}{3}}$. So $y'' > 0$ on $(-\infty, 0)$, $y = x^{\frac{1}{3}}$ is concave on $(-\infty, 0)$; $y'' < 0$ on $(0, +\infty)$, $y = x^{\frac{1}{3}}$ is convex on $(0, +\infty)$.

Example 3 Judge the concavity and convexity of $y = 1 + x^2 + \frac{x^3}{3} - x^4$.

Solution $y' = 2x + x^2 - 4x^3$, $y'' = 2 + 2x - 12x^2 = 2(1+3x)(1-2x)$.

Let $y'' = 0$, we can obtain that $x = -\frac{1}{3}$ and $x = \frac{1}{2}$, listed as follows:

Chapter 4 Applications of derivatives

x	$\left(-\infty, -\frac{1}{3}\right)$	$-\frac{1}{3}$	$\left(-\frac{1}{3}, \frac{1}{2}\right)$	$\frac{1}{2}$	$\left(\frac{1}{2}, +\infty\right)$
y''	$-$	0	$+$	0	$-$
	convex		concave		convex

Around $x=0$ in Example 2 and $x=-\frac{1}{3}$, $x=\frac{1}{2}$ in Example 3, the concavity and convexity of function has changed. The dividing points $(x_0, f(x_0))$ of convex and concave interval are inflection point of curve.

Definition 2 If the point C on continuous curve is the dividing point of concave arc and convex arc, the point C is called as inflection point of the curve.

Note From the definition of inflection point, the inflection point $M_0(x_0, f(x_0))$ is a point on the curve. It is different from the definition of extreme point.

The point $(0, 0)$ in Example 2 and $\left(-\frac{1}{3}, y\big|_{x=-\frac{1}{3}}\right)$, $\left(\frac{1}{2}, y\big|_{x=\frac{1}{2}}\right)$ in Example 3 are inflection points of curve. The curve corresponding to Example 1 is convex on the whole interval, so there is no inflection point. How to find the inflection point of $y=f(x)$? If the sign of $f''(x)$ is different on either side, the point $(x_0, f(x_0))$ is inflection point. To the second derivable function, the following is necessary condition of the existence of inflection point:

Theorem 2 Suppose the second derivative of $y=f(x)$ exists. If the point (x_0, y_0) is the inflection point of $y=f(x)$, $f''(x_0)=0$.

Theorem 2 can be proved easily through the definition of inflection point and Theorem 1.

Theorem 2 points out the scope for finding inflection point. The abscissa of inflection points of the curve $y=f(x)$ which has the second derivative $f''(x)$ only need to be found from the point at which $f''(x)=0$ and the point at which $f''(x)$ doesn't exist. Generally, the steps of finding inflection points are as follows:

(1) Find $f''(x)$;

(2) Let $f''(x)=0$, solve the real roots of this equation in the domain and the point at which $f''(x)$ doesn't exist in the domain;

(3) To each root solved in (2) and the point x_0 at which $f''(x)$ doesn't exist, considering the sign of $f''(x)$ near the point x_0. If the signs are opposite, the point $(x_0, f(x_0))$ is the inflection point. If the signs are same, the point $(x_0, f(x_0))$ is not the inflection point.

Example 4 Find the inflection point of $y=e^{-x^2}$.

Solution
$$y' = -2xe^{-x^2},$$
$$y'' = 2(2x^2-1)e^{-x^2}.$$

Let $y''=0$, we can obtain
$$x = \pm\frac{1}{\sqrt{2}}.$$

When $x > \frac{1}{\sqrt{2}}$ or $x < -\frac{1}{\sqrt{2}}$, $y'' > 0$; when $-\frac{1}{\sqrt{2}} < x < \frac{1}{\sqrt{2}}$, $y'' < 0$. So the curve is concave in $\left(-\infty, -\frac{1}{\sqrt{2}}\right)$ and $\left(\frac{1}{\sqrt{2}}, +\infty\right)$; the curve is convex on $\left(-\frac{1}{\sqrt{2}}, \frac{1}{\sqrt{2}}\right)$. Thus, the points $\left(\frac{1}{\sqrt{2}}, e^{-\frac{1}{2}}\right)$ and $\left(-\frac{1}{\sqrt{2}}, e^{-\frac{1}{2}}\right)$ are inflection points.

Example 5 To prove the following inequalities by applying the concavity and convexity: suppose $a > 0$, $b > 0$, $a \neq b$, then
$$a^p + b^p > 2^{1-p}(a+b)^p \quad (p>1).$$

Proof Observing the function $f(x) = x^p$ on $x > 0$,
$$f'(x) = px^{p-1},$$
$$f''(x) = p(p-1)x^{p-2}.$$

When $p > 1$, $f''(x) > 0$ (to arbitrary $x > 0$), so the curve $y = f(x)$ is concave on $(0, +\infty)$. From Theorem 1, to arbitrary $a > 0$, $b > 0$, $a \neq b$, then
$$\frac{f(a)+f(b)}{2} > f\left(\frac{a+b}{2}\right),$$

which is
$$\frac{1}{2}(a^p + b^p) > \left(\frac{a+b}{2}\right)^p,$$

so
$$a^p + b^p > 2^{1-p}(a+b)^p \quad (p>1).$$

Example 6 (Inflection points and stock market) Figure 4-14 is a hypothetical Dow Jones Industrial (stock) Average curve. It reflects comprehensively the index of rising and falling trend of American industrial stock price. The Dow Jones Industrial Average is an index of the stock market which can capture the overall growth of the stock market that has partial rises and falls.

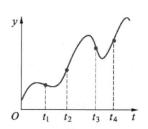

Figure 4-14

One way to invest in the stock market is to buy index fund stocks. It buys different stocks in turn to track the index. The main goal of index fund is undoubtedly buying in low price (buy at local minimum) and selling in high price (sell at local maximum). However, because it is impossible to predict the extreme value of stock market, the time of stock market is unpredictable. When investors realize the stock market is really rising, the minimum has already been over.

The inflection point provides investors with the way to predict it before the reverse trend, because the inflection point indicates radical changes of the rate of growth of the function. Buying stocks at (or near) the inflection point can make investors stay in longer-term upward trend (the inflection point warns the change of trend). Investors want to reduce the impact on stock market and this method captures the upward trend for a long time.

Exercises 4-3

1. Judge the concavity and convexity of function and point out the inflection points.

   (1) $y = 2x^3 - 3x^2 - 36x + 25$;   (2) $y = x + \dfrac{1}{x}$;

   (3) $y = \ln(x^2 + 1)$;   (4) $y = x^2 + \dfrac{1}{x}$.

2. When a is what, the point $(1, 1)$ is the inflection point of the curve $y = x^3 + a\ln x$.

3. If the point $(1, 3)$ is the inflection point of the curve $y = ax^3 + bx^2$, a, b should be what?

4. To prove inequalities by the concavity and convexity of curve.

   (1) $2\cos \dfrac{x+y}{2} > \cos x + \cos y$, $(x, y) \in \left(-\dfrac{\pi}{2}, \dfrac{\pi}{2}\right)$;

   (2) $(x+y)\ln \dfrac{x+y}{2} < x\ln x + y\ln y$, $0 < x < y$;

   (3) $e^{\frac{x+y}{2}} < \dfrac{e^x + e^y}{2}$.

5. Prove that there are 3 inflection points of $y = \dfrac{x+1}{x^2+1}$ are on the same line.

4.4 Description of the function graph

4.4.1 Asymptote of curve

The asymptote of curve is used to depict the curve at infinity. Definitions of three kinds of asymptotes and the method to find them will be given.

1) Horizontal asymptote and vertical asymptote

If $\lim\limits_{x \to \infty} f(x) = A$ (A is constant), the straight line $y = A$ is called the horizontal asymptote of the curve $y = f(x)$. If $\lim\limits_{x \to x_0} f(x) = \infty$, the straight line $x = x_0$ is the vertical asymptote of the curve $y = f(x)$. Such as the curve $y = \dfrac{1}{x}$, because $\lim\limits_{x \to \infty} \dfrac{1}{x} = 0$, the straight line $y = 0$ (x-axis) is the horizontal asymptote of it; because $\lim\limits_{x \to 0} \dfrac{1}{x} = \infty$, the straight line $x = 0$ (y-axis) is the vertical asymptote. This shows that the curves are infinitely close to two axes at infinity, which means when x tends to be ∞ or 0, the curve are more and more close to x-axis or y-axis.

Example 1 Find the asymptote of the curve $y = \dfrac{1}{\sqrt{2\pi}} e^{-\frac{x^2}{2}}$.

Solution Because $\lim\limits_{x\to+\infty} f(x) = \lim\limits_{x\to+\infty} \dfrac{1}{\sqrt{2\pi}} e^{-\frac{x^2}{2}} = 0$, $y=0$ is the horizontal asymptote of the curve.

Because $0 \leqslant y = \dfrac{1}{\sqrt{2\pi}} e^{-\frac{x^2}{2}} \leqslant \dfrac{1}{\sqrt{2\pi}}$, there is no vertical asymptote.

It should be noted that not all curves have asymptote.

2) Slanting asymptote

If there is a straight line $y = kx + b$ which can make $\lim\limits_{x\to\infty}[f(x) - kx - b] = 0$, the straight line $y = kx + b$ is called as the slanting asymptote of $y = f(x)$. This indicates that the curve tends to be a straight line when $x \to \infty$.

It is easily to be proved that the sufficient and necessary condition of the straight line $y = kx + b$ being slanting asymptote of the curve $y = f(x)$ is

$$k = \lim_{x\to\infty} \dfrac{f(x)}{x}, \quad b = \lim_{x\to\infty}[f(x) - kx].$$

Note As long as one of the two extreme values, we can conclude that $y = f(x)$ does not have asymptote.

Example 2 Find the asymptote of $y = \dfrac{x^2}{x+1}$.

Solution Because $\lim\limits_{x\to -1} f(x) = \lim\limits_{x\to -1} \dfrac{x^2}{x+1} = \infty$, the straight line $x = -1$ is vertical asymptote of the curve.

Because

$$\lim_{x\to\infty}\dfrac{f(x)}{x} = \lim_{x\to\infty}\dfrac{x^2}{x(x+1)} = 1,$$

$$\lim_{x\to\infty}[f(x) - kx] = \lim_{x\to\infty}\left(\dfrac{x^2}{x+1} - x\right) = -1,$$

the straight line $y = x - 1$ is the slanting asymptote of $y = \dfrac{x^2}{x+1}$.

4.4.2 Description of the function graph

By using the monotonicity, extreme value of function and concavity, convexity, inflection point of curve discussed in this chapter and the method of finding asymptote given in 4.4.1, and ploting some points selectively, the graph of function can be described accurately. Steps of describing the graph of function are as follows:

Firstly, determine the domain of $f(x)$ and some characteristics such as parity, periodicity of function.

Secondly, find the first and the second derivative of function. Find the points at which $f'(x)$, $f''(x)$ are equal to 0 and at which derivative does not exist, and divide the domain into several intervals by these points.

Thirdly, determine the sign of $f'(x)$, $f''(x)$ on the divided intervals, make a list to discuss the monotonicity, extreme value, concavity and convexity, inflection point of the function.

Fourthly, determine whether the function has asymptote or not.

Fifthly, choose some points on the graph and describe.

Example 3 Draw the graph of $y=f(x)=\dfrac{x^3}{3}-2x^2+3x+1$.

Solution (1) The domain is $(-\infty, +\infty)$.

(2) $f'(x)=x^2-4x+3=(x-1)(x-3)$, $f''(x)=2x-4=2(x-2)$.

Let $f'(x)=0$, $f''(x)=0$, we can obtain that $x=1,2,3$. They divide $(-\infty, +\infty)$ into $(-\infty, 1)$, $(1, 2)$, $(2, 3)$, $(3, +\infty)$, and we can obtain that $f(1)=\dfrac{7}{3}, f(2)=\dfrac{5}{3}, f(3)=1$.

(3) The characteristics of function graph: monotonicity, extreme value, concavity and convexity, inflection point and so on can be given by list:

x	$(-\infty, 1)$	1	$(1,2)$	2	$(2,3)$	3	$(3,+\infty)$
$f'(x)$	+	0	−	−	−	0	+
$f''(x)$	−	−	−	0	+	+	+
The graph of $y=f(x)$	↗	local maximum $\dfrac{7}{3}$	↘	inflection point $\left(2,\dfrac{5}{3}\right)$	↘	local minimum 1	↗

"↗" means increasing monotonically and convexity; "↗" means increasing monotonically and concavity; "↘" means decreasing monotonically and convexity; "↘" means decreasing monotonically and concavity.

(4) There is not asymptote in this exercise.

(5) Choose some points. By $f(0)=1$, $f(-1)=-\dfrac{13}{3}$, $f(4)=\dfrac{7}{3}$, we can determine points $(0, 1)$, $\left(-1,-\dfrac{13}{3}\right)$, $\left(4,\dfrac{7}{3}\right)$. Draw the graph (Figure 4-15).

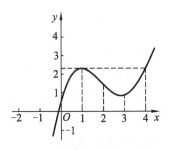

Figure 4-15

Example 4 Draw the graph of $y=e^{-x^2}$.

Solution (1) The domain is $(-\infty, +\infty)$, and $f(x)$ is parity. Its graph is symmetry about y-axis. So we only need to discuss the graph of function on $(0, +\infty)$, the other half can be obtained by symmetry.

(2) $f'(x)=-2xe^{-x^2}$, $f''(x)=-2(e^{-x^2}-2x^2e^{-x^2})=-2e^{-x^2}(1-2x^2)$.

Let $f'(x)=0$, $f''(x)=0$, we can obtain that $x=0$, $\pm\dfrac{1}{\sqrt{2}}$. And $f(0)=1$, $f\left(\pm\dfrac{1}{\sqrt{2}}\right)=e^{-\frac{1}{2}}$.

(3) Make a list and give the monotocity, extreme value, concavity and convexity, inflection points of function.

x	$\left(-\infty,-\dfrac{1}{\sqrt{2}}\right)$	$-\dfrac{1}{\sqrt{2}}$	$\left(-\dfrac{1}{\sqrt{2}},0\right)$	0	$\left(0,\dfrac{1}{\sqrt{2}}\right)$	$\dfrac{1}{\sqrt{2}}$	$\left(\dfrac{1}{\sqrt{2}},+\infty\right)$
$f'(x)$	+	+	+	0	−	−	−
$f''(x)$	+	0	−	−	−	0	+
The graph of $y=f(x)$	↗	inflection point	↗	maximal point	↘	inflection point	↘

157

(4) Because $\lim\limits_{x\to\infty} e^{-x^2}=0$, the straight line $y=0$ is the horizontal asymptote.

(5) Choose the points $(0,1)$, $\left(\pm\dfrac{1}{\sqrt{2}}, e^{-\frac{1}{2}}\right)$, $(\pm 1, e^{-1})$, and draw the graph (Figure 4-16).

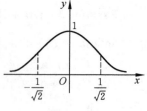

Figure 4-16

Example 5 Draw the graph of $y=\dfrac{(x-1)^3}{x^2}$.

Solution (1) The domain is $(-\infty,0)\cup(0,+\infty)$.

(2) $f'(x)=\dfrac{(x-1)^2(x+2)}{x^3}$, $f''(x)=\dfrac{6(x-1)}{x^4}$.

Let $f'(x)=0$, $f''(x)=0$, we can obtain that $x=-2, 1$. Divide the domain into $(-\infty, -2)$, $(-2,0)$, $(0,1)$, $(1,+\infty)$ and we can obtain that $f(-2)=-\dfrac{27}{4}$, $f(1)=0$.

(3) Make a list and give the monotocity, extreme value, concavity and convexity, inflection points of function.

x	$(-\infty,-2)$	-2	$(-2,0)$	$(0, 1)$	1	$(1,+\infty)$
$f'(x)$	$+$	0	$-$	$+$	0	$+$
$f''(x)$	$-$	$-\dfrac{9}{8}$	$-$	$-$	0	$+$
The graph of $y=f(x)$	↗	local maximum $-\dfrac{27}{4}$	↘	↗	inflection point $(1,0)$	↗

(4) Because $\lim\limits_{x\to 0} f(x)=-\infty$, the straight line $x=0$ is vertical asymptote.

Because $\lim\limits_{x\to\infty}\dfrac{f(x)}{x}=1$, $\lim\limits_{x\to\infty}[f(x)-x]=-3$. The straight line $y=x-3$ is the slanting asymptote of $y=\dfrac{(x-1)^3}{x^2}$.

(5) Draw the graph (Figure 4-17).

Figure 4-17

Exercises 4-4

1. Draw the graphs of the following functions following the steps:

(1) $y=x^3+6x^2-15x-20$;

(2) $y=\dfrac{x^3}{2(1+x)^2}$;

(3) $y=\dfrac{x^2}{x+1}$;

(4) $y=x-2\arctan x$;

(5) $y=xe^{-x}$;

(6) $y=\ln\dfrac{1+x}{1-x}$;

(7) $y = \dfrac{(x-3)^2}{4(x-1)}$;

(8) $y = \dfrac{1}{\sqrt{2\pi}} e^{-\frac{(x-a)^2}{2\sigma^2}}$ $(a, \sigma > 0)$.

4.5 Curvature

In practical problems, we often need to study the degree of the curvature of the curve. For example, when the train is turning, the rail needs a proper curve so that the train can move smoothly into the curve; the bending degree of the steel beam needs to be discussed. Therefore, how to calculate the bending degree is necessary.

4.5.1 Differentiation of an arc

To study the curving degree of curve, the concept of arc differential will be introduced first. So the definition of arc will be given first:

Suppose the function $y = f(x)$ has continuous derivative on (a, b), take the fixed point $M_0(x_0, y_0)$ on the curve $y = f(x)$ as the base point of measuring the arc length (Figure 4-18), and stipulate that the direction in which increases is positive. To the any variable point $M(x, y)$, we use s denote the value of the arc $\overparen{M_0 M}$ with direction: the absolute value of s is equal to the length of $\overparen{M_0 M}$. If the direction of $\overparen{M_0 M}$ is the same with positive direction of the curve, the length of curve $s > 0$; on the contrary, the length of curve $s < 0$. So the length of arc s is a function of x, which is $s = s(x)$. And it is the monotonically increasing function of x. The following is to find the differential of $s(x)$.

Suppose x, $x + \Delta x$ are two adjacent points on (a, b), their corresponding points on the curve $y = f(x)$ are M, M' (Figure 4-18), and suppose the increment of x is Δx, the increment of the length of arc s is Δs, so

$$\Delta s = \overparen{M_0 M'} - \overparen{M_0 M} = \overparen{MM'}.$$

In the above formula, we use the mark of directed arc, such as $\overparen{M_0 M'}$ to express the value of this directed arc, so

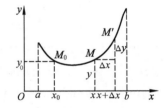

Figure 4-18

$$\left(\dfrac{\Delta s}{\Delta x}\right)^2 = \left(\dfrac{\overparen{MM'}}{\Delta x}\right)^2 = \left(\dfrac{\overparen{MM'}}{|MM'|}\right)^2 \cdot \dfrac{|MM'|^2}{(\Delta x)^2}$$

$$= \left(\dfrac{\overparen{MM'}}{|MM'|}\right)^2 \cdot \dfrac{(\Delta x)^2 + (\Delta y)^2}{(\Delta x)^2}$$

$$= \left(\dfrac{\overparen{MM'}}{|MM'|}\right)^2 \left[1 + \left(\dfrac{\Delta y}{\Delta x}\right)^2\right],$$

then

$$\dfrac{\Delta s}{\Delta x} = \pm \sqrt{\left[\dfrac{\overparen{MM'}}{|MM'|}\right]^2 \cdot \left[1 + \left(\dfrac{\Delta y}{\Delta x}\right)^2\right]}.$$

Because $s = s(x)$ is monotonically increasing function of x, Δs and Δx have the same sign, which means $\dfrac{\Delta s}{\Delta x} > 0$, so it is positive in the above formula. Thus,

$$\dfrac{\Delta s}{\Delta x} = \dfrac{\overparen{MM'}}{|MM'|} \sqrt{1 + \left(\dfrac{\Delta y}{\Delta x}\right)^2}.$$

Let $\Delta x \to 0$, so $M' \to M$. And because

$$\lim_{M' \to M} \left| \frac{\widehat{MM'}}{|MM'|} \right| = 1 \text{ and } \lim_{\Delta x \to 0} \frac{\Delta y}{\Delta x} = y',$$

According to the definition of derivative,

$$\frac{ds}{dx} = \lim_{\Delta x \to 0} \frac{\Delta s}{\Delta x} = \lim_{\Delta x \to 0} \sqrt{1 + \left(\frac{\Delta y}{\Delta x}\right)^2} \cdot \lim_{M' \to M} \frac{\widehat{MM'}}{|MM'|} = \sqrt{1 + y'^2},$$

Which is $\qquad ds = \sqrt{1 + y'^2} dx \quad \text{or} \quad ds = \sqrt{(dx)^2 + (dy)^2}. \qquad (1)$

The above formula is called as arc differential calculation formula. By using arc differential calculation formula, calculation formula of the quantity that describes the bending degree of the curve-curvature.

4.5.2 Calculation formula of curvature

In the following, we will introduce the concept of curvature. Straight line does not bend, the circle of smaller radius bends more obviously than the circle of bigger radius, but other curves have different curvature at different part.

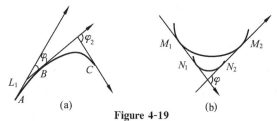

Figure 4-19

As shown in Figure 4-19a, take points A, B, C separately on the curve L_1, and $\widehat{AB} = \widehat{BC} = \Delta s$. When the moving point moves continuous along the curve along the curve from the point A to B, the tangent angle of the curve is φ_1. When the moving point moves from B to C, its tangent angle is φ_2, from the Figure 4-19a, we know that $\varphi_2 > \varphi_1$. The reason is that the curvature of $\widehat{BC}$ is bigger than $\widehat{AB}$. But the change of tangent angle can not reflect the degree of curvature. For example Figure 4-19b indicates that the bending degree of the curve is related to the length of arc. Thus, to the same length $\widehat{MM'} = \Delta s$, its bending degree is related to the tangent angle of $\widehat{MM'}$ at two endpoints $\varphi = \Delta \alpha$. If $\Delta \alpha$ is bigger, the bending degree is bigger. But to the same tangent angle, the length of Δs is longer, the bending degree is smaller. so:

(1) use $\left| \dfrac{\Delta \alpha}{\Delta s} \right|$ to describe the average bending degree of Δs;

(2) use $\lim\limits_{\Delta s \to 0} \left| \dfrac{\Delta \alpha}{\Delta s} \right|$ to describe the bending degree of curve at the point M.

Suppose the curve L has continuously rotating tangent. Choose a point M_0 as the base point of measuring the length of arc s on the curve L. Suppose the point M on the curve corresponds to the length of arc s, the tangent angle at the point M is α, the other point M' corresponds to the length of arc $s + \Delta s$, the tangent angle here is $\alpha + \Delta \alpha$ (Figure 4-20). The length of $\widehat{MM'}$ is $|\Delta s|$. When the moving point moves from M

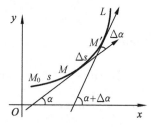

Figure 4-20

to M', the angle tangent rotated is $|\Delta\alpha|$.

The ratio $\dfrac{|\Delta\alpha|}{|\Delta s|}$ expresses the size of tangent angle on unit arc. And we describe the average bending degree of $\widehat{MM'}$ by the ratio. We call it as average curvature of $\widehat{MM'}$, and record as $\overline{K}$, which means

$$\overline{K}=\dfrac{|\Delta\alpha|}{|\Delta s|}.$$

When $\Delta s \to 0$, which is $M \to M'$, the limit of average curvature $\overline{K}$ is called as the curvature of the curve L at the point M, and is recorded as, which is

$$K=\lim_{\Delta s\to 0}\dfrac{|\Delta\alpha|}{|\Delta s|}.$$

If $\lim\limits_{\Delta s\to 0}\dfrac{\Delta\alpha}{\Delta s}=\dfrac{\mathrm{d}\alpha}{\mathrm{d}s}$ exists,

$$K=\left|\dfrac{\mathrm{d}\alpha}{\mathrm{d}s}\right|. \tag{2}$$

The curvature at one point describes the bending degree of curve at this point. The calculating formula of curvature is given as follows:

Suppose the rectangular coordinate equation of $y=f(x)$, and $f(x)$ has the second derivative. Because $y'=\tan\alpha$, $\alpha=\arctan y'$, So

$$\dfrac{\mathrm{d}\alpha}{\mathrm{d}x}=\dfrac{y''}{1+y'^2},$$

then

$$\mathrm{d}\alpha=\left(\dfrac{\mathrm{d}\alpha}{\mathrm{d}x}\right)\mathrm{d}x=\dfrac{y''}{1+y'^2}\mathrm{d}x.$$

And by arc differential in this chapter,

$$\mathrm{d}s=\sqrt{1+y'^2}\,\mathrm{d}x.$$

So from the definition of curvature, we can obtain

$$K=\left|\dfrac{\mathrm{d}\alpha}{\mathrm{d}s}\right|=\dfrac{|y''|}{(1+y'^2)^{\frac{3}{2}}}. \tag{3}$$

The above formula is the calculation formula of curvature.

Example 1 Calculate curvature of the straight line $y=ax+b$ at arbitrary point.

Solution Because $y'=a$, $y''=0$, from curvature formula we can obtain that $K=0$.

It means that the curvature at arbitrary point on straight line is equal to zero. This is consistent with "straight line does not bend".

Example 2 Calculate the curvature of the circle $x^2+y^2=R^2$.

Solution Because $y=\pm\sqrt{R^2-x^2}$,

$$y'=\dfrac{x}{\mp\sqrt{R^2-x^2}},\quad y''=\dfrac{R^2}{(R^2-x^2)^{\frac{3}{2}}}.$$

So

$$K=\dfrac{|y''|}{(1+y'^2)^{\frac{3}{2}}}=\dfrac{1}{R}.$$

This conclusion indicates: (1) the bending degree at each point on circle is the same;

(2) the radius of circle is bigger, the bending degree is larger.

Suppose the curve is given by parametric equation $\begin{cases} x=\varphi(t), \\ y=\psi(t) \end{cases}$ we can find the y'_x and y''_x by the method of finding derivative of function determined by parametric equation. Plug (3), and we can get

$$K = \frac{|\varphi'(t)\psi''(t) - \varphi''(t)\psi'(t)|}{[\varphi'^2(t) + \psi'^2(t)]^{\frac{3}{2}}}. \tag{4}$$

Such as in Example 2, we plug parametric equation $\begin{cases} x = R\cos t, \\ y = R\sin t \end{cases}$ into (4), and we can obtain $K = \frac{1}{R}$.

4.5.3 Curvature circle

The slope of tangent has been used as change rate of function, which is geometric interpretation of derivatives. The radius of the circle is used as geometric scale of curvature, which means in order to represent curvature vividly, the concept of curvature circle is introduced.

Suppose the curvature of $y = f(x)$ is $K(K \neq 0)$ at the point A, lead a normal through the point A. Take a point C on the normal on the concave side of, let $|AC| = \frac{1}{K}$, and then take C as the center, $|AC|$ as radius to make a circle (Figure 4-21). We can call this circle as curvature circle of the curve on the point A, its radius is called as curvature radius, the center is curvature center.

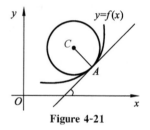

Figure 4-21

Generally, if $K \neq 0$, the reciprocal of the curvature at one point is called as curvature radius of this point, and written as ρ, which means $\rho = \frac{1}{K}$; if $K = 0$, the curvature radius is specified as $+\infty$.

Obviously, curvature circle and curve are tangent at the point A, and they have the same direction of curvature. Because the curvature of curvature circle is equal to the reciprocal of radius, it and the curve have the same curvature at the point A. Thus, in all circles passing through the point, the shape curvature circle is the most similar to the curve near the point A. In the practical problems concerning the concavity and convexity and curvature of curve, this is the theoretical basis of replacing a segment of circular arc of curvature circle near the point A with curve arc.

The concept of curvature circle not only gives geometric interpretation of curvature, but also uses curvature circle or curvature radius in practical applications.

Example 3 Calculate the curvature and curvature radius of $y = \sin x$ at the point $\left(\frac{\pi}{2}, 1\right)$.

Solution $y' = \cos x$, $y'' = -\sin x$, and $y'\big|_{x=\frac{\pi}{2}} = 0$, $y''\big|_{x=\frac{\pi}{2}} = -1$, so

$$K=\frac{|y''|}{(1+y'^2)^{\frac{3}{2}}}\bigg|_{x=\frac{\pi}{2}}=1, \rho=\frac{1}{K}=1,$$

which means the curvature radius is $\rho=1$.

Example 4 Figure 4-22 is a railway line that turns from straight rail BO to the point $A(2,2)$. Because the curve of train is usually an arc of large radius, but straight rail can not be connected with arc when the train turns from straight rail to the curve, it must be turned into a circle arc by a transition curve. The key is that the curvature should be the same at bonding point, so that the train will not oscillate at this point.

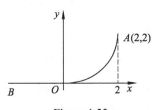

Figure 4-22

From physics we can know that when a train runs at a curve track at a speed of v, the centrifugal force is

$$\frac{mv^2}{r}=mv^2 k,$$

here m is the weight of train, r is curvature radius at any point, k is curvature. When m, v are constants, centrifugal force changes with curvature k. When k does not change continuously, centrifugal force will change discontinuously, so that the train will oscillate. Therefore, only when the curvature k changes continuously, centrifugal force also changes continuously, violent vibration will not be caused.

Because curvature of straight line is 0, so we should choose the simplest curve $y=f(x)$ as the arc $\overset{\frown}{OA}$, which will make BOA be a curve which is continuous, smooth and curvature changing continuously. Especially, it should meet following conditions at joint point O:

(1) $f(0)=0$;

(2) $f'(0)=0$;

(3) $f''(0)=0$, which means $k=0$.

(4) $f(2)=2$, which means that the curve should pass the point $A(2,2)$.

Obviously, the most simplest curve which meet the above conditions is cubic polynomial

$$f(x)=ax^3+bx^2+cx+d,$$

a, b, c, d are undetermined constants. These four constants can be determined by the above four conditions, which means the equation of the arc $\overset{\frown}{OA}$ should be cubic parabola $y=\frac{1}{4}x^3$. It is not a curve which connects with an arc directly as imaginary, but needs to be transformed into an arc through a transition curve.

Exercises 4-5

1. Find the curvature and curvature radius of following curves at specified point:
   (1) $y=x^3+x$ at the point $(0,0)$;
   (2) $y=\cos x$ when $x=0$;
   (3) $y=x^2-4x+3$ at the vertex;
   (4) $y=x+\ln x$ at the point $(1,1)$.

2. Find the curvature of $y=\frac{a}{4}x^4$ ($a>0$) at any point, and find the points at which the curvature will be maximum and minimum.

3. Find the curvature radius of $y=a\ln\left(1-\frac{x^2}{a^2}\right)$ ($a>0$) at any point, and find the point at which curvature radius is minimum.

4. A plane dives along the parabola $y=\frac{x^2}{10\ 000}$ (the y-axis is vertical upward, the unit is m), the speed of plane$=200$ m/s at the coordinate origin. The weight of pilot $G=70$ kg, find the effect of the seat on the pilot when the plane dives to the lowest point which is origin. (Notice: the centrifugal force of an object that makes circular motion at the uniform velocity $F=\frac{mv^2}{r}$, m is the weight of object, v is its speed, r is radius of circle.)

*4.6 Approximation of equation

In science, technology and engineering, we often encounter the problem of solving equation $f(x)=0$. In general, it is difficult to calculate exact root of equation, so the problem of finding approximation of real root of equation is given. In order to find approximation of $f(x)=0$, we should find approximate range of root first, which means we should determine an interval $[a, b]$ to make the real root which we find the only one in this interval. So we call $[a, b]$ as an isolated interval of the real root we find. Then we narrow the isolated interval and gradually improve the accuracy of approximate roots until the approximation which meets accuracy requirement is obtained. Two common methods of finding approximation will be introduced in this chapter.

4.6.1 Bisection method

Suppose $f(x)$ is continuous on $[a, b]$, $f(a) \cdot f(b)<0$, and $f(x)=0$ only has one real root ξ on (a, b), so the interval $[a, b]$ is an isolated interval of the real root ξ.

The basic idea of bisection method is: take the midpoint $\xi_1=\frac{a+b}{2}$ of the interval $[a, b]$, and calculate $f(\xi_1)$. There are two possibilities: (1) If $f(\xi_1)=0$, $\xi=\xi_1$. (2) If $f(\xi_1)\neq 0$, $f(\xi_1)$ has the same sign of one of $f(a)$ and $f(b)$. If $f(\xi_1)$ and $f(a)$ have the same sign, $f(\xi_1)$ and $f(b)$ have contrary sign, so $\xi_1<\xi<b$ and $b-\xi_1=\frac{1}{2}(b-a)$. If $f(\xi_1)$ has the same sign with $f(b)$, $f(\xi_1)$ and $f(a)$ have contrary sign, so $a<\xi<\xi_1$ and $\xi_1-a=\frac{1}{2}(b-a)$. If ξ_1 is approximation of the root of equation, the error is less than $\frac{1}{2}(b-a)$.

If the error cannot fulfill the requirement, we can repeat the above process on new interval $[\xi_1, b]$ (or $[a, \xi_1]$) to find ξ_2 $\left(\xi_2=\frac{\xi_1+b}{2}\ \text{or}\ \xi_2=\frac{a+\xi_1}{2}\right)$, when $\xi_2\neq\xi$, repeat again until the required accuracy is obtained.

Example 1 Find an approximate root of $x^3-2x^2-4x-7=0$ on $[3, 4]$, and the error is less than 0.1.

Solution Suppose $f(x)=x^3-2x^2-4x-7$, $f(x)$ is continuous on $[3,4]$. Because $f(3)=-10<0$, $f(4)=9>0$, there is one real root of $f(x)=0$ on $(3,4)$ at least. Because $f'(x)=3x^2-4x-4$, let $f'(x)=0$, we can obtain $x_1=-\frac{2}{3}, x_2=2$. When $x>2$, $f'(x)>0$, $f(x)$ increases monotonically, so there is only one real root of $f(x)=0$ on $(3, 4)$, $[3, 4]$ is the only isolated interval of this real root.

Let $\xi_1=3.5, f(3.5)=-2.625<0, f(4)=9>0$, the new isolated interval is $[3.5, 4]$.

Let $\xi_2=3.75, f(3.75)=2.609>0, f(3.5)=-2.625<0$, the new isolated interval is $[3.5, 3.75]$.

Let $\xi_3=3.63, f(3.63)=-0.042<0, f(3.75)=2.609>0$. Because,
$$\frac{1}{2}(3.75-3.63)=0.06<0.1,$$

$\xi=3.63$ is an approximate root of the equation and the error is less than 0.1.

4.6.2 Tangent method

Suppose $f(x)$ has the second derivative on $[a,b]$, $f(a)\cdot f(b)<0$, $f'(x)$ and $f''(x)$ keep the same sign on $[a, b]$, there is only one real root ξ of $f(x)=0$ on (a, b), the interval $[a, b]$ is an isolated interval of ξ. There are four kinds of the graph of $y=f(x)$ on $[a,b]$ $\overset{\frown}{AB}$ (Figure 4-23): a. be concave and increase monotonically; b. be concave and decrease monotonically; c. be convex and increase monotonically; d. be convex and decrease monotonically.

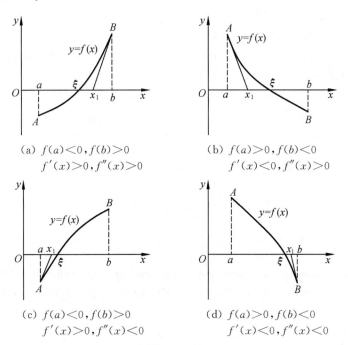

Figure 4-23

The basic idea of tangent method is: replace curve arc $\overset{\frown}{AB}$ with the tangent of one end of $\overparen{AB}$ on $[a,b]$, so the abscissa of intersection point of this tangent and x-axis is approximation of the root of equation. From Figure 4-26, make tangent of the curve at the endpoint $B(b, f(b))$ at which the ordinate $f(x)$ and $f''(x)$ have the same sign, the equation of this tangent is

$$y - f(b) = f'(b)(x - b).$$

Let $y = 0$, the abscissa of the intersection point of tangent and x-axis

$$x_1 = b - \frac{f(b)}{f'(b)}.$$

It is closer to the root ξ than b.

Then replace $[a, b]$ with $[a, x_1]$, make tangent at the point $(x_1, f(x_1))$, and we can obtain the second approximation of ξ

$$x_2 = x_1 - \frac{f(x_1)}{f'(x_1)},$$

x_2 is closer to the root ξ than x_1. So continue, we can obtain a series of approximation of ξ

$$b > x_1 > x_2 > \cdots > x_n > \cdots > \xi.$$

Because the sequence x_n decreases monotonically, and it has lower bound ξ, it must have limit. This limit is ξ. When

$$|x_{n-1} - x_{n-2}| < \delta \quad \text{and} \quad |x_n - x_{n-1}| < \delta$$

x_n can be approximation of ξ and the error is less than δ.

The other three kinds can also be considered, it should be noticed that the tangent is always made at the endpoint at which $f(x)$ and $f''(x)$ have the same sign.

Example 2 Find approximation of real root of $x^3 - 2x^2 - 4x - 7 = 0$ on $(3, 4)$ by using tangent method and the error is less than 0.01.

Solution From Example 1 we can know that $[3, 4]$ is the isolated interval of the real root. Let $f(x) = x^3 - 2x^2 - 4x - 7$ on $[3, 4]$,

$$f'(x) = 3x^2 - 4x - 4 > 0, f''(x) = 6x - 4 > 0.$$

But $f(3) = -10 < 0, f(4) = 9 > 0$, so from the right endpoint $b = 4$ of interval, we can obtain that

$$x_1 = 4 - \frac{f(4)}{f'(4)} = 4 - \frac{9}{28} \approx 4 - 0.321 = 3.679,$$

$$x_2 = 3.679 - \frac{f(3.679)}{f'(3.679)} \approx 3.679 - 0.046 = 3.633,$$

$$x_3 = 3.633 - \frac{f(3.633)}{f'(3.633)} \approx 3.633 - 0.001 = 3.632.$$

We can find two digits after decimal point of x_3 and x_2, and $f(3.632) > 0$, $f(3.631) < 0$, which means $3.631 < \xi < 3.632$. Let 3.631 be the approximation of ξ, its error is less than 0.01.

Exercises 4-6

1. Prove there is only one real root of $x^3 - 3x^2 + 6x - 1 = 0$ on $(0, 1)$, and find the

approximation by dichotomy and the error is less than 0.01.

2. Find the real root of $x^5+x+1=0$ by tangent method and the error is less than 0.001.

3. Find the approximation of $x=0.538\sin x+1$.

4.7 The application of extreme value of function in economic management

4.7.1 Demand analysis

Derivatives can be applied to analyze how does the demand change with the price. A cubic demand function is observed as follows:
$$D=-ap^3+bp^2-cp+d \quad (a>0,b>0,c>0,d>0).$$
Its first and second derivatives are:
$$\frac{dD}{dp}=-3ap^2+2bp-c, \quad \frac{d^2D}{dp^2}=-6ap+2b.$$
Therefore, the following analysis can be made:

(1) When $p=0, D=d$.

(2) Making $\frac{d^2D}{dp^2}=0$, the answer is $p=\frac{b}{3a}$. When $p<\frac{b}{3a}$, $\frac{d^2D}{dp^2}>0$, the curvature is concave-down; when $p>\frac{b}{3a}$, $\frac{d^2D}{dp^2}<0$, the curvature is concave-up. Thus, $p=\frac{b}{3a}$ is corresponded to the inflection point of demand curvature.

The demand functions of many durable consumer goods often belong to this kind of function which can be met frequently. When the price is falling down, the sales volume will rise rapidly. However, when the price is dropped closely to $\frac{b}{3a}$, market demand takes on a phenomenon of saturation that means the demand changes a little. But if the price plummets and is far away from $\frac{b}{3a}$, the sales volume will go up sharply again (Figure 4-24).

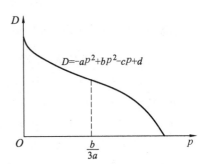

Figure 4-24

Then we discuss Engel function and Tornquist function.

Generally, the demand for goods changes not only with price, but also with many other factors. If we regard income of consumers as the main factor and the others is changeless, the functional relationship that demand for goods changes with consumers' income is called the Engel function, written as
$$Q=f(x).$$
In this formula, Q is the demand and x is the income of consumers.

Economically, if the Engel function of a goods is strictly increasing, then this goods is normal; if the function is strictly decreasing, this goods is inferior. Therefore, if $Q=f(x)$ is differentiable, the goods is normal when $f'(x)>0$ and the goods is low-quality when $f'(x)<0$.

Supposing that the Engel function of a goods is a power function:
$$f(x)=Ax^b, \ A>0.$$
Due to $f'(x)=Abx^{b-1}$, if this goods is normal, $b>0$ is the requirement and the elasticity that demand corresponds to income is
$$\varepsilon_{fx}=\frac{\mathrm{d}f}{\mathrm{d}x}\frac{x}{f}=bAx^{b-1}\frac{x}{Ax^b}=b.$$
Thus the elasticity that demand for this goods corresponds to income is b.

The following forms of the Engel function are put forward by Tornquist and named after him:
$$f(x)=\frac{ax}{x+b}, \ a,b>0;$$
$$g(x)=\frac{a(x-c)}{x+b}, \ a,b,c>0;$$
$$h(x)=\frac{ax(x-c)}{x+b}, \ a>0, b,c\geq 0.$$

The derivative of $f(x)$ is
$$f'(x)=\frac{ab}{(x+b)^2}, \ f(0)=0, \ \lim_{x\to\infty}f(x)=a.$$

This illustrates that with the increase of consumers' income, demand for goods $f(x)$ is strictly increasing to the saturated demand a; then $f''(x)=\frac{-2ab}{(x+b)^3}<0$, so the curvature $f(x)$ is up-convex(Figure 4-25).

Function $g(x)$ is similar to $f(x)$ that $g(x)$ is strictly increasing to the saturation point a, but the function value is negative within the interval $0<x<c$. So in this interval, the function makes no sense in economics. It explains that only if the income is greater than c, the market will have requirements for this goods.

Function $h(x)$ has no saturated demand, it can be easily demonstrated : this curvature is a down-convex incremental curvature and $\lim_{x\to\infty}h(x)=\infty$(Figure 4-26).

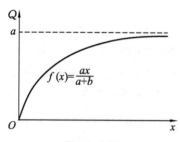

Figure 4-25

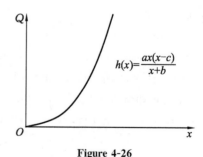

Figure 4-26

A kind of Engel function in the exponential form is analyzed as follows:
$$Q(x) = A e^{\frac{b}{x}}, \quad A > 0.$$
The first and second derivatives are
$$Q'(x) = -\frac{Ab}{x^2} e^{\frac{b}{x}}, \quad Q''(x) = \frac{Abe^{\frac{b}{x}}}{x^3}\left(2 + \frac{b}{x}\right).$$

It can be seen easily that when $b < 0$, $Q'(x) > 0$, therefore for normal goods, it is advisable to select $b < 0$. Due to $Q''\left(-\frac{b}{2}\right) = 0$ and the second derivative that changes symbol at $x = -\frac{b}{2}$, so the concavity and convexity also have changes. Thus the curvature $Q(x)$ has an inflection point at $x = -\frac{b}{2}$. Then because of $\lim_{x \to \infty} A e^{\frac{b}{x}} = A$, so the saturated demand of this goods is A. The income elasticity of demand is
$$\varepsilon_{Qx} = \frac{dQ}{dx}\frac{x}{Q} = -\frac{Ab}{x^2} e^{\frac{b}{x}} \frac{x}{A e^{\frac{b}{x}}} = -\frac{b}{x}.$$

Therefore, when $b < 0$, the elasticity is strictly decreasing. According to the analysis above, the figure of the Engel function is shown as Figure 4-27.

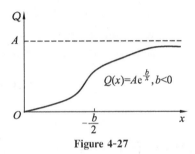

Figure 4-27

In economic problems, many practical problems that include the minimum value of cost, the maximum value of income and the maximum value of profit have to be solved often under certain conditions. In Mathematics, these problems are to obtain the maximum and minimum value of the function.

The next introduces the maximum and minimum value of some common functions in economic analysis.

4.7.2 Problems of minimum average cost

Providing that $C(Q)$ is the total cost function, then the average cost function is
$$\overline{C(Q)} = \frac{C(Q)}{Q}.$$
And Q is the output. During the process of production, enterprises are concerned about the minimum of average cost, which is the minimum average cost. Make,
$$\overline{C(Q)}' = \frac{QC'(Q) - C(Q)}{Q^2} = \frac{1}{Q}\left[C'(Q) - \frac{C(Q)}{Q}\right] = \frac{1}{Q}[C'(Q) - \overline{C(Q)}] = 0,$$
it can get the answer
$$C'(Q) = \overline{C(Q)},$$
which means the marginal cost is equal to the average cost.

This is one important conclusion in economy that making average cost be the minimum production means the marginal cost is equal to the production of average cost.

When the average cost is at lowest, the average cost function $\overline{C(Q)}$ and the total cost function $C(Q)$ also have certain relations with the elasticity of output Q.

When the average cost is the minimum, $C'(Q)=\overline{C(Q)}$, which is
$$C'(Q)=\frac{C(Q)}{Q}, \text{ so } Q=\frac{C(Q)}{C'(Q)}.$$
If the elasticity that the average cost $\overline{C(Q)}$ to output Q is written as $E_{\bar{c}}$, then
$$E_{\bar{c}}=\overline{C(Q)}'\frac{Q}{\overline{C(Q)}}=\frac{1}{Q}[C'(Q)-\overline{C(Q)}]\cdot\frac{Q}{\overline{C(Q)}}=0.$$

It means that when the average cost is the minimum, the elasticity that the average cost to output is zero.

If the elasticity that the total cost function to $C(Q)$ output Q is written as E_c, then
$$E_c=C'(Q)\frac{Q}{C(Q)}=\frac{C'(Q)}{C(Q)}\cdot\frac{C(Q)}{C'(Q)}=1.$$

This illustrates that when the average cost is the minimum, the elasticity that the total cost to output is one.

Example 1 Supposing the total cost function that makes Q products is
$$C(Q)=0.002Q^2+40Q+18\,000 \text{ Yuan}.$$
How many products should be made that the average cost is the minimum? And what is the minimum average cost?

Solution The average cost function is
$$\overline{C(Q)}=\frac{C(Q)}{Q}=0.002Q+40+\frac{18\,000}{Q},$$
$$\overline{C(Q)}'=0.002-\frac{18\,000}{Q^2}.$$
Make $\overline{C(Q)}'=0$, so $Q=3\,000$ (rounding $-3\,000$). Due to
$$\overline{C(3\,000)}''=\frac{36\,000}{Q^3}\Big|_{Q=3\,000}=\frac{4}{3\times10^6}>0,$$
when making 3 000 products, the average cost is the minimum and the minimum average cost $\overline{C(3\,000)}=52$ (Yuan/PC).

Example 2 By raw material, someone has to make five storage cabinets every day. Providing the transportation cost of woods from outside is 6 000 Yuan and the cost of storing one unit material is 8 Yuan, how many raw materials does he order for one time and how long does he make an order that can ensure the average daily cost within a production cycle between two transportations is the minimum?

Solution Supposing that someone makes an order every x days, the order must purchase $5x$ materials within one cycle of transportation and the average storage is approximately half of the transportation quantity, which is $\frac{5x}{2}$. Thus,

the cost of every cycle = the transportation cost + the storage cost $=6\,000+\frac{5x}{2}\cdot x\cdot 8$,

average cost $\overline{C(x)}=\frac{\text{cost of every cycle}}{x}=\frac{6\,000}{x}+20x$, $x>0$.

With $\overline{C(x)}'=-\frac{6\,000}{x^2}+20$, solving equation $\overline{C(x)}'=0$ will get stagnation points

$x_1 = 10\sqrt{3} \approx 17.32$, $x_2 = -10\sqrt{3} \approx -17.32$ (rounding).

Because of $\overline{C(x)}'' = \dfrac{12\,000}{x^3}$, then $\overline{C(x)}'' > 0$, so there is the minimum at the point of $x_1 = 10\sqrt{3} \approx 17.32$. The one who makes storage cabinets should convey units of woods what is $5 \times 17 = 85$ unit material outside every 17 days.

4.7.3 Problems of minimizing inventory cost

Retail stores in business pay attention to the inventory cost. Supposing that one store can sell 360 calculators annually, this store may ensure normal business by ordering all calculators for one time. But on the other hand, the storekeeper has to confront with the production cost that stores all calculators (such as insurance, floor space and etc.). Then he may divide the amount into several small orders. For example, the amount is divided into 6 groups, so the maximum that must be stored is 60. However, the storekeeper also has to pay for cost including clerical work, delivery expenses and labors when he makes an order again. Therefore, it seems that an equilibrium point exists between inventory cost and re-order cost. The following will demonstrate how does the differential calculus help us to confirm the equilibrium point. We minimize the function below:

total inventory cost = (annual production cost) + (annual reorder cost).

Batch refers to the maximum of goods in every re-order cycle. If is the order quantity of each cycle, the storage on hand is an integer from 0 to x within that time. In order to obtain an formula about storage on hand at every moment of that cycle, it is advisable to adopt average quantity $\dfrac{x}{2}$ to express this year's average inventory in the corresponding period.

Refer to Figure 4-28. If this batch is 360, the storage on hand is in one position between 0 and 360 within the period of two orders. The average inventory of goods on hand is $\dfrac{360}{2}$, 180 units. If this batch is 180, the storage on hand is in one position between 0 and 180 within the period of two orders. The average inventory of goods on hand is $\dfrac{180}{2}$, 90 units.

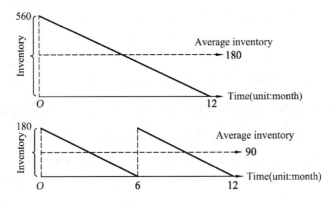

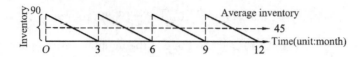
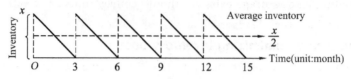

Figure 4-28

Example 3 One calculator store sells 360 units of calculator annually. The cost of storing one calculator for one year is 8 Yuan. For reordering, it should pay the fixed cost 10 Yuan and add 8 Yuan to every calculator. In order to minimize the inventory cost, how many times does the store order calculators every year? What's the number of every batch?

Solution Providing that x refers to the batch. The inventory cost refers to
$$C(x) = (\text{annual production cost}) + (\text{annual reorder cost}).$$
We discuss respectively the annual production cost and annual reorder cost.

The average inventory is $\frac{x}{2}$ and every calculator costs 8 Yuan to store. Therefore,

annual production cost $=$ (annual cost of one calculator) $\times$ (average number) $= 8 \cdot \frac{x}{2} = 4x$.

It is known that x refers to the batch, and then supposing the times of reordering annually is n, so $nx = 360 \Rightarrow n = \frac{360}{x}$. Thus

annual reorder cost $=$ (order cost of one time) $\times$ (times of reordering)
$$= (10 + 8x)\frac{360}{x} = \frac{3\,600}{x} + 2\,880.$$

Therefore, $$C(x) = 4x + \frac{3\,600}{x} + 2\,880.$$

Make $C'(x) = 4 - \frac{3\,600}{x^2} = 0$, it can obtain the stagnation points $x = \pm 30$.

But $$C''(x) = \frac{7\,200}{x^3} > 0.$$

Because there is only one stagnation point that $x = 30$ in the interval $[1, 360]$, so the minimum exists at the point of $x = 30$ (Figure 4-29).

Therefore, in order to minimize inventory cost, the store should annually order goods $\frac{360}{30} = 12$ (Time).

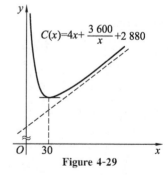

Figure 4-29

In this series of problems, what will happen if the answer is not an integral? For these functions, we can consider the two integrals that is closed to the answer. Then putting them into $C(x)$, and the smaller one of $C(x)$ is its batch.

Example 4 Discuss Example 3 again. Apart from changing the inventory cost from 8

Yuan to 9 Yuan, all data that provided from Example 3 is identical. For minimizing inventory cost, how large the batch is that the store plans to reorder calculators and how many times should it order annually?

Solution Compared with Example 3, its inventory cost changes into
$$C(x) = 9 \cdot \frac{x}{2} + (10 + 9x)\frac{360}{x} = \frac{9x}{2} + \frac{3\,600}{x} + 3\,240.$$

Then let $C'(x)$ be equal to 0 to work out x:
$$C'(x) = \frac{9}{2} - \frac{3\,600}{x^2} = 0 \Rightarrow x = \sqrt{800} \approx 28.3.$$

Because reordering 28.3 units for one time has no sense, so we can consider the two integrals beside 28.3, which are 28 and 29. Now we can get
$$C(28) \approx 3\,494.57(\text{Yuan}) \quad \text{and} \quad C(29) \approx 3\,494.64(\text{Yuan}).$$

Above all, the batch that minimizes inventory cost is 28. Although there is a gap of 0.07 between two numbers, this is not important. (Notice: this procedure cannot be applied to all kinds of functions, but it is applicative in the function we discuss.) Due to the time of reordering is $\frac{360}{28} \approx 13$, so it still refers to some approximate value.

The batch that minimizes total inventory cost is often called economic order quantity. To confirm the economic order quantity by methods above, it has to make three assumptions. Firstly, in the whole year, time from sending out indents to taking over cargo is consistent every time. Secondly, demand for productions is consistent in the whole year. It may be reasonable for calculators but unreasonable for seasonal goods such as clothing or skis. Finally, various cost involved (such as storage, transportation expenses) is invariable. Though calculating these cost by predicting and utilizing average cost, it may be unreasonable during inflation or deflation. In a word, the methods described above are still useful and make us utilize differential calculus to analyze some problems which seem difficult.

4.7.4 Problems of maximal profit

The revenue R of selling some goods is equal to the result that the unit price of goods P multiplies by quantity of sales x, $R = P \cdot x$. Then the selling profits equal to the result that revenue R minus cost C, $L = R - C$.

Example 5 For selling pieces of clothing, one garment company Ltd. confirms that its unit price should be $P = 150 - 0.5x$. Meanwhile, it also confirms the total cost of producing x pieces of clothing that can be expressed as $C(x) = 4\,000 + 0.25x^2$.

(1) Find general income $R(x)$;

(2) Find gross profit $L(x)$.

(3) For maximizing the profit, how many pieces of clothing should the company produce and sell?

(4) What is the maximal profit?

(5) For the sake of the maximal profit, how much should the unit price be?

Solution (1) General income.

$R(x)=$(pieces of clothing) $\cdot$ (unit price) $= x \cdot p = x(150-0.5x) = 150x - 0.5x^2$.

(2) Gross profit.
$$L(x) = R(x) - C(x) = (150x - 0.5x^2) - (4\,000 + 0.25x^2)$$
$$= -0.75x^2 + 150x - 4\,000.$$

(3) For calculating the maximum value of $L(x)$, firstly calculate,
$$L'(x) = -1.5x + 150,$$
then solve the equation $L'(x) = 0$ and get the answer $x = 100$.

Note $L''(x) = -1.5 < 0$. Because there is only one stagnation point, so $L(100)$ is the maximum value.

(4) The maximal profit is
$$L(100) = -0.75 \times 100^2 + 150 \times 100 - 4\,000 = 3\,500 (\text{Yuan}).$$

Thus the company must produce and sell 100 pieces of clothing to realize the maximal profit 3 500 Yuan.

(5) The unit price to realize maximum profit is
$$P = 150 - 0.5 \times 100 = 100 (\text{Yuan}).$$

Gross profit function and its related functions are normally inspected. Figure 4-30 displays an example about total cost function and general income function.

According to observation, it can estimate that the maximal profit is the widest gap between $R(x)$ and $C(x)$, $C_0 R_0$ Point B_1 and B_2 are the break-even points.

Figure 4-31 displays an example about gross profit function. Notice that when the production is too low ($< x_0$), there will be in a loss that caused by high fixed cost or high initial cost and low revenue. When the production is too high ($> x_2$), there will also be in a loss that caused by high marginal cost and low marginal profit (Figure 4-32).

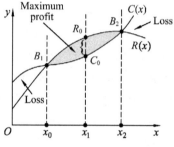

Figure 4-30

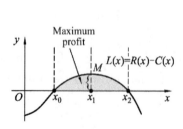

Figure 4-31

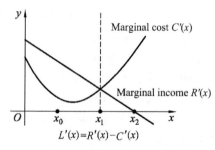

Figure 4-32

Business are made at every profitable offering between x_0 and x_2. Notice that the maximal profit exists at $L(x)$'s stagnation point x_1. Assuming that $L'(x)$ always exists to all the x in one interval (usually $[0, \infty)$), so the stagnation point is one number x that makes

$$L'(x)=0 \text{ and } L''(x)<0.$$

Because $L(x)=R(x)-C(x)$, it can obtain that
$$L'(x)=R'(x)-C'(x) \text{ and } L''(x)=R''(x)-C''(x).$$

Therefore, the maximal profit exists one point that makes
$$L'(x)=R'(x)-C'(x)=0 \text{ and } L''(x)=R''(x)-C''(x)<0$$

or
$$R'(x)=C'(x) \text{ and } R''(x)<C''(x).$$

Above all, we can get the following theorem.

Theorem When marginal revenue is equal to marginal cost and the rate of change of marginal revenue is less than that of marginal cost,
$$R'(x)=C'(x) \text{ and } R''(x)<C''(x)$$
the maximum profit can be made.

Example 6 One university is trying to price football tickets. If one ticket is 6 Yuan, then there will have 70 000 spectators in every match. 1 Yuan the price raise each time, 10 000 spectators will lose from the average number. The average cost of every spectator spent on concessional rate is 1.5 Yuan. For maximizing the profit, how much should one ticket be? According to this price, how many spectators will come to watch matches?

Solution Assuming that the sum raised of one ticket is x (if x is negative, the price of ticket will reduce). Firstly, express general income R as function of x.

$R(x)=$(revenue of ticket)$+$(revenue of concessional rate)
$\quad\quad=$(number of spectators)$\times$(ticket price)$+1.5\times$(number of spectators)
$\quad\quad=(70\ 000-10\ 000x)(6+x)+1.5(70\ 000-10\ 000x)$
$\quad\quad=-10\ 000x^2-5\ 000x+525\ 000.$

For working out that x maximizes $R(x)$, calculating $R'(x)$:
$$R'(x)=-20\ 000x-5\ 000.$$

After solving equation $R'(x)=0$, it can obtain that $x=-0.25$(Yuan).

Because $R''(x)=-20\ 000<0$ and this is the only stagnation point, the maximum is $R(-0.25)$.

Therefore, in order to maximize the revenue, the football ticket price should be
$$6-0.25=5.75(\text{Yuan})$$

In other words, the price reduced will appeal to more spectators to watch matches that can make the maximal profit. The number is
$$70\ 000-10\ 000\times(-0.25)=72\ 500.$$

4.7.5 Problems of maximal revenue

Providing that price is P, quantity demanded is Q, then revenue function is $R=PQ$.

(1) How to confirm the price of commodity that maximizes the revenue.

When researching problems of revenue, Q is the function of P. Due to

$$R' = Q + PQ',$$

and $R' = 0$ is known from requirement that extreme value exists, it can obtain that

$$P = -\frac{Q}{Q'}.$$

Because the stagnation point is unique, so the revenue is the maximum value when price $P = -\frac{Q}{Q'}$.

When the revenue is the maximum, elasticity of demand E_d also conforms to certain relations:

$$E_d = \frac{dQ}{dP} \cdot \frac{P}{Q} = Q' \cdot \frac{-Q}{Q'Q} = -1.$$

It illustrates that when the revenue is the maximum value, elasticity of demand is equal to -1.

(2) When discussing the relations between sales volume and revenue, how to confirm the sales volume of goods that maximizes the revenue.

P is the function of Q. Because $R = P(Q)Q$, therefore $R' = P + QP'$. Make $R' = 0$, it can obtain that $Q = -\frac{P}{P'}$. Thus when $Q = -\frac{P}{P'}$, the revenue is the maximum value.

Example 7 Assuming that sales volume of one commodity is $Q = 50 - 5P$, and the unit of price P is: ten-thousand Yuan/t. How many tons the sales volume are that maximizes revenue? What is the maximal revenue?

Solution Through the equation $Q = 50 - 5P$, it can obtain that $P = 10 - \frac{Q}{5}$. Thus

$$R = P \cdot Q = 10Q - \frac{1}{5}Q^2.$$

Make $R' = 10 - \frac{2}{5}Q = 0$, and then obtain $Q = 25$.

Because $R''(25) = -\frac{2}{5} < 0$, so the revenue is the maximum value when the sales volume are 25 t. The maximal revenue is

$$R(25) = 10 \times 25 - \frac{25^2}{5} = 125 \text{ (ten-thousand Yuan)}.$$

4.7.6 Problems of maximal tax

Example 8 The average cost of one commodity is $\overline{C(Q)} = 2$. The price function is $P(Q) = 20 - 40Q$ (Q refers to quantity of commodity). The tax that levied by nation from each enterprise is t.

(1) How many commodities produced are that can maximize profit?

(2) Under the condition the enterprise makes maximal profit, what is t that maximizes the total tax?

Solution (1) The total cost is $C(Q) = \overline{C(Q)} \cdot Q = 2Q$, so there will have

$$R(Q) = QP = 20Q - 40Q^2,$$

$$T(Q)=tQ,$$
$$L(Q)=R(Q)-C(Q)-T(Q)=(18-t)Q-40Q^2.$$

Mak $L'(Q)=18-t-80Q=0$ and obtain $Q=\dfrac{18-t}{80}$. Due to $L''(Q)=-80<0$, therefore the maximal profit is
$$L\left(\dfrac{18-t}{80}\right)=\dfrac{(18-t)^2}{160}.$$

(2) At the moment that profit is the maximum value, the tax is
$$T=tQ=\dfrac{18t-t^2}{80}.$$

Make $T'=\dfrac{9-t}{40}=0$ and obtain $t=9$.

Due to $T''=-\dfrac{1}{40}<0$, so when $t=9$, the total tax has its maximum value
$$T(9)=\dfrac{9(18-9)}{80}=\dfrac{81}{80}.$$

Meanwhile, the gross profit is
$$L=\dfrac{(18-9)^2}{160}=\dfrac{81}{160}.$$

4.7.7 Problems of compound interest

Generally when making a loan from banks, enterprises have to pay for interest that paid according to stated profit and deadline.

Providing that principal is P, annual profit is r and the deadline of loan is x. The interest can be calculated through two methods, which are simple interest and compound interest. The formula is expressed as follows:

amount calculated by simple interest $=P(1+xr)$,

amount calculated by compound interest $=P(1+r)^x$.

Now if we divide one year into periods to calculate compound interest, amount paid after making a loan for years should be
$$S=P\left(1+\dfrac{r}{t}\right)^{tx}.$$

By utilizing binomial
$$(1+x)^p=1+px+\dfrac{p(p-1)}{2}x^2+\cdots$$

to unfold the formula above and noticing that $\left(1+\dfrac{r}{t}\right)^t>1+r$, so there will be
$$P\left(1+\dfrac{r}{t}\right)^{tx}>P(1+r)^x.$$

That is to say, in one year, compound amount calculated for t times is much larger than one calculated for one time. And the more times of calculating compound interest are, the larger the compound amount calculated is. However, the amount will also not infinitely increase. Because

$$\lim_{t\to\infty} P\left(1+\frac{r}{t}\right) = \lim_{t\to\infty} P\left(1+\frac{r}{t}\right)^{\frac{t}{r}\cdot rx} = Pe^{rx}.$$

Therefore, calculating compound interest continually on the basis of principal P and nominal annual rate r, the amount after x years will be $S = Pe^{rx}$.

The limit above refers as continuous compound interest formula and x in this formula refers as continuous variable. Thus it is only a theoretical formula that not used in practice. Only for the long deposit term, it will be used as an approximate evaluation.

The following discusses problems that are contrary to the condition above. In order to achieve amount S at the end of x years, what is the principal P deposited in the bank?

Assuming that annual rate is r and calculate compound interest for one time a year, so P refers to the formula

$$P(1+r)^x = S,$$

which is

$$P = \frac{S}{(1+r)^x}.$$

Obviously, if nominal annual rate is r and calculate compound interest for t times a year, therefore

$$P = \frac{S}{\left(1+\frac{r}{t}\right)^{tx}}.$$

Furthermore, if the formula is calculated by continuous compound interest, it can obtain $Pe^{rx} = S$, which is equal to $P = Se^{-rx}$.

From the discussion above it is known that the amount of money after several years is related to rate r and time t of calculating compound interest. The more rate and time are, the less the amount of money is.

Example 9 On the basis of technologies and resources, one company intended to obtain x Yuan this year and y Yuan the next year by investment. Between y and x, there exists a relation: $y = \varphi(x)$ (called conversion curve or line of investment opportunity). Supposing that annual rate is r and calculate compound interest for one time a year, how to adjust the investment that maximizes current value?

Solution Providing W refers to current value including the amount x obtained this year and the amount y obtained the next year, therefore

$$W = x + \frac{\varphi(x)}{1+r}.$$

According to the question, for maximizing W, it just needs to find out that satisfies

$$\frac{dW}{dx} = 0 \text{ and } \frac{d^2W}{dx^2} < 0.$$

That is to say,

$$1 + \frac{\varphi'(x)}{1+r} = 0 \text{ and } \frac{\varphi''(x)}{1+r} < 0.$$

Due to $1+r > 0$, so when $\varphi(x)$ is up-concave function that $\varphi''(x) < 0$, adjustable investments make revenue x obtained this year meet the relation $\varphi'(x) = -(1+r)$.

Meanwhile, revenue y obtained the next year can also calculated through $y=\varphi(x)$. Thus after such an adjustment, the current value will be the maximum value.

Exercises 4-7

1. The profit producing units of one commodity is $L(x)=5\ 000+x-0.000\ 01x^2$ (Yuan). How many units are produced that maximize the profit?

2. The demand function of one product is $P=10-3D$ and the average cost is $\bar{c}=D$. How large is the quantity demanded that maximizes the profit. And calculate the maximal profit.

3. One factory produces a product whose annual output is x(Unit: hundred Pcs) and total cost is C(Unit: ten-thousand Yuan) including fixed cost twenty thousand Yuan. Each one hundred Pcs are produced, the cost will increase ten thousand Yuan. On the market, this product can be annually sold 400 Pcs. The gross income from sales R is the function of x

$$R=R(x)=\begin{cases} 4x-\dfrac{1}{2}x^2, & 0\leqslant x\leqslant 4, \\ 8, & x>4. \end{cases}$$

For maximizing gross profit L, how many products should be produced annually?

4. Assuming that total cost function of one commodity is $C=1\ 000+3D$ and demand function is $D=-100P+1\ 000$. The unit price of the commodity is P. What is that can maximize the profit?

5. The cost function of one commodity is $C=15Q-6Q^2+Q^3$ (Q refers to production).

(1) How large the production is that minimizes average cost?

(2) Calculate marginal cost and demonstrate that marginal cost equals average cost when the average cost is the minimum value.

6. The annual sales of commodity produced by one factory is one million. Each batch has to increase 1 000 Yuan for production arrangements and the storage fee of one is 0.05 Yuan. If annual sales rate is homogeneous (At this time, average inventory of the commodity is half of the batch). For minimizing the sum of production arrangements and storage fee, how many times should the batch be divided into for production?

7. One company annually sells 5 000 units of one commodity whose unit price is 200 Yuan. The purchase expenses for one time are 40 Yuan and the rate of annual storage fee is 20%. Calculate the economic order quantity (Optimal order quantity).

8. One material demanded for production a year is 5 170 t. The subscription cost each time is 570 Yuan. The unit price of this material is 600 Yuan per ton and the rate of storage fee is 14.2%.

Calculate: (1) optimal order quantity; (2) optimal order batch; (3) optimal order cycle; (4) minimal total cost.

9. Sports shop annually sells 100 billiard tables. The fee for storing a billiard table a

year is 20 Yuan. For recovering, it needs to pay fixed cost 40 Yuan and add 16 Yuan to each table. In order to minimize storage cost, how many times should the shop order billiard tables a year? How large the batch is for one time?

10. The unit price of one commodity is $P=7-0.2x$ (ten-thousand Yuan/t). x refers to the offtake and total cost function is
$$C=3x+1 (\text{ten-thousand Yuan})$$

(1) If one ton of the commodity is sold out, the government will levy t (ten-thousand Yuan). Calculate sales volume when the merchant earns maximal profit.

(2) Under the condition that the merchant earns maximal profit, what is t which maximizes total tax?

Summary

1. Main contents

The main content of this chapter is to calculate the infinitive and the research function's form by the derivative.

(1) L'Hospital's Rule.

The indeterminate form $\frac{0}{0}$, $\frac{\infty}{\infty}$, its limit can be calculated by using the L'Hospital's Rule. There are types of indeterminate forms: $0 \cdot \infty$, $\infty - \infty$, 0^0, 1^∞, ∞^0. These indeterminate forms can be calculated by being turned into type $\frac{0}{0}$ or $\frac{\infty}{\infty}$, then using the L'Hospital's Rule to calculate the limit.

Find the limit by L'Hospital's Rule, the following points should be noted:

① L'Hospital's Rule is to take the derivative of the numerator and the denominator and take the limit instead of taking the derivative of the whole quotient.

② The indeterminate form is not $\frac{0}{0}$, $\frac{\infty}{\infty}$, it cannot use the L'Hospital's Rule.

③ Sequences' limits cannot directly use L'Hospital's Rule.

④ As far as possible, the limit method is combined with the limit method (important limit formula, equivalent infinitesimal substitution, etc.) to simplify the calculation of the limit.

(2) The steps of finding the monotone interval and extreme of function:

① Defining the domain of the continuous function $f(x)$.

② Find the $f'(x)$, assume $f'(x)=0$ (stagnation point) and $f'(x)$ that does not exist, then dividing a domain into a number of parts.

③ The monotonic increase and subtraction of the function $f(x)$ can be determined by judging the symbols of $f'(x)$ in each part of the interval.

④ Investigate the monotonicity increasing and decreasing properties of the functions near and near each stagnation point and $f'(x)$. The turning point of monotonicity increasing and decreasing is the extreme point. Determine whether it is a maximum point or a minimum point, and find the extreme point and the extreme value.

(3) The steps of the concave and convex interval of the curve and the step of the inflection point and the monotone interval of the function are similar to those of the extreme. Only the $f'(x)$ is changed to $f''(x)$.

(4) Maximum and minimum.

The continuous function on closed interval $[a, b]$, $f(x)$ can get the maximum and minimum value on $[a, b]$. It only needs all stationary points in (a, b). There is no point function value and $f(a)$, $f(b)$ in the first derivative. The maximum (or minimum) value is the maximum (minimum) value of the function.

The steps of finding the maximum value and the minimum value of the practical application problem:

① Understand the meaning of practical problems, find out the problems to be solved.

② To establish a mathematical model of a practical problem and to determine the domain of definition, that is, the actual problem is converted to the maximum and minimum value problem of a continuous function on an interval.

③ Finding the end point of function in the defined domain and the interval end where the function changes, it is the suspect point of the maximum (minimum) value, and the maximum (minimum) value of the corresponding function value is obtained.

(5) The steps of drawing the graph of the function $y=f(x)$.

① Determine the domain of the function $y=f(x)$ and some features of the function, such as parity, periodicity, etc..

② Find the first and two derivatives of the function, find out the points of $f'(x)=0$, $f''(x)=0$ and the points are not existent, and divide the domain into several intervals with these points.

③ Determine the positive and negative numbers of $f'(x)$, $f''(x)$ in the partitions, set up a table, determine the monotone range, extreme point, concave and convex interval and inflection point of the function.

④ Check the graphics function have or without asymptote.

⑤ A number of points on the graph are properly fixed, drawing the graph.

(6) Arc differential and curvature.

Arc differential: $ds = \sqrt{1+y'^2}\,dx$ or $ds = \sqrt{(dx)^2+(dy)^2}$.

Curvature: $K = \left| \dfrac{d\alpha}{ds} \right| = \dfrac{|y''|}{(1+y'^2)^{\frac{3}{2}}}$.

(7) Demand analysis, minimum average cost problem, inventory cost minimization problem, maximum profit problem, maximum income problem, maximum tax problem, compound interest problem.

2. Basic requirements

(1) Familiar with the limit of indeterminate form L' Hospital's Rule.

(2) Understand the extreme value concept of function, grasp the monotonicity and extreme value method of function with derivative, and solve the application problem of simpler maximum and minimum. We will use the monotonicity of function and the maximum (minimum) to prove inequality.

(3) Master the judgment of convex function of graphics and method of inflection point by derivative, horizontal and vertical asymptote will find the curve, will describe the function of the graphics.

(4) To understand the concept of arc differential, to understand the concept of curvature and radius of curvature and to calculate the radius of curvature.

(5) The problem of maximum and minimum in economic management is solved by using the method of finding the extreme value and the most value of the function.

Quiz

1. Find the limits of following exercises:

(1) $\lim\limits_{x \to 0} \dfrac{e^{x^2}-1}{\sin^2 x}$;

(2) $\lim\limits_{x \to +\infty} \dfrac{\ln(x-1)}{\ln(x^2-1)}$;

(3) $\lim\limits_{x \to 0} \dfrac{\sin x}{\sqrt{1-\cos x}}$.

2. Try to solve the following exercises:

(1) Discuss the monotonicity of $y=(x+1)^3$ on the $(-1,2)$.

(2) Find the local minimum of $y=(x-1)^2$ on the $(-\infty,+\infty)$.

(3) How many inflection point of curve $y=e^{-x^2}$? If so, how many inflection point.

(4) Find the horizontal asymptote of $y=2\ln\dfrac{x+3}{x}-3$.

(5) Find the maxima of function $y=x^2+1$ on the $(-1, 1)$.

3. Suppose $f(x)=\begin{cases} -x^2, & x<0, \\ x\arctan x, & x\geqslant 0. \end{cases}$

(1) Find the $f'(0)$;

(2) Discuss the monotonicity of $f(x)$ (Writing the calculation).

4. Discuss the function $y=e^{|x|}$ at the point $x=0$ whether or not derivable? Whether or not have extreme value. If the function has extreme value, find it.

5. A, B both factories and docks are located on the same side of a straight line river, A is off the dock is 10 km. B is north of the dock, and off the dock 4 km. A highway is to be built between the two factories. If built road along the river, the cost is 3 000 Yuan/km. Otherwise, the cost is 5 000 Yuan/km. Find how long the highway has been built along the river from A, can make road building total cost the most province.

6. Prove: $x>\ln(1+x^2)$, where $x>0$.

Chapter 4　Applications of derivatives

7. Find the curvature and radius of curvature on the curve $y=\dfrac{1}{3}x^3$ ($x>0$) at each point, and point with the smallest radius of curvature.

8. Suppose the total cost of producing a product is $C(x)=10\ 000+50x+x^2$ (x is production), the average cost of each product is the lowest when how many it produces.

9. Suppose price function is $P=15e^{-\frac{x}{3}}$ (x is production). For the maximizing the income find the production, price, and income.

10. Suppose the product's demand and Q price P function is $Q=12\ 000-80\ P$. The product's total cost C and demand Q function is $C=25\ 000+50Q$, the unit tax is 2. Try to maximize the price of the product and the maximum profit.

11. A stereo radio receiver factory to determine, to sell a new stereo radio sets, a set's price must be $P=800-x$. And, the product x sets' total costs is $C(x)=2\ 000+10x$.

(1) Find the general income $R(x)$.

(2) Find the total profit $L(x)$.

(3) For the maximizing the profit, how many the factory must product and sell?

(4) Find the maximum profit.

(5) For the maximum profit, how much every set's price?

Exercises

1. Find the following limits:

(1) $\lim\limits_{x\to 0}\dfrac{a^x-a^{-x}}{1-x-\log_a(a-x)}$;

(2) $\lim\limits_{x\to 0}\dfrac{xe^{2x}+xe^x-2e^{2x}+2e^x}{(e^x-1)^3}$;

(3) $\lim\limits_{x\to\infty}\dfrac{x^8}{8^x}$;

(4) $\lim\limits_{x\to 1^-}\dfrac{1}{\cos\dfrac{\pi x}{2}\ln(1-x)}$;

(5) $\lim\limits_{x\to\frac{\pi}{6}}\sin\left(\dfrac{\pi}{6}-x\right)\tan 3x$;

(6) $\lim\limits_{x\to 1}\left(\dfrac{\pi}{2}-\arctan x\right)^{\frac{1}{x}}$.

2. If a constant C, suppose $\lim\limits_{x\to+\infty}\dfrac{\ln(1+Ce^x)}{\sqrt{1+Cx^2}}=4$, find C.

3. Determine the monotonicity intervals of the following function:

(1) $y=\dfrac{1-x+x^2}{1+x+x^2}$;

(2) $y=x\sqrt{ax-x^2}$.

4. Find the extreme value, maximum, minimum of the function $y=x\cos x-\sin x$ ($0\leqslant x\leqslant 2\pi$).

5. Prove:

(1) $\ln x>\dfrac{2(x-1)}{x+1}$, $x>1$;

(2) $1+x\ln(x+\sqrt{1+x^2})>\sqrt{1+x^2}$, $x>1$.

6. For a inscribed in the radius of R ball's cone, when it has the largest volume, find the high.

7. Parallel to the tangent point A at the circle of the circular chord BC, the area of $\triangle ABC$ is the largest, find the distance between A and BC.

8. Suppose the curve $y=x^3+ax^2+bx+c$ has inflection point $(1,-1)$, and at the $x=0$, it has local maximum, find these constant a, b, c.

9. Draw the graph of the function $y=x^2 e^{-x}$.

10. Suppose the function $f(x)$ on the $[a, +\infty)$ has second-order derivative, and (1) $f(a)>0$; (2) $f'(a)<0$; (3) $f''(x)\leqslant 0$, when $x>a$. Try to prove: the function $f(x)=0$ has only one real root on the $(a, +\infty)$.

11. Suppose the function $y=f(x)$ at $x=x_0$'s one neighborhood has continuous derivatives of order three. If $f''(x_0)=0$, but $f'''(x_0)\neq 0$. Prove $(x_0, f(x_0))$ is inflection point of the curve $y=f(x)$.

12. The company has an estimated cost function for the product is
$$C(x)=2\,600+2x+0.001x^2,$$
how much output, the average cost can reach the minimum, find the minimum average cost.

13. The fixed cost of one production is 10 000 Yuan, The variable cost is proportional to the cube of the product's daily output of x t, when the daily output is 20 t, the total cost is 10 320 Yuan. How many daily output should be made that the average cost is the minimum? And what is the minimum average cost? (Assuming the maximum output of 100 tons per day)

14. A home appliance factory is producing a new refrigerator, in order to sell the x refrigerator, its unit price is $P=280-0.4x$. At the same time, it is also determined that the total cost of producing the x refrigerator is
$$C(x)=5\,000+0.6x^2.$$
(1) Find the total income $R(x)$.
(2) Find the total profit $L(x)$.
(3) For the maximizing the profit, how many the factory must product and sell?
(4) Find the maximum profit.
(5) For the maximum profit, how much every set's price?

15. Busy salesman rides a thin line between profit and loss especially in determining the price. For example, an agreement on the fare of a cinema. According to a continuous record of a theater, determination, if the ticket is 20 Yuan, the average number of 1 000 for the cinema viewing, but the price is increase1 Yuan, the theater the average number of lost 100 customers, each customer in the price on the average cost of 1.8 Yuan, in order to maximize the total profit, the theater should determine how much is the fare ticket?

16. A retail appliance store annual sales of 2 500 TV sets. A TV inventory one year in the store, the store need to spend 10 Yuan, in order to re order, pay a fixed cost of 20

Yuan, and each other to pay 9 Yuan. In order to minimize the inventory cost, how many the store should order and how often should be ordered a year?

17. Discussion of the Q16, in addition to changing the cost of the inventory to 10 Yuan to 20 Yuan, all the data given in the above question are adopted. In order to minimize the inventory cost, how many the store should order and how often should be ordered a year?

Chapter 5　Indefinite integrals

In the differential calculus of unary functions, we have studied how to find the derivative of a function. But in many practical problems, it always need to solve the opposite problems that we recover from its known derivative. It is one of basic problems in integral calculus—indefinite integral.

5.1　Indefinite integrals

5.1.1　Antiderivative

In practice, it often consider the following problems: calculate $f(x)$ through the given derivative $f'(x)$ of one function $f(x)$. For example, the speed of point M which makes rectilinear motion at any time is $v=v(t)$. In order to find the law of motion of point M, it means finding that the relation of distances of the point M moves and time is $s=f(t)$. Mathematically, the substance of the opposite problem is: seek out a function $s=f(t)$ that makes the derivative $f'(t)$ of function $f(t)$ equal to the given function $v(t)$, $f'(t)=v(t)$. Because the problem has universal meaning, therefore, we introduce the conception of antiderivatives.

Definition 1　A function $F(x)$ is called an antiderivative of the functionon $f(x)$ an interval I if $F'(x)=f(x)$ or $\mathrm{d}F(x)=f(x)\mathrm{d}x$ for all x in I.

For example, due to $(x^2)'=2x$ on $(-\infty,+\infty)$, so x^2 is an antiderivative of $2x$ on $(-\infty,+\infty)$. For another example, due to $(\sqrt{x})'=\dfrac{1}{2\sqrt{x}}$ on $x\in(0,+\infty)$, so $\sqrt{x}$ is an antiderivative of $\dfrac{1}{2\sqrt{x}}$ on $x\in(0,+\infty)$.

As for antiderivatives, firstly, we put forward such a question: what the function is that exists the antiderivative?

Theorem 1 (Existence theorem of antiderivatives)　If a function $f(x)$ is continuous on the interval I, then there exists a differentiable function $F(x)$ on the interval that makes $F'(x)=f(x)$ for all $x\in I$.

Briefly, continuous functions have antiderivatives.

Because elementary functions in their domains are continuous, so all of these are antiderivatives. Is the antiderivatives of $f(x)$ unique? The answer is no. Known from the conception of antiderivativs, if $F'(x)=f(x)$, for any constant C there all exists:
$$[F(x)+C]'=f(x).$$
It can be seen that when $F(x)$ is the antiderivatives of $f(x)$, $F(x)+C$ is also the antiderivativs of $f(x)$. Thus if for a given function $f(x)$ that exists the antiderivatives, $f(x)$ will have infinite antiderivativs.

Theorem 2 If $F(x)$ is one antiderivative of $f(x)$, function $f(x)$ has infinite antiderivativs and any antiderivative can refer to the form $F(x)+C$.

Proof Prove that $\Phi(x)$ is one antiderivative of $f(x)$ and make
$$y=\Phi(x)-F(x).$$
Due to $\qquad F'(x)=f(x),\ \Phi'(x)=f(x).$
Therefore $\qquad y'=\Phi'(x)-F'(x)=f(x)-f(x)\equiv 0.$
Thereupon $\Phi(x)-F(x)=C$, thus $\Phi(x)=F(x)+C.$

According to Theorem 2, there is only one constant between any two original functions of a function.

5.1.2 The conception of indefinite integral

Definition 2 On the interval I, function $f(x)$'s antiderivatives which have any constant refer to indefinite integral of $f(x)$ (or $f(x)dx$) on the interval I, denoted by $\int f(x)dx$,
$$\int f(x)dx = F(x)+C,$$
where C is an arbitrary constant and $F(x)$ is one antiderivative of $f(x)$, $f(x)$ is called the integrand, $f(x)dx$ is called the expression under the integrant sign, $\int$ is the integrant sign and x is the variable of integration.

It can be seen that the indefinite integral of one function is not an exact number or a function but a family of functions instead.

For example, due to $(\sin x)'=\cos x$, so $\int \cos x dx = \sin x + C.$

Then due to $(x^2)'=2x$, so $\int 2x dx = x^2 + C.$

Significance of indefinite integral in geometry: Suppose that $F(x)$ is one antiderivative of $f(x)$, in the plane $y=F(x)$ is a curve which is titled as the integral curve of $f(x)$. And $\int f(x)dx=F(x)+C$ refers to a set of curves obtained by $y=F(x)$ moving along y-axis. The tangent line's slope of any curve at the same abscissa x is $f(x)$.

Example 1 Suppose that the curve passes through point$(2,5)$, and the tangent slope in an arbitrary point (x,y) of the curve is $2x$. Find the curve.

Solution Suppose that the curve is $y=F(x)$. According to the question:
$$F'(x)=2x.$$
Thus $$F(x) = \int 2x dx = x^2 + C.$$

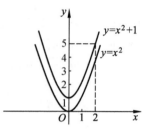

Because the curve passes through point $(2,5)$, it can obtain that $F(2)=5$,
$$5=2^2+C.$$
Thereupon, $C=1$. Therefore the curve is $y=x^2+1$(Figure 5-1).

Figure 5-1

According to the definition of indefinite integral, we can know the following relations:

(1) $\left[\int f(x)dx\right]' = f(x)$ or $d\left[\int f(x)dx\right] = f(x)dx.$

In fact, assumed that $F(x)$ is one antiderivative of $f(x)$, $F'(x)=f(x)$, then,
$$\int f(x)dx = F(x)+C,$$
therefore
$$\left(\int f(x)dx\right)' = (F(x)+C)' = F'(x) = f(x).$$

(2) $\int F'(x)dx = F(x)+C$ or $\int dF(x) = F(x)+C.$

Actually, because $F'(x)$ is the antiderivative of $F(x)$, so there is
$$\int F'(x)dx = F(x)+C.$$

The method to calculate indefinite integral is titled as integration. If regard integration as an operation, it can say that integration is an inversion operation of differential method. Thus we can obtain basic integral formulas by using derivative formulas.

5.1.3 Table of basic indefinite integrals

We observe that integration and differentiation are mutually inverse operations. The inverse relation can be seen clearly from the following operation formulas. Then we give the following table of integrals of the simplest functions which is called table of basic indefinite integrals.

(1) $\int k dx = kx + C$ (k is a constant);

(2) $\int x^{\alpha} dx = \dfrac{1}{\alpha+1}x^{\alpha+1} + C$ ($\alpha \neq -1$);

(3) $\int \dfrac{1}{x} dx = \ln|x| + C;$

(4) $\int \dfrac{dx}{1+x^2} = \arctan x + C$;

(5) $\int \dfrac{dx}{\sqrt{1-x^2}} = \arcsin x + C$;

(6) $\int \cos x \, dx = \sin x + C$;

(7) $\int \sin x \, dx = -\cos x + C$;

(8) $\int \sec^2 x \, dx = \tan x + C$;

(9) $\int \csc^2 x \, dx = -\cot x + C$;

(10) $\int \sec x \tan x \, dx = \sec x + C$;

(11) $\int \csc x \cot x \, dx = -\csc x + C$;

(12) $\int e^x \, dx = e^x + C$;

(13) $\int a^x \, dx = \dfrac{a^x}{\ln a} + C \ (a>0,\ a\neq 1)$;

(14) $\int \operatorname{sh} x \, dx = \operatorname{ch} x + C$;

(15) $\int \operatorname{ch} x \, dx = \operatorname{sh} x + C$.

As for the formula $\int \dfrac{1}{x} dx = \ln|x| + C$, demonstrate as follows:

① When $x>0$, $(\ln x)' = \dfrac{1}{x}$, so $\int \dfrac{1}{x} dx = \ln x + C$.

② When $x<0$, $[\ln(-x)]' = \dfrac{1}{x}$, so $\int \dfrac{1}{x} dx = \ln(-x) + C$.

Thus for any one that $x \neq 0$, $\int \dfrac{1}{x} dx = \ln|x| + C$.

Example 2 Find $\int \dfrac{1}{x^4} dx$.

Solution $\int \dfrac{1}{x^4} dx = \int x^{-4} dx = \dfrac{x^{-4+1}}{-4+1} + C = -\dfrac{1}{3x^3} + C$.

Example 3 Find $\int x^2 \sqrt[3]{x} \, dx$.

Solution $\int x^2 \sqrt[3]{x} \, dx = \int x^{\frac{7}{3}} dx = \dfrac{x^{\frac{7}{3}+1}}{\frac{7}{3}+1} + C = \dfrac{3}{10} x^{\frac{10}{3}} + C$.

After, the two integrands above are power functions. Under this condition, firstly, change integrand into the form x^a, then calculate indefinite integral by integral formulas of power function.

These formulas are the foundations to compute indefinite integrals. However, integrals that we shall encounter in the future are usually more complicated than these in the table. We have to make some deformation for the integrand of an integral so that we can obtain the result by looking up the table.

5.1.4 Properties of the indefinite integral

According to the definition of indefinite integral, it can easily demonstrate the following integral algorithms. In the concrete, let the antiderivatives of $f(x)$, $g(x)$ both exist on the interval I.

Property 1 $\int kf(x)\mathrm{d}x = k\int f(x)\mathrm{d}x$, k is a constant ($k \neq 0$).

Property 1 means the constant of integrand can be moved out of the sign of integration.

In fact, $\left[k\int f(x)\mathrm{d}x\right]' = kf(x)$, which is $\int kf(x)\mathrm{d}x = k\int f(x)\mathrm{d}x$.

Property 2 $\int [f(x) \pm g(x)]\mathrm{d}x = \int f(x)\mathrm{d}x \pm \int g(x)\mathrm{d}x$.

Actually, $\left[\int f(x)\mathrm{d}x \pm \int g(x)\mathrm{d}x\right]' = \left[\int f(x)\mathrm{d}x\right]' \pm \left[\int g(x)\mathrm{d}x\right]' = f(x) \pm g(x)$,

Thus $\int [f(x) \pm g(x)]\mathrm{d}x = \int f(x)\mathrm{d}x \pm \int g(x)\mathrm{d}x$.

The one that combines Property 1 and Property 2 together is titled as linear property of integrals

$$\int [k_1 f(x) \pm k_2 g(x)]\mathrm{d}x = k_1 \int f(x)\mathrm{d}x \pm k_2 \int g(x)\mathrm{d}x,$$

k_1, k_2 are both the nonzero constants.

This algorithm can be generalized to the condition that have limited number of functions

$$\int [k_1 f_1(x) \pm k_2 f_2(x) \pm \cdots \pm k_m f_m(x)]\mathrm{d}x$$
$$= k_1 \int f_1(x)\mathrm{d}x \pm k_2 \int f_2(x)\mathrm{d}x \pm \cdots \pm k_m \int f_m(x)\mathrm{d}x,$$

k_1, k_2, $\cdots$, k_m are all the nonzero constant.

Example 4 Find $\int (3^x e^x + 2x^{-2})\mathrm{d}x$.

Solution $\int (3^x e^x + 2x^{-2})\mathrm{d}x = \int (3e)^x \mathrm{d}x + 2\int x^{-2}\mathrm{d}x$

$$= \frac{(3e)^x}{\ln(3e)} - 2\frac{1}{x} + C = \frac{3^x e^x}{1 + \ln 3} - \frac{2}{x} + C.$$

Example 5 Find $\int \left(1 - \frac{1}{x^2}\right)\sqrt{x\sqrt{x}}\,\mathrm{d}x$.

Solution $\int \left(1 - \frac{1}{x^2}\right)\sqrt{x\sqrt{x}}\,\mathrm{d}x = \int \sqrt{x\sqrt{x}}\,\mathrm{d}x - \int \frac{1}{x^2}\sqrt{x\sqrt{x}}\,\mathrm{d}x$

$$= \int x^{\frac{3}{4}} dx - \int x^{-\frac{5}{4}} dx$$

$$= \frac{4}{7} x^{\frac{7}{4}} + 4x^{-\frac{1}{4}} + C.$$

Example 6 Find $\int \left(5\cos x + 2 - 3x^2 + \frac{1}{x} - \frac{4}{1+x^2} \right) dx$.

Solution $\int \left(5\cos x + 2 - 3x^2 + \frac{1}{x} - \frac{4}{1+x^2} \right) dx$

$$= 5 \int \cos x \, dx + 2 \int dx - 3 \int x^2 \, dx + \int \frac{1}{x} dx - 4 \int \frac{1}{1+x^2} dx$$

$$= 5\sin x + 2x - x^3 + \ln|x| - 4\arctan x + C.$$

Example 7 Find $\int (10^x + 3\sin x + \sqrt{x}) dx$.

Solution $\int (10^x + 3\sin x + \sqrt{x}) dx = \int 10^x dx + 3 \int \sin x \, dx + \int \sqrt{x} dx$

$$= \frac{10^x}{\ln 10} - 3\cos x + \frac{1}{\frac{1}{2}+1} x^{\frac{1}{2}+1} + C$$

$$= \frac{10^x}{\ln 10} - 3\cos x + \frac{2}{3} x^{\frac{3}{2}} + C.$$

Example 8 Find $\int \frac{\cos 2x}{\cos x + \sin x} dx$.

Solution $\int \frac{\cos 2x}{\cos x + \sin x} dx = \int \frac{\cos^2 x - \sin^2 x}{\cos x + \sin x} dx = \int (\cos x - \sin x) dx$

$$= \int \cos x \, dx - \int \sin x \, dx = \sin x + \cos x + C.$$

Example 9 Find $\int \frac{x^4}{1+x^2} dx$.

Solution $\int \frac{x^4}{1+x^2} dx = \int \frac{x^4 - 1 + 1}{1+x^2} dx = \int \frac{(x^2-1)(x^2+1)+1}{1+x^2} dx$

$$= \int \left(x^2 - 1 + \frac{1}{1+x^2} \right) dx = \int x^2 dx - \int dx + \int \frac{1}{1+x^2} dx$$

$$= \frac{x^3}{3} - x + \arctan x + C.$$

Example 10 Find $\int \frac{1}{\sin^2 \frac{x}{2} \cos^2 \frac{x}{2}} dx$.

Solution $\int \frac{1}{\sin^2 \frac{x}{2} \cos^2 \frac{x}{2}} dx = \int \frac{1}{\left(\frac{\sin x}{2} \right)^2} dx = 4 \int \csc^2 x \, dx$

$$= -4\cot x + C.$$

Example 11 Find $\int \cos^2 \frac{x}{2} dx$.

Solution $\int \cos^2 \frac{x}{2} dx = \int \frac{1+\cos x}{2} dx = \frac{1}{2} \left(\int dx + \int \cos x \, dx \right)$

$$= \frac{1}{2}(x + \sin x) + C.$$

The examples above all calculate indefinite integrals through basic means such as identical transformation, add or subtract items and so on. The aim is to simplify complex integrands into indefinite integrals in the basic integration table or forms of indefinite integrals that are given.

Exercises 5-1

1. Find the following indefinite integrals:

(1) $\int (2^x + x^2) dx$;

(2) $\int 2^x e^x dx$;

(3) $\int \frac{1 + 2x^2}{x^2(1 + x^2)} dx$;

(4) $\int \sin^2\left(\frac{x}{2}\right) dx$;

(5) $\int \frac{1}{1 - \cos^2 x} dx$;

(6) $\int \frac{1}{x^2(1 + x^2)} dx$;

(7) $\int \frac{x^2 + 1}{\sqrt{x}} dx$;

(8) $\int \frac{x^2}{1 + x^2} dx$;

(9) $\int \tan^2 x\, dx$;

(10) $\int \frac{2 \cdot 3^x - 5 \cdot 2^x}{3^x} dx$;

(11) $\int (\sqrt{x} + 1)(\sqrt{x^3} - 1) dx$;

(12) $\int \frac{1 + \cos^2 x}{1 + \cos 2x} dx$;

(13) $\int \frac{x^6}{1 + x^2} dx$;

(14) $\int \frac{e^{3x} + 1}{e^x + 1} dx$;

(15) $\int (x^2 + x)^2 dx$;

(16) $\int \frac{x^3 - 27}{x - 3} dx$;

(17) $\int \frac{3x^2}{1 + x^2} dx$;

(18) $\int \frac{1}{1 - \cos 2x} dx$;

(19) $\int \frac{\cos 2x}{\cos x - \sin x} dx$;

(20) $\int (1 + \sqrt{x}) \sqrt{x \sqrt{x}}\, dx$;

(21) $\int \frac{\cos 2x}{\cos^2 x \sin^2 x} dx$;

(22) $\int \frac{1}{\cos^2 x \sin^2 x} dx.$

2. Suppose that the curve passes through point $(e^2, 3)$, and the tangent equals to the reciprocal of its abscissa. Find the curve.

3. Suppose that the curve is $y = f(x)$, and the tangent slope in an arbitrary point (x, y) of the curve is $-2 + 2ax + 3x^2$, a is a constant. The abscissa of its inflection point is $-\frac{1}{3}$. Find the curve.

4. Prove that $\int x f(x) dx = \arccos x + C$. Find $f(x)$.

5.2 Integration by substitution

Using characters and basic formulas of indefinite integral, we can find indefinite integrals of some simple functions. But for finding integral problems of some complex functions, it also needs to learn related methods. The aim of integration by substitution is to simplify some integrals into those in table of basic indefinite integrals through proper variable substitution.

5.2.1 Integration by the first substitution (Gather differential together)

Theorem 1 Suppose $u = \varphi(x)$ is differentiable and $F(u)$ is an antiderivative of function $f(u)$, we have

$$\int f[\varphi(x)]\varphi'(x)\mathrm{d}x = F[\varphi(x)] + C. \tag{1}$$

Proof Supposed that $F(u)$ is one antiderivative of $f(u)$, so

$$F'(u) = f(u), \int f(u)\mathrm{d}u = F(u) + C,$$

where u is variable and $u = \varphi(x)$ is differentiable. According to the differential method of composite function, so there is

$$\mathrm{d}F[\varphi(x)] = f[\varphi(x)]\varphi'(x)\mathrm{d}x.$$

Due to definition of indefinite integral:

$$\int f[\varphi(x)]\varphi'(x)\mathrm{d}x = F[\varphi(x)] + C.$$

The basic idea of this method is to pick up $\varphi'(x)$ from integrand $f[\varphi(x)]\varphi'(x)$ to combine with $\mathrm{d}x$ to make $\mathrm{d}\varphi(x)$, therefore

$$\int f[\varphi(x)]\varphi'(x)\mathrm{d}x = \int f[\varphi(x)]\mathrm{d}\varphi(x)$$

$$\xrightarrow{\varphi(x)=u} \int f(u)\mathrm{d}u = F(u) + C$$

$$\xrightarrow{u=\varphi(x)} F[\varphi(x)] + C.$$

Thus integrals calculated can transformed into indefinite integrals in table of basic indefinite integrals. We title the Formula (1) as the first substitution method of indefinite integrals (also as gathering differential together).

Example 1 Find the indefinite integral of $\int \sin(2x-1)\mathrm{d}x$.

Solution Make $\varphi(x) = 2x - 1$, $\varphi'(x) = 2$. The expression of the integrant is

$$\sin(2x-1)\mathrm{d}x = \frac{1}{2}\sin\varphi(x)\mathrm{d}\varphi(x).$$

Regarding $\varphi(x) = 2x - 1$ as a variable u and $\sin u$ has an antiderivative $-\cos u$, it can obtain by Theorem 1

$$\int \sin(2x-1)\mathrm{d}x = \frac{1}{2}\int \sin(2x-1)\mathrm{d}(2x-1)$$

$$\xrightarrow{2x-1=u} \frac{1}{2}\int \sin u\,\mathrm{d}u = -\frac{1}{2}\cos u + C$$

$$\xrightarrow{u=2x-1} -\frac{1}{2}\cos(2x-1) + C.$$

Example 2 Find the indefinite integral of $\int x^2 \sqrt{x^3+2}\,dx$.

Solution Make $\varphi(x)=x^3+2$, $\varphi'(x)=3x^2$. The integrand is

$$x^2\sqrt{x^3+2}\,dx = \frac{1}{3}[\varphi(x)]^{\frac{1}{2}}d\varphi(x).$$

Regarding $\varphi(x)=x^3+2$ as a variable u and $u^{\frac{1}{2}}$ has an antiderivative $\frac{2}{3}u^{\frac{3}{2}}$, it can obtain by Theorem 1

$$\int x^2\sqrt{x^3+2}\,dx = \frac{1}{3}\int(x^3+2)^{\frac{1}{2}}d(x^3+2)$$

$$\xequal{x^3+2=u} \frac{1}{3}\int u^{\frac{1}{2}}du = \frac{2}{9}u^{\frac{3}{2}}+C$$

$$\xequal{u=x^3+2} \frac{2}{9}(x^3+2)^{\frac{3}{2}}+C.$$

After using integration by the first substitution, we can leave out the process of integration by substitution.

Example 3 Find the indefinite integral of $\int \tan x\,dx$.

Solution Because $\int \tan x\,dx = \int \frac{\sin x}{\cos x}dx = -\int \frac{1}{\cos x}d(\cos x) = -\ln|\cos x|+C.$

Similarly,
$$\int \cot x\,dx = \ln|\sin x|+C.$$

Example 4 Find the indefinite integral of $\int \frac{dx}{x(1+\ln x)}$.

Solution $\int \frac{dx}{x(1+\ln x)} = \int \frac{1}{1+\ln x}d(1+\ln x) = \ln|1+\ln x|+C.$

Example 5 Find the indefinite integral of $\int \frac{a^{\frac{1}{x}}}{x^2}dx\ (a>0)$.

Solution $\int \frac{a^{\frac{1}{x}}}{x^2}dx = -\int a^{\frac{1}{x}}d\left(\frac{1}{x}\right) = -\frac{a^{\frac{1}{x}}}{\ln a}+C.$

Example 6 Find the indefinite integral of $\int \frac{1}{(\arcsin x)^3\sqrt{1-x^2}}dx$.

Solution $\int \frac{1}{(\arcsin x)^3\sqrt{1-x^2}}dx = \int(\arcsin x)^{-3}d(\arcsin x)$

$$= \frac{1}{-3+1}(\arcsin x)^{-3+1}+C$$

$$= -\frac{1}{2(\arcsin x)^2}+C.$$

Example 7 Find the indefinite integral of $\int \sec x\,dx$.

Solution Because

$$\sec x\,dx = \frac{\sec x(\sec x+\tan x)}{\sec x+\tan x}dx = \frac{\sec^2 x+\sec x\tan x}{\sec x+\tan x}dx = \frac{d(\sec x+\tan x)}{\sec x+\tan x},$$

Hence
$$\int \sec x \, dx = \int \frac{d(\sec x + \tan x)}{\sec x + \tan x} = \ln|\sec x + \tan x| + C.$$

Similarly,
$$\int \csc x \, dx = \ln|\csc x - \cot x| + C.$$

Example 8 Find the indefinite integral of $\int \sin^3 x \, dx$.

Solution
$$\int \sin^3 x \, dx = \int \sin^2 x \sin x \, dx = -\int (1 - \cos^2 x) d(\cos x)$$
$$= -\int d(\cos x) + \int \cos^2 x \, d(\cos x)$$
$$= -\cos x + \frac{1}{3}\cos^3 x + C.$$

Example 9 Find the indefinite integral of $\int \sec^6 x \, dx$.

Solution
$$\int \sec^6 x \, dx = \int (\sec^2 x)^2 \sec^2 x \, dx = \int (1 + \tan^2 x)^2 d(\tan x)$$
$$= \int (1 + 2\tan^2 x + \tan^4 x) d(\tan x)$$
$$= \tan x + \frac{2}{3}\tan^3 x + \frac{1}{5}\tan^5 x + C.$$

Example 10 Find the indefinite integral of $\int \frac{\cos^3 x}{\sqrt{\sin x}} dx$.

Solution
$$\int \frac{\cos^3 x}{\sqrt{\sin x}} dx = \int \frac{\cos^2 x \cos x}{\sqrt{\sin x}} dx = \int (1 - \sin^2 x) \sin^{-\frac{1}{2}} x \, d(\sin x)$$
$$= \int \sin^{-\frac{1}{2}} x \, d(\sin x) - \int \sin^{\frac{3}{2}} x \, d(\sin x)$$
$$= 2\sin^{\frac{1}{2}} x - \frac{2}{5}\sin^{\frac{5}{2}} x + C.$$

Example 11 Find the indefinite integral of $\int \frac{e^{\sqrt[3]{x}}}{\sqrt{x}} dx$.

Solution
$$\int \frac{e^{\sqrt[3]{x}}}{\sqrt{x}} dx = \frac{2}{3}\int e^{\sqrt[3]{x}} d(3\sqrt{x}) = \frac{2}{3} e^{\sqrt[3]{x}} + C.$$

Example 12 Find the indefinite integral of $\int \frac{1}{\sqrt{a^2 - x^2}} dx \ (a > 0)$.

Solution
$$\int \frac{1}{\sqrt{a^2 - x^2}} dx = \int \frac{1}{a\sqrt{1 - \left(\frac{x}{a}\right)^2}} dx = \int \frac{1}{\sqrt{1 - \left(\frac{x}{a}\right)^2}} d\left(\frac{x}{a}\right)$$
$$= \arcsin \frac{x}{a} + C.$$

Example 13 Find the indefinite integral of $\int \frac{1}{a^2 + x^2} dx$.

Solution
$$\int \frac{1}{a^2 + x^2} dx = \int \frac{1}{a^2 \left[1 + \left(\frac{x}{a}\right)^2\right]} dx = \frac{1}{a} \int \frac{1}{1 + \left(\frac{x}{a}\right)^2} d\left(\frac{x}{a}\right)$$

$$= \frac{1}{a}\arctan\frac{x}{a} + C.$$

Example 14 Find the indefinite integral of $\int \frac{1}{a^2 - x^2} dx$.

Solution Because $\frac{1}{a^2 - x^2} = \frac{1}{2a}\left(\frac{1}{a+x} + \frac{1}{a-x}\right)$,

$$\int \frac{dx}{a^2 - x^2} = \frac{1}{2a}\left(\int \frac{1}{a+x} dx + \int \frac{1}{a-x} dx\right)$$

$$= \frac{1}{2a}\left[\int \frac{1}{a+x} d(a+x) - \int \frac{1}{a-x} d(a-x)\right]$$

$$= \frac{1}{2a}\ln\left|\frac{a+x}{a-x}\right| + C.$$

Example 15 Find the indefinite integral of $\int \frac{x}{2+x^4} dx$.

Solution $\int \frac{x}{2+x^4} dx = \frac{1}{2}\int \frac{1}{(\sqrt{2})^2 + (x^2)^2} dx^2 = \frac{1}{2\sqrt{2}}\arctan\frac{x^2}{\sqrt{2}} + C.$

Example 16 Find the indefinite integral of $\int \frac{\ln\tan\frac{x}{2}}{\sin x} dx$.

Solution $\int \frac{\ln\tan\frac{x}{2}}{\sin x} dx = \int \frac{\ln\tan\frac{x}{2}}{2\sin\frac{x}{2}\cos\frac{x}{2}} dx = \int \frac{\ln\tan\frac{x}{2}}{\tan\frac{x}{2}\cos^2\frac{x}{2}} d\left(\frac{x}{2}\right)$

$$= \int \frac{\ln\tan\frac{x}{2}}{\tan\frac{x}{2}} d\left(\tan\frac{x}{2}\right) = \int \ln\tan\frac{x}{2} d\left(\ln\tan\frac{x}{2}\right)$$

$$= \frac{1}{2}\ln^2\left(\tan\frac{x}{2}\right) + C.$$

Regarding conclusions in Example 3, 7, 12~14 as new integral formulas that

(16) $\int \tan x dx = -\ln|\cos x| + C;$

(17) $\int \cot x dx = \ln|\sin x| + C;$

(18) $\int \sec x dx = \ln|\sec x + \tan x| + C;$

(19) $\int \csc x dx = \ln|\csc x - \cot x| + C;$

(20) $\int \frac{1}{\sqrt{a^2 - x^2}} dx = \arcsin\frac{x}{a} + C;$

(21) $\int \frac{1}{a^2 + x^2} dx = \frac{1}{a}\arctan\frac{x}{a} + C;$

(22) $\int \frac{1}{a^2 - x^2} dx = \frac{1}{2a}\ln\left|\frac{a+x}{a-x}\right| + C.$

The key to use integration by the first substitution is to remember basic integral formulas firmly. After learning basic methods about calculating indefinite integral, we can directly transform integrands into the form of $\int f[\varphi(x)]\mathrm{d}\varphi(x)$ to find indefinite integrals and regard function $\varphi(x)$ as a variable, then apply the basic integral formulas. Thus this is titled as "gather differentials together".

5.2.2 Integration by the second substitution

In finding indefinite integrals, it often exists problems in contrast to conditions above that the indefinite integral $\int f(x)\mathrm{d}x$ is not complex but hard to work out directly. However, through proper variable substitution $x=\varphi(t)$, it can transform integral $\int f(x)\mathrm{d}x$ into the form $\int f[\varphi(t)]\varphi'(t)\mathrm{d}t$ which is easy to find. This is another method of calculating indefinite integrals by variable substitution, which is titled as integration by the second substitution.

Theorem 2 (The second substitution method) Assume that $x=\varphi(t)$ is a monotonic and differentiable function and $\varphi'(t)\neq 0$. Then providing that $f[\varphi(t)]\varphi'(t)$ has antiderivativs, therefore there is the formula for integration by substitution

$$\int f(x)\mathrm{d}x \xrightarrow{x=\varphi(t)} \left\{\int f[\varphi(t)]\varphi'(t)\mathrm{d}t\right\}_{t=\varphi^{-1}(x)},$$

where $t=\varphi^{-1}(x)$ is the inverse function of $x=\varphi(t)$.

Proof Given that $f[\varphi(t)]\varphi'(t)$ has antiderivativs, then assume that its antiderivative is $F(t)$. It just needs to prove

$$\{F[\varphi^{-1}(x)]+C\}'=f(x).$$

Because $\{F[\varphi^{-1}(x)]+C\}'=F'(t)\dfrac{\mathrm{d}t}{\mathrm{d}x}=f[\varphi(t)]\varphi'(t)\dfrac{1}{\varphi'(t)}=f[\varphi(t)]=f(x).$

Integration by the second substitution is to substitute proper function $x=\varphi(t)$ for variable x and then transform it into the form below:

$$\int f(x)\mathrm{d}x = \int f[\varphi(t)]\varphi'(t)\mathrm{d}t.$$

After finding the integral of right part of the upper formula, use $x=\varphi(t)$'s inverse function $t=\varphi^{-1}(x)$ to substitute back.

The establishment of this formula for integration by substitution requires certain conditions. Because it is necessary to use $x=\varphi(t)$'s inverse function $t=\varphi^{-1}(x)$ to substitute back after finding the integral $\int g[\varphi(t)]\varphi'(t)\mathrm{d}t$, so in the condition of the theorem, we require that the function $x=\varphi(t)$ in one interval of t (This interval corresponds to integral interval of x considered) is monotonic, derived and $\varphi'(t)\neq 0$ ensures that the inverse function $t=\varphi^{-1}(x)$ exists.

Example 17 Find the indefinite integral of $\int \dfrac{dx}{1+\sqrt{1+x}}$.

Solution For removing the surd of integrand's denominator, assumed that $\sqrt{1+x}=t$, so $x=t^2-1$, $dx=2tdt$.

$$\int \dfrac{dx}{1+\sqrt{1+x}} = \int \dfrac{2t}{1+t}dt = 2\int \dfrac{t+1-1}{1+t}dt$$

$$= 2\left(\int dt - \int \dfrac{1}{1+t}dt\right) = 2(t - \ln|1+t|) + C$$

$$\xlongequal{t=\sqrt{1+x}} 2(\sqrt{1+x} - \ln|1+\sqrt{1+x}|) + C.$$

When applying the second substitution method, notice that substitute the variable back after integrating.

Example 18 Find the indefinite integral of $\int \sqrt{a^2-x^2}\,dx$ $(a>0)$.

Solution As shown in Figure 5-2, $|x|\leqslant a$. Assume that $x=a\sin t$, when $-\dfrac{\pi}{2}\leqslant t\leqslant \dfrac{\pi}{2}$, $x=a\sin t$ has single valued inverse function $t=\arcsin \dfrac{x}{a}$, and $|\cos t|=\cos t$, $dx=a\cos t dt$. Thus

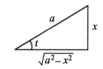

Figure 5-2

$$\int \sqrt{a^2-x^2}\,dx = \int \sqrt{a^2-a^2\sin^2 t}\cdot a\cos t dt = a^2\int |\cos t|\cos t dt$$

$$= a^2 \int \cos^2 t dt = \dfrac{a^2}{2}\int (1+\cos 2t) dt = \dfrac{a^2}{2}\left(\int dt + \int \cos 2t dt\right)$$

$$= \dfrac{a^2}{2}\left(t+\dfrac{1}{2}\sin 2t\right) + C = \dfrac{a^2}{2}t + \dfrac{a^2}{4}\sin 2t + C.$$

Due to $t=\arcsin \dfrac{x}{a}$, so $\sin 2t = 2\sin t\cos t = 2\dfrac{x}{a^2}\sqrt{a^2-x^2}$. Thereupon,

$$\int \sqrt{a^2-x^2}\,dx = \dfrac{a^2}{2}\arcsin \dfrac{x}{a} + \dfrac{x}{2}\sqrt{a^2-x^2} + C.$$

Example 19 Find the indefinite integral of $\int \dfrac{1}{\sqrt{x^2+a^2}}dx$ $(a>0)$.

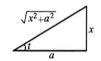

Figure 5-3

Solution As shown in Figure 5-3, assume that $x=a\tan t$, so $dx=a\sec^2 t dt$. When $-\dfrac{\pi}{2}<t<\dfrac{\pi}{2}$, $x=a\tan t$ has single valued inverse function and $|\sec t|=\sec t$. Thereupon,

$$\int \dfrac{dx}{\sqrt{x^2+a^2}} = \int \dfrac{a\sec^2 t}{a\sqrt{1+\tan^2 t}}dt = \int \sec t dt = \ln|\sec t + \tan t| + C.$$

In order to transform $\sec t$ and $\tan t$ to the function of x, we can plot an right triangle (Figure 5-3). By Figure 5-3, we know that $\tan t=\dfrac{x}{a}$, so we have $\sec t=\dfrac{\sqrt{x^2+a^2}}{a}$.

Therefore,

$$\int \dfrac{dx}{\sqrt{x^2+a^2}} = \ln\left|\dfrac{\sqrt{a^2+x^2}}{a} + \dfrac{x}{a}\right| + C_1 = \ln|\sqrt{a^2+x^2}+x| + C.$$

$C = C_1 - \ln a$ is still an arbitrary constant.

Example 20 Find the indefinite integral of $\int \dfrac{dx}{\sqrt{x^2 - a^2}}$ $(a > 0)$.

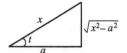

Figure 5-4

Solution Assume that $x = a\sec t$, so $dx = a\sec t \tan t\, dt$ (Figure 5-4).

When $0 < t < \dfrac{\pi}{2}$ or $\dfrac{\pi}{2} < t < \pi$, $x = a\sec t$, which has inverse function. This just discusses the condition that $0 < t < \dfrac{\pi}{2}$, it can discuss the condition that $\dfrac{\pi}{2} < t < \pi$ in the same way.

When $0 < t < \dfrac{\pi}{2}$, $|\tan t| = \tan t$. Thus

$$\int \frac{dx}{\sqrt{x^2 - a^2}} = \int \frac{a\sec t \tan t\, dt}{a \tan t} = \int \sec t\, dt = \ln|\sec t + \tan t| + C.$$

Due to $\sec t = \dfrac{x}{a}$, so $\tan t = \dfrac{\sqrt{x^2 - a^2}}{a}$.

Thereupon,

$$\int \frac{dx}{\sqrt{x^2 - a^2}} = \ln\left|\frac{x}{a} + \frac{\sqrt{x^2 - a^2}}{a}\right| + C_1 = \ln|x + \sqrt{x^2 - a^2}| + C.$$

Example 18~20 are also common integral formulas.

(23) $\int \sqrt{a^2 - x^2}\, dx = \dfrac{a^2}{2}\arcsin\dfrac{x}{a} + \dfrac{x}{2}\sqrt{a^2 - x^2} + C;$

(24) $\int \dfrac{dx}{\sqrt{x^2 + a^2}} = \ln|\sqrt{a^2 + x^2} + x| + C;$

(25) $\int \dfrac{dx}{\sqrt{x^2 - a^2}} = \ln|x + \sqrt{x^2 - a^2}| + C.$

The second substitution method is mainly used in two kinds of indefinite integrals.

One is: the surd in the integrand cannot be solved by the first substitution method, this time use the second substitution method to remove the surd and then find. It can be seen in Example 17.

The other is: the integrand has factors including surds $\sqrt{a^2 - x^2}$, $\sqrt{a^2 + x^2}$, $\sqrt{x^2 - a^2}$ that cannot be calculated through the first substitution method. It can be seen in Example 18~20.

Example 21 Find the indefinite integral of $\int \dfrac{dx}{x^2\sqrt{1 + x^2}}$.

Solution Assume that $x = \tan t$, so $dx = \sec^2 t\, dt$. Thereupon,

$$\int \frac{dx}{x^2\sqrt{1 + x^2}} = \int \frac{\sec^2 t}{\tan^2 t \sec t}\, dt = \int \frac{\sec t}{\tan^2 t}\, dt$$

$$= \int \frac{dt}{\dfrac{\sin^2 t}{\cos^2 t} \cdot \cos t} = \int \frac{dt}{\sin t \cdot \dfrac{\sin t}{\cos t}}$$

$$= \int \csc t \cot t \, dt = -\csc t + C = -\frac{\sqrt{1+x^2}}{x} + C.$$

Note These examples above prove that the main aim of the second substitution method is to rationalize surds in integrands.

Exercises 5-2

1. Find the following indefinite integrals:

(1) $\int \dfrac{1}{2x+1} dx$;

(2) $\int x\sqrt{1-x^2}\, dx$;

(3) $\int \dfrac{1}{1+9x^2} dx$;

(4) $\int \dfrac{1}{x}\sqrt{\ln x}\, dx$;

(5) $\int \cos x \sin^2 x\, dx$;

(6) $\int \dfrac{(\arcsin x)^{-2}}{\sqrt{1-x^2}} dx$;

(7) $\int x^2 e^{-x^3}\, dx$;

(8) $\int \dfrac{x}{3-2x^2} dx$;

(9) $\int \dfrac{2x-3}{x^2-3x+8} dx$;

(10) $\int \dfrac{\sqrt{1+x^2}+\sqrt{1-x^2}}{\sqrt{1-x^4}} dx$;

(11) $\int \dfrac{\sin x}{2+\cos^2 x} dx$;

(12) $\int \dfrac{1}{\cos^2 x(1+\tan x)} dx$;

(13) $\int \dfrac{e^x}{4+e^{2x}} dx$;

(14) $\int \dfrac{1}{\sqrt{x}\sin^2 \sqrt{x}} dx$;

(15) $\int \dfrac{dx}{\sqrt{x}(1+2x)}$;

(16) $\int \dfrac{1}{e^x+e^{-x}} dx$;

(17) $\int \dfrac{1+x}{\sqrt{1-x^2}} dx$;

(18) $\int \dfrac{1}{3+2x-x^2} dx$;

(19) $\int \dfrac{dx}{x \ln x \ln(\ln x)}$;

(20) $\int x \cos x^2\, dx$;

(21) $\int \dfrac{\cos^2 x}{\sin 2x} dx$.

2. Find the following indefinite integrals:

(1) $\int (1-x^2)^{-\frac{3}{2}} dx$;

(2) $\int \dfrac{x^2}{\sqrt{a^2-x^2}} dx$;

(3) $\int \dfrac{dx}{x^2\sqrt{x^2-9}}$;

(4) $\int \dfrac{\sin 2x}{\sqrt{2-\cos^4 x}} dx$;

(5) $\int \dfrac{dx}{x\sqrt{x^2-a^2}}$;

(6) $\int \dfrac{dx}{\sqrt{(x^2+1)^3}}$;

(7) $\int \dfrac{x^2}{\sqrt{x-x^2}} dx$;

(8) $\int \dfrac{x^2+x+1}{(x+1)^{50}} dx$;

(9) $\int \tan^4 x\, dx$;

(10) $\int x^3 \sqrt{1+x^2}\, dx$;

(11) $\int \dfrac{x}{1+\sqrt{1+x^2}} dx$.

5.3 Integration by parts

Corresponding to the multiplication formula for differentiation, we also have a powerful method of integration, named integration by parts.

Let $u=u(x)$ and $v=v(x)$ are differentiable functions of x and their antiderivatives are continuous. The derivative of the product $u(x)v(x)$ is found from the following formula:
$$(uv)' = u'v + uv' \quad \text{or} \quad uv' = (uv)' - u'v.$$
Integrating both sides, we have
$$\int uv' \, dx = \int (uv)' \, dx - \int u'v \, dx,$$
which is
$$\int uv' \, dx = uv - \int u'v \, dx \tag{1}$$
or
$$\int u \, dv = uv - \int v \, du. \tag{2}$$

The two formulas above are titled as formula of integration by parts.

If it is easier to find $\int u'v \, dx$ than to find $\int uv' \, dx$, the formula for integration by parts plays an important role. The next examples illustrate the technique.

Example 1 Find the indefinite integral of $\int x \sin x \, dx$.

Solution The expression of the integrant $x \sin x \, dx$ can be written to the form $x \, d(-\cos x)$ and $u = x$, $v = -\cos x$. Through the formula of integration by parts, it can obtain that
$$\int x \sin x \, dx = \int x \, d(-\cos x) = x(-\cos x) - \int (-\cos x) \, dx$$
$$= -x \cos x + \int \cos x \, dx = -x \cos x + \sin x + C.$$

Note If in this example, make $u = \sin x$, $v = \dfrac{x^2}{2}$, therefore
$$\int x \sin x \, dx = \int \sin x \, d\left(\frac{1}{2} x^2\right) = \frac{x^2}{2} \sin x - \int \frac{x^2}{2} \cos x \, dx.$$

But integral $\int \dfrac{x^2}{2} \cos x \, dx$ is more complex than the former integral. It can be seen that it important to choose correct and when using integration by parts.

Generally, we consider two aspects in choosing u and dv.

(1) It is easy to find v.

(2) It is easier to find $\int v \, du$ than to find $\int u \, dv$.

Example 2 Find the indefinite integral of $\int xe^x \, dx$..

Solution The expression of the integrant $xe^x \, dx$ is written to the form $x \, de^x$ and this time $u = x$, $v = e^x$. Through the formula of integration by parts, it can obtain that
$$\int xe^x \, dx = \int x \, de^x = xe^x - \int e^x \, dx = xe^x - e^x + C.$$

Example 3 Find the indefinite integral of $\int x^2 \ln x \, dx$.

Solution Because $x^2 \ln x \, dx$ cannot be found in table of basic indefinite integrals, the expression of the integrant $x^2 \ln x \, dx$ must be written to the form $\ln x \, d\left(\frac{1}{3}x^3\right)$. At this moment $u = \ln x$, $v = \frac{1}{3}x^3$. Through the formula of integration by parts, it can obtain that

$$\int x^2 \ln x \, dx = \int \ln x \, d\left(\frac{1}{3}x^3\right) = \left(\frac{1}{3}x^3\right)\ln x - \int \frac{1}{3}x^3 \cdot d(\ln x)$$

$$= \frac{1}{3}x^3 \ln x - \frac{1}{3}\int x^3 \cdot \frac{1}{x}dx = \frac{1}{3}x^3 \ln x - \frac{1}{9}x^3 + C.$$

Example 4 Find the indefinite integral of $\int x \arctan x \, dx$.

Solution The expression of the integrant $x \arctan x \, dx$ can be written to the form $\arctan x \, d\left(\frac{1}{2}x^2\right)$ and this time $u = \arctan x$, $v = \frac{1}{2}x^2$. Through the formula of integration by parts, it can obtain that

$$\int x \arctan x \, dx = \int \arctan x \, d\left(\frac{1}{2}x^2\right)$$

$$= \frac{1}{2}x^2 \cdot \arctan x - \int \frac{1}{2}x^2 \, d(\arctan x)$$

$$= \frac{1}{2}x^2 \arctan x - \frac{1}{2}\int \frac{x^2}{1+x^2}dx$$

$$= \frac{1}{2}x^2 \arctan x - \frac{1}{2}\int \frac{x^2+1-1}{1+x^2}dx$$

$$= \frac{1}{2}x^2 \arctan x - \frac{1}{2}\left(\int dx - \int \frac{1}{1+x^2}dx\right)$$

$$= \frac{1}{2}x^2 \arctan x - \frac{1}{2}x + \frac{1}{2}\arctan x + C.$$

Example 5 Find the indefinite integral of $\int e^x \cos x \, dx$.

Solution $\int e^x \cos x \, dx = \int \cos x \, d(e^x) = e^x \cos x + \int e^x \sin x \, dx.$

Using integration by parts again (still choose $e^x dx$ to make $d(e^x)$)

$$\int e^x \cos x \, dx = e^x \cos x + \int \sin x \, d(e^x)$$

$$= e^x \cos x + e^x \sin x - \int e^x \cos x \, dx.$$

Transposing, divided it by 2 and add arbitrary constant C, then obtain that

$$\int e^x \cos x \, dx = \frac{e^x}{2}(\cos x + \sin x) + C.$$

Example 6 Find the indefinite integral of $\int \sin(\ln x) \, dx$.

Solution This time assumed that $u(x) = \sin(\ln x)$, $dx = dv$, then

$$\int \sin(\ln x)\,dx = x\sin(\ln x) - \int x\,d[\sin(\ln x)]$$
$$= x\sin(\ln x) - \int x\cos(\ln x)\frac{1}{x}\,dx$$
$$= x[\sin(\ln x) - \cos(\ln x)] - \int \sin(\ln x)\,dx.$$

Thus $\int \sin(\ln x)\,dx = \frac{1}{2}x[\sin(\ln x) - \cos(\ln x)] + C.$

Note (1) Integration by parts is mainly applicable to the following five kinds of integrants $f(x)$:

① $f(x)$ is the product of polynomial $P_n(x)$ and trigonometric function. Then select $u(x) = P_n(x)$ and $v'(x)$ is trigonometric function.

② $f(x)$ is the product of $P_n(x)$ and exponential function. Then select $u(x) = P_n(x)$ and $v'(x)$ is exponential function.

③ $f(x)$ is the product of $P_n(x)$ and logarithmic function. Then select $v'(x) = P_n(x)$ and $u(x)$ is logarithmic function.

④ $f(x)$ is the product of $P_n(x)$ and inverse trigonometric function. Then select $v'(x) = P_n(x)$ and $u(x)$ is inverse trigonometric function.

⑤ $f(x)$ is the product of exponential function and trigonometric function. Then select $u(x)$ = exponential function, $v'(x)$ is trigonometric function or $u(x)$ = trigonometric function, $v'(x)$ is exponential function.

(2) When calculating indefinite integrals, firstly consider using the first substitution method to work out. Notice that even if use the second substitution method or integration by parts to solve problems, it also use the first substitution method at the same time.

(3) Applying integration by parts, it can also obtain some useful recursion formulas.

Example 7 Find the indefinite integral of $J_n = \int \dfrac{1}{(a^2+x^2)^n}\,dx \ (a \neq 0).$

Solution
$$J_n = \int \frac{1}{(a^2+x^2)^n}\,dx = \frac{x}{(x^2+a^2)^n} + 2n\int \frac{x^2}{(x^2+a^2)^{n+1}}\,dx$$
$$= \frac{x}{(x^2+a^2)^n} + 2n\int \frac{(x^2+a^2) - a^2}{(x^2+a^2)^{n+1}}\,dx$$
$$= \frac{x}{(x^2+a^2)^n} + 2n\int \frac{1}{(x^2+a^2)^n}\,dx - 2na^2\int \frac{1}{(x^2+a^2)^{n+1}}\,dx$$
$$= \frac{x}{(x^2+a^2)^n} + 2nJ_n - 2na^2 J_{n+1}.$$

Then
$$J_{n+1} = \frac{1}{2na^2}\frac{x}{(x^2+a^2)^n} + \frac{2n-1}{2n}\cdot\frac{1}{a^2}J_n \ (n\geq 1).$$

The formula is titled as the recursion formula of J_n.

When $n=1$, $J_1 = \int \dfrac{1}{x^2+a^2}\,dx = \dfrac{1}{a}\arctan\dfrac{x}{a} + C.$

Example 8 Find the indefinite integral of $I_n = \int (\ln x)^n\,dx.$

Solution $I_n = \int (\ln x)^n dx = x(\ln x)^n - n\int (\ln x)^{n-1} dx$
$= x(\ln x)^n - nI_{n-1},$

hence $I_n = x(\ln x)^n - nI_{n-1} \ (n \geq 2).$

When $n=1$, $I_1 = \int \ln x dx = x\ln x - x + C.$

When finding integrals, it sometimes has to combine integration by parts with integration by substitution together to use.

Example 9 Find the indefinite integral of $\int \sec^3 x dx$.

Solution $\int \sec^3 x dx = \int \sec x \sec^2 x dx = \int \sec x d(\tan x) = \sec x \tan x - \int \tan x d(\sec x)$
$= \sec x \tan x - \int \tan^2 x \sec x dx = \sec x \tan x - \int (\sec^2 x - 1)\sec x dx$
$= \sec x \tan x - \int \sec^3 x dx + \int \sec x dx.$

Therefore
$$\int \sec^3 x dx = \frac{1}{2}\left(\sec x \tan x + \int \sec x dx\right)$$
$$= \frac{1}{2}(\sec x \tan x + \ln|\sec x + \tan x|) + C.$$

Example 10 Find the indefinite integral of $\int e^{\sqrt{x}} dx$.

Solution Make $\sqrt{x} = t$, then $x = t^2, dx = 2tdt$. Hence,
$$\int e^{\sqrt{x}} dx = 2\int te^t dt = 2e^t(t-1) + C = 2e^{\sqrt{x}}(\sqrt{x} - 1) + C.$$

Example 11 Find the indefinite integral of $\int \cos\sqrt{1+x} dx$.

Solution Make $\sqrt{1+x} = t$, then $x = t^2 - 1, dx = 2tdt$.
Thereupon
$$\int \cos\sqrt{1+x} dx = \int \cos t \cdot 2t dt = 2\int t d(\sin t) = 2t\sin t - 2\int \sin t dt$$
$$= 2t\sin t + 2\cos t + C$$
$$= 2\sqrt{1+x}\sin\sqrt{1+x} + 2\cos\sqrt{1+x} + C.$$

Example 12 Find the indefinite integral of $\int \frac{\arcsin x}{x^2} dx$.

Solution $\int \frac{\arcsin x}{x^2} dx = -\int \arcsin x d\left(\frac{1}{x}\right) = -\frac{1}{x}\arcsin x + \int \frac{1}{x} d(\arcsin x)$
$= -\frac{1}{x}\arcsin x + \int \frac{dx}{x\sqrt{1-x^2}}.$

Integrate the second one in the right of the upper formula, make $x = \sin t$, hence
$$\int \frac{dx}{x\sqrt{1-x^2}} = \int \frac{\cos t dt}{\sin t \cos t} = \int \frac{1}{\sin t} dt = \ln|\csc t - \cot t| + C$$

$$= \ln\left|\frac{1}{x} - \frac{\sqrt{1-x^2}}{x}\right| + C.$$

Thus $\int \frac{\arcsin x}{x^2} dx = -\frac{1}{x}\arcsin x + \ln\left|\frac{1-\sqrt{1-x^2}}{x}\right| + C.$

Exercises 5-3

1. Find the following indefinite integrals:

   (1) $\int x\sin 2x\,dx$;

   (2) $\int x^2 \cos x\,dx$;

   (3) $\int e^{-x}\sin x\,dx$;

   (4) $\int \frac{\ln x}{x^2}\,dx$;

   (5) $\int \arcsin x\,dx$;

   (6) $\int \ln(x + \sqrt{1+x^2})\,dx$;

   (7) $\int x^n \ln x\,dx, n \neq -1$;

   (8) $\int x^2 \arctan x\,dx$;

   (9) $\int x^2 \ln(1+x)\,dx$;

   (10) $\int \frac{\ln^3 x}{x^2}\,dx$;

   (11) $\int (\arcsin x)^2\,dx$;

   (12) $\int x^2 \cos \omega x\,dx, \omega \neq 0$;

   (13) $\int \frac{\ln(1+x)}{\sqrt{x}}\,dx$.

2. Find the following indefinite integrals:

   (1) $\int x\tan x\sec^4 x\,dx$;

   (2) $\int \frac{2x+3}{\sqrt{1+x^2}}\,dx$;

   (3) $\int [\tan^2 x - x\sin(x^2+1)]\,dx$;

   (4) $\int \arctan \sqrt{x}\,dx$;

   (5) $\int e^{2x}\sin^2 x\,dx$;

   (6) $\int \cos(\ln x)\,dx$.

3. Given that one antiderivative of $f(x)$ is e^{-x^2}, Find $\int xf'(x)\,dx$.

5.4 Integration of rational functions

The basic methods of finding indefinite integrals are integration by substitution and integration by parts. Through the two methods, we can find indefinite integrals of many functions but not ensure that finding out primitive functions of integrands such as $\int \frac{\sin x}{x}\,dx$, $\int e^{x^2}\,dx$, that cannot be expressed by elementary functions.

Next we introduce kinds of integrals that can find their primitive functions.

5.4.1 Integration of rational functions

A rational function is such a function that is represented by a quotient or ratio of two polynomials:

$$R(x) = \frac{P(x)}{Q(x)} = \frac{a_0 x^n + a_1 x^{n-1} + \cdots + a_n}{b_0 x^m + b_1 x^{m-1} + \cdots + b_m}, \tag{1}$$

where, m, n are nonnegative integers; a_0, a_1, $\cdots$, a_n, b_0, b_1, $\cdots$, b_m are constants and $a_0 \neq 0$, $b_0 \neq 0$.

When $n \geq m$, $R(x)$ is titled as improper rational fraction; when $n < m$, $R(x)$ is titled as proper rational fraction.

An improper fraction can be decomposed into a sum of a polynomial and a proper fraction by division of polynomials. Then it is easy to work out the indefinite integral of polynomial. Thus finding indefinite integral of rational function is mainly to work out the indefinite integral of the proper rational fraction.

Because the sum of two proper fractions is still a proper fraction, therefore, we hope that transform one proper fraction into the sum of several simple proper fractions.

As for proper fractions, there are the following conclusions in algebra:

Assumed that the Formula (1) is the proper rational fraction ($n < m$), make $b_0 = 1$, then $Q(x)$ can be always divided into the product of the expression of liner factor and quadratic prime factor that

$$Q(x) = (x-a)^\alpha \cdots (x-b)^\beta (x^2+px+q)^\mu \cdots (x^2+rx+s)^v, \tag{2}$$

where $p^2 - 4q < 0, \cdots, r^2 - 4s < 0$.

According to the algebra theorem, when $Q(x)$ can be expressed to the Formula (2), the proper rational fraction (1) can be uniquely written to the sum of the following fractions:

$$\frac{P(x)}{Q(x)} = \frac{A_1}{x-a} + \frac{A_2}{(x-a)^2} + \cdots + \frac{A_\alpha}{(x-a)^\alpha} + \cdots + \frac{B_1}{x-b} + \frac{B_2}{(x-b)^2} + \cdots +$$
$$\frac{B_\beta}{(x-b)^\beta} + \frac{M_1 x + N_1}{x^2+px+q} + \frac{M_2 x + N_2}{(x^2+px+q)^2} + \cdots + \frac{M_\mu x + N_\mu}{(x^2+px+q)^\mu} + \cdots +$$
$$\frac{U_1 x + V_1}{x^2+rx+s} + \frac{U_2 x + V_2}{(x^2+rx+s)^2} + \cdots + \frac{U_v x + V_v}{(x^2+rx+x)^v},$$

where $A_1, \cdots, A_\alpha, B_1, \cdots, B_\beta, M_1, \cdots, M_\mu, N_1, \cdots, N_\mu, U_1, \cdots, U_v, V_1, \cdots, V_v$ are all constants.

Known from the theorem above, every proper rational fraction can be divided into the sum of fractions such as $\dfrac{A}{x-a}$, $\dfrac{B}{(x-a)^k}$, $\dfrac{Cx+D}{x^2+px+q}$, $\dfrac{Ex+F}{(x^2+px+q)^k}$. A, B, C, D, E, F are real numbers, natural number $k > 1$ and equation $x^2 + px + q = 0$ does not have real roots ($p^2 - 4q < 0$). Thus problems about integrals of rational expressions can be concluded in the following integrals of four simple fractions:

$$\int \frac{A}{x-a} dx; \quad \int \frac{B}{(x-a)^k} dx; \quad \int \frac{Cx+D}{x^2+px+q} dx; \quad \int \frac{Ex+F}{(x^2+px+q)^k} dx \ (k > 1).$$

Now discuss the indefinite integrals of four kinds of functions above. The problems about integrals of rational expressions can be solved by just finding the indefinite integrals of these kinds.

(1) $\displaystyle\int \frac{A}{x-a} dx = A \ln |x-a| + C$.

(2) $\displaystyle\int \frac{B}{(x-a)^k} dx = B \int \frac{1}{(x-a)^k} d(x-a) = \frac{B}{-k+1} (x-a)^{-k+1} + C$.

(3) $\int \dfrac{Cx+D}{x^2+px+q}dx \quad (p^2-4q<0)$.

$$\int \dfrac{Cx+D}{x^2+px+q}dx = \int \dfrac{Cx+\dfrac{Cp}{2}-\dfrac{Cp}{2}+D}{x^2+px+q}dx$$

$$= \dfrac{C}{2}\int \dfrac{2x+p}{x^2+px+q}dx + \left(D-\dfrac{Cp}{2}\right)\int \dfrac{dx}{\left(x+\dfrac{p}{2}\right)^2+q-\dfrac{p^2}{4}}$$

$$= \dfrac{C}{2}\int \dfrac{d(x^2+px+q)}{x^2+px+q} + \left(D-\dfrac{Cp}{2}\right)\int \dfrac{d\left(x+\dfrac{p}{2}\right)}{\left(x+\dfrac{p}{2}\right)^2+\left(\sqrt{q-\dfrac{p^2}{4}}\right)^2}$$

$$= \dfrac{C}{2}\ln|x^2+px+q| + \dfrac{2D-Cp}{\sqrt{4q-p^2}}\arctan \dfrac{2x+p}{\sqrt{4q-p^2}}+C.$$

(4) $\int \dfrac{Ex+F}{(x^2+px+q)^k}dx \quad (p^2-4q<0,\ k>1)$.

Make $t=x+\dfrac{p}{2}$, therefore

$$\int \dfrac{Ex+F}{(x^2+px+q)^k}dx = \dfrac{E}{2}\int \dfrac{2t}{(t^2+a^2)^k}dt + \left(F-\dfrac{Ep}{2}\right)\int \dfrac{1}{(t^2+a^2)^k}dt,$$

where $a^2 = q - \dfrac{p^2}{4}, \int \dfrac{2t}{(t^2+a^2)^k}dt = \dfrac{1}{1-k}\cdot\dfrac{1}{(t^2+a^2)^{k-1}}+C$.

The second integral can use the conclusion in 5.3 Example 7. Thus integral problems like the kind of (4) can be solved.

Then take several examples about integrals of rational expressions.

Example 1 Find the indefinite integral of $\int \dfrac{x}{(x^2+1)(x-1)}dx$.

Solution Assumed that

$$\dfrac{x}{(x^2+1)(x-1)} = \dfrac{Ax+B}{x^2+1}+\dfrac{C}{x-1},$$

Comparing the numerators in the two sides, it obtains that
$$x=(Ax+B)(x-1)+C(x^2+1)=(A+C)x^2+(B-A)x+(C-B).$$
Then comparing the coefficients of which has the same power in the two sides:
$$\begin{cases} A+C=0, \\ -A+B=1, \\ -B+C=0. \end{cases}$$

It can be solved that: $A=-\dfrac{1}{2}$, $B=\dfrac{1}{2}$, $C=\dfrac{1}{2}$. Therefore

$$\int \dfrac{x}{(x^2+1)(x-1)}dx = -\dfrac{1}{2}\int \dfrac{x-1}{x^2+1}dx + \dfrac{1}{2}\int \dfrac{1}{x-1}dx$$

$$= -\dfrac{1}{4}\ln(1+x^2)+\dfrac{1}{2}\arctan x + \dfrac{1}{2}\ln|x-1|+C.$$

Example 2 Find the indefinite integral of $\int \dfrac{1}{x^2+2x+3}dx$.

Solution The equation $x^2+2x+3=0$ has no real roots ($p^2-4q=4-12=-8<0$). Then completing the square:

$$x^2+2x+3=(x+1)^2+(\sqrt{2})^2,$$

Thus
$$\int \frac{1}{x^2+2x+3}dx = \int \frac{1}{(x+1)^2+(\sqrt{2})^2}d(x+1)$$
$$= \frac{1}{\sqrt{2}}\arctan \frac{x+1}{\sqrt{2}}+C.$$

Example 3 Find the indefinite integral of $\int \frac{3x+5}{x^2+2x+2}dx$.

Solution
$$\int \frac{3x+5}{x^2+2x+2}dx = \frac{3}{2}\int \frac{d(x^2+2x+2)}{x^2+2x+2}+2\int \frac{d(x+1)}{(x+1)^2+1}$$
$$= \frac{3}{2}\ln|x^2+2x+2|+2\arctan(x+1)+C.$$

Example 4 Find the indefinite integral of $\int \frac{x^2+1}{x(x-1)^2}dx$.

Solution Due to $\frac{x^2+1}{x(x-1)^2}=\frac{1}{x}+\frac{2}{(x-1)^2}$, so

$$\int \frac{x^2+1}{x(x-1)^2}dx = \int \left[\frac{1}{x}+\frac{2}{(x-1)^2}\right]dx$$
$$= \ln|x|-\frac{2}{x-1}+C.$$

Example 5 Find the indefinite integral of $\int \frac{x^2+2x-1}{(x-1)^2(x^2-x+1)}dx$.

Solution Due to $\frac{x^2+2x-1}{(x-1)^2(x^2-x+1)}=\frac{2}{x-1}+\frac{2}{(x-1)^2}-\frac{2x+1}{x^2-x+1}$, so

$$\int \frac{x^2+2x-1}{(x-1)^2(x^2-x+1)}dx = \int\left[\frac{2}{x-1}+\frac{2}{(x-1)^2}-\frac{2x+1}{x^2-x+1}\right]dx$$
$$= \int \frac{2}{x-1}dx+\int \frac{2}{(x-1)^2}dx-\int \frac{2\left(x-\frac{1}{2}\right)+2}{\left(x-\frac{1}{2}\right)^2+\frac{3}{4}}dx$$
$$= 2\ln|x-1|-\frac{2}{x-1}-\ln|x^2-x+1|-\frac{4}{\sqrt{3}}\arctan \frac{2x-1}{\sqrt{3}}+C.$$

Note There is the common method above for the indefinite integrals of rational functions. However, this may not be the easiest method of finding in details. Thus when solving the integral of rational function, it should consider other integral methods according to characters of integrands.

Example 6 Find the indefinite integral of $\int \frac{1}{x(x^7+2)}dx$.

Solution Make $x=\frac{1}{t}, dx=-\frac{1}{t^2}dt$, therefore

$$\int \frac{1}{x(x^7+2)}dx = \int \frac{t}{\left(\frac{1}{t}\right)^7+2}\left(-\frac{1}{t^2}\right)dt = -\int \frac{t^6}{1+2t^7}dt$$

$$= -\frac{1}{14}\ln|1+2t^7|+C = -\frac{1}{14}\ln|2+x^7|+\frac{1}{2}\ln|x|+C.$$

Note It often substitutes $x=\frac{1}{t}$ when the order of denominator in rational function is high.

In the following two chapters, we will discuss several integral problems that integrands can transform into rational functions.

5.4.2 Integration of rational trigonometric functions

Rational trigonometric function refers to the function composed by constants and trigonometric functions after limited times of rational operation (arithmetic is also titled as rational operation). This is one kind that can transform into rational function integral through conversion. Because tangent, cotangent, secant and cosecant in trigonometric function can be all expressed by $\sin x$ and $\cos x$, so we write the rational trigonometric function as $R(\sin x, \cos x)$, its indefinite integral as

$$\int R(\sin x, \cos x)dx.$$

1) Universal substitution

Assumed that $\tan\frac{x}{2}=t$ $(-\pi<x<\pi)$ or $x=2\arctan t$, so $dx=\frac{2}{1+t^2}dt$.

Due to,
$$\sin x = \frac{2\tan\frac{x}{2}}{1+\tan^2\frac{x}{2}} = \frac{2t}{1+t^2},$$

$$\cos x = \frac{1-\tan^2\frac{x}{2}}{1+\tan^2\frac{x}{2}} = \frac{1-t^2}{1+t^2}.$$

Therefore
$$\int R(\sin x, \cos x)dx = \int R\left(\frac{2t}{1+t^2}, \frac{1-t^2}{1+t^2}\right) \cdot \frac{2}{1+t^2}dt.$$

In this way, it can transform the integration of a rational trigonometric function into the integration of rational function. Because the integration of rational function can be integrated, the integration of a rational trigonometric function can also be found. Conversion $\tan\frac{x}{2}=t$ is titled as the universal substitution of the rational trigonometric function $R(\sin x, \cos x)$.

Example 7 Find the indefinite integral of $\int \frac{dx}{\sin x(1+\cos x)}$.

Solution Assumed that $\tan\frac{x}{2}=t$, then $x=2\arctan t$, $dx=\frac{2}{1+t^2}dt$, $\sin x=\frac{2t}{1+t^2}$, $\cos x=\frac{1-t^2}{1+t^2}$. Thereupon

$$\int \frac{dx}{\sin x(1+\cos x)} = \int \frac{1}{\frac{2t}{1+t^2}\left(1+\frac{1-t^2}{1+t^2}\right)} \cdot \frac{2}{1+t^2}dt = \frac{1}{2}\int \left(t+\frac{1}{t}\right)dt$$

$$= \frac{1}{2}\left(\frac{t^2}{2}+\ln|t|\right)+C=\frac{1}{4}\tan^2\frac{x}{2}+\frac{1}{2}\ln\left|\tan\frac{x}{2}\right|+C.$$

2) The kind is $R(\sin x, \cos x) = R(-\sin x, -\cos x)$

Example 8 Find the indefinite integral of $\int \frac{1}{\sin^4 x \cos^2 x} dx$.

Solution This kind is $R(\sin x, \cos x) = R(-\sin x, -\cos x)$. Make substitution $t = \tan x$, then

$$\int \frac{1}{\sin^4 x \cos^2 x} dx = \int \frac{(1+t^2)^2}{t^4} dt$$

$$= \tan x - 2\cot x - \frac{1}{3}\cot^3 x + C.$$

(3) If $R(\sin x, \cos x) = \sin ax\cos bx$ or $\sin ax\sin bx$ or $\cos ax\cos bx$

It is easy to obtain the antiderivative by applying the product to sum formula of trigonometric function.

Example 9 Find the indefinite integral of $\int \sin 9x \sin x dx$.

Solution
$$\int \sin 9x \sin x dx = \frac{1}{2}\int(\cos 8x - \cos 10x)dx$$

$$= \frac{1}{2}\left(\int \cos 8x dx - \int \cos 10x dx\right)$$

$$= \frac{1}{16}\int \cos 8x d(8x) - \frac{1}{20}\int \cos 10x d(10x)$$

$$= \frac{1}{16}\sin 8x - \frac{1}{20}\sin 10x + C.$$

Note In theorem, the method of universal substitution is important and also common method in finding the integration of a rational trigonometrical function. Due to the large found amount in application, firstly consider simple methods when finding the integration of rational trigonometric function. Using the universal substitution unless it is necessary.

5.4.3 Integration of simply irrational functions

Simply irrational function refers to one kind of functions that can transform into rational function through conversion.

Integrals like $\int R(x, \sqrt[n]{ax+b})dx$ can make variable substitution that $\sqrt[n]{ax+b}=t$; integrals like $\int R\left(x, \sqrt[n]{\frac{ax+b}{cx+d}}\right)dx$ can make variable substitution $t=\sqrt[n]{\frac{ax+b}{cx+d}}$ that rationalizes the integrand $R\left(x, \sqrt[n]{\frac{ax+b}{cx+d}}\right)$.

Example 10 Find the indefinite integral of $\int \frac{\sqrt{x+1}}{x+2}dx$.

Solution Make $\sqrt{x+1}=t$, then $x=t^2-1$, so $dx=2tdt$. Thereupon

$$\int \frac{\sqrt{x+1}}{x+2}dx = \int \frac{t}{t^2+1}2tdt = 2\int\left(1-\frac{1}{1+t^2}\right)dt = 2(t-\arctan t)+C$$

$$=2(\sqrt{x+1}-\arctan\sqrt{x+1})+C.$$

Example 11 Find the indefinite integral of $\int \dfrac{dx}{\sqrt{x}(1+\sqrt[3]{x})}$.

Solution In order to rationalize all the surds $\sqrt{x}$ and $\sqrt[3]{x}$ of the integrand, make $\sqrt[6]{x}=t$ or $x=t^6$, then $dx=6t^5 dt$, therefore

$$\int \frac{dx}{\sqrt{x}(1+\sqrt[3]{x})} = \int \frac{6t^5}{t^3(1+t^2)}dt = 6\int \frac{t^2}{1+t^2}dt = 6(t-\arctan t)+C$$
$$=6(\sqrt[6]{x}-\arctan\sqrt[6]{x})+C.$$

Example 12 Find the indefinite integral of $\int \dfrac{1}{\sqrt[3]{(x+1)^2(x-1)^4}}dx$.

Solution Due to

$$\frac{1}{\sqrt[3]{(x+1)^2(x-1)^4}} = \frac{1}{(x-1)(x+1)}\sqrt[3]{\frac{x+1}{x-1}},$$

assumed that $\sqrt[3]{\dfrac{x+1}{x-1}}=t$, then

$$x=\frac{t^3+1}{t^3-1},\ dx=\frac{-6t^2}{(t^3-1)^2}dt,$$

therefore the original integral is

$$\int \frac{1}{\sqrt[3]{(x+1)^2(x-1)^4}}dx = \int \frac{1}{\dfrac{2}{t^3-1}\cdot\dfrac{2t^3}{t^3-1}}\cdot t \cdot \frac{-6t^2}{(t^3-1)^2}dt$$

$$=-\frac{3}{2}\int dt = -\frac{3}{2}t+C = -\frac{3}{2}\sqrt[3]{\frac{x+1}{x-1}}+C.$$

Note In practice, the integrals may be complex. In order to apply expediently, people have built the integration tables for finding. In the appendix of this book, there is a simple integration table which is arranged by the kind of integrand. When finding integrals, it can look up the result you need from the table directly or through simple transformation according to the kind of integrand. It will be easier to work out indefinite integrals if people apply the general mathematical software.

Exercises 5-4

1. Find indefinite integrals of the following rational functions

(1) $\int \dfrac{x}{(x+1)(x+2)(x+3)}dx$;

(2) $\int \dfrac{x-5}{x^3-3x^2+4}dx$;

(3) $\int \dfrac{dx}{(x-1)(x+1)^2}$;

(4) $\int \dfrac{2x-1}{x^2-5x+6}dx$;

(5) $\int \dfrac{x^5+x^4-8}{x^3-x}dx$;

(6) $\int \dfrac{dx}{x(x^2+1)}$;

(7) $\int \dfrac{x^3-3x+2}{x(x^2+2x+1)}dx$;

(8) $\int \dfrac{x^6+x^4-4x^2-2}{x^3(x^2+1)^2}dx$;

(9) $\int \dfrac{5x+1}{x^2+2x+3}dx$;   (10) $\int \dfrac{x}{x^8-1}dx$.

2. Find the following indefinite integrals:

(1) $\int \dfrac{1}{3+5\cos x}dx$;   (2) $\int \dfrac{1-\cos x}{1+\cos x}dx$;

(3) $\int \dfrac{\cos x}{1+\cos x}dx$;   (4) $\int \dfrac{\sin^2 x+1}{\cos^4 x}dx$;

(5) $\int \dfrac{\cot x}{\sin x+\cos x+1}dx$;   (6) $\int \dfrac{\sin x}{1+\sin x}dx$.

3. Find the following indefinite integrals:

(1) $\int \dfrac{2+x}{3\sqrt{3-x}}dx$;   (2) $\int x\sqrt{\dfrac{x-1}{x+1}}dx$;

(3) $\int \sqrt{\dfrac{x}{1-\sqrt{x}}}dx$;   (4) $\int \dfrac{\sqrt{x+1}-1}{\sqrt{x+1}+1}dx$;

(5) $\int \dfrac{dx}{x\sqrt{x^2+4x+3}}$;   (6) $\int \dfrac{dx}{\sqrt{1+\sqrt[3]{x^2}}}$.

5.5 Application of indefinite integral in economy

Known from the marginal analysis in Chapter 2, for a given economic function $F(x)$, its marginal function is the derived function $F'(x)$. As the inverse operation of differential coefficient, if calculating indefinite integral $\int F'(x)dx$ of the given marginal function $F'(x)$, it will work out the antiderivative

$$\int F'(x)dx + C.$$

The integration constant C is determined by actual condition of $F(0)=F_0$.

For example, given that the marginal cost function of producing one commodity is $C'(x)$ and x refers to the output of this commodity. Thus the total cost function of producing this commodity is

$$C(x) = \int C'(x)dx + C_0.$$

The economic significance of $C_0 = C(0)$ always stands up.

In the same way, given that marginal product, marginal revenue, marginal profit, marginal demand and so on, it can work out the production function, revenue function, profit function and demand function through indefinite integral. Then take examples to demonstrate.

Example 1 If the manufacturer's marginal cost is the following function of gross output: $C'(x)=2e^{0.2x}$ and the fixed cost is $C_F=90$. Find total cost function.

Solution Integrate

$$\int 2e^{0.2x}dx = 2\dfrac{1}{0.2}e^{0.2x} + C = 10e^{0.2x} + C.$$

Regarding $C_F = 90$ as the initial condition that determines constant C. When $x=0$, the total cost only includes C_F. Thus it can obtain: $10e^0 + C = 90$, then $C = 80$. Total cost function is
$$C(x) = 10e^{0.2x} + 80.$$

Example 2 Assumed that the marginal revenue of one enterprise is
$$R'(x) = 100 - 0.01x$$
(x refers to the output of the product) and revenue is $R=0$ when the output is $x=0$. Find revenue function $R(x)$ and average revenue function.

Solution Given that the marginal revenue is
$$R'(x) = 100 - 0.01x,$$
Therefore integrate both sides of the formula above, it obtains
$$R(x) = \int (100 - 0.01x) dx = 100x - 0.005x^2 + C.$$
Substitute $x=0$, $R=0$ into the formula, then $C=0$.

Thus revenue function is
$$R(x) = 100x - 0.005x^2,$$
and average revenue function is
$$\overline{R(x)} = 100 - 0.005x.$$

Example 3 The marginal cost of one enterprise's product is $3x+20$ and its marginal revenue is $44-5x$. When producing and then selling 80 units of this product, the cost is 11 400 Yuan. Find:

(1) The best level of production.

(2) Profit function.

(3) When the production is at the best level, will the enterprise make a profit or lose money?

Solution (1) Because the condition that satisfies the best level of production is
Marginal cost = Marginal revenue.
So it can solve that $x=3$, according to the equation $3x+20 = 44-5x$.

(2) Cost function is
$$C(x) = \int (3x+20) dx = \frac{3}{2}x^2 + 20x + C.$$
Substitute the given condition $x=80$, $C(80) = 11\ 400$ into the formula above, then it can obtain that $C=200$.

Thus the cost function is
$$C(x) = \frac{3}{2}x^2 + 20x + 200.$$

Revenue function is
$$R(x) = \int (44-5x) dx = 44x - \frac{5}{2}x^2 + C.$$
Substitute the given condition $x=0$, $R(0)=0$, into the formula, it can obtain that $C=0$.

Thus the revenue function is
$$R(x)=44x-\frac{5}{2}x^2.$$
Therefore profit function is
$$L(x)=R(x)-C(x)=24x-4x^2-200.$$
(3) Substitute the best level of production $x=3$ known from (1) into the profit function, then
$$L(3)=24\times 3-4\times 3^2-200=-164.$$
Thus when the production is at the best level, the enterprise will have a loss of 164 Yuan.

Exercises 5-5

1. If the marginal cost of total cost $C(x)$ (ten-thousand Yuan) that producing units of one product is $C'(x)=5+\dfrac{6}{\sqrt{x}}$ and the fixed cost is 10 (ten-thousand Yuan). Find total cost function and average cost function.

2. If the marginal propensity to save is the following function of income:
$$S'(x)=0.3-0.1x^{-\frac{1}{2}}.$$
When income $x=81$, the gross saving is $S=0$. Find savings function.

3. The demand Q of one commodity is the function of price P. Given that the demand's rate of change is
$$Q'(P)=-1\,000\ln 3\left(\frac{1}{3}\right)^P$$
and the maximum demand of this commodity is 1 000. Find the demand function of this commodity.

Summary

1. Main contents

(1) The concepts and characters of indefinite integral.

① If $f(x)$ is continuous on the interval I, then $f(x)$ exists the antiderivative $F(x)$ on the interval I. Thus all elementary functions exist antiderivatives in their domains.

② Assumed that $F(x)$ is one antiderivative of $f(x)$, then $F'(x)=f(x)$ or $dF(x)=f(x)dx$. Thus $f(x)$ has infinite antiderivativs and the difference between any two antiderivativs is only a constant. The general expression of function $f(x)$'s antiderivativs is $F(x)+C$ (C is an arbitrary constant) which is titled as the indefinite integral of $f(x)$,
$$\int f(x)dx = F(x)+C.$$
③ $\int [af(x)\pm bg(x)]dx = a\int f(x)dx \pm b\int g(x)dx.$

④ $\dfrac{d}{dx}\left[\int f(x)dx\right]= f(x).$

⑤ $\int f'(x)dx = f(x) + C$.

(2) Find indefinite integrals.

① Immediate integration.

Immediate integration is the most basic integration method that using algebra or trigonometric identical transformation and then integrating through the properties of indefinite integral and basic integral formulas. However, notice that the transformation of integrands in integral are all identical transformation.

② The first substitution method

The first substitution method is also titled as gathering differential together. Its key is how to transform the expression of integrant $g(x)dx$ in the integral $\int g(x)dx$ into the given form $f(u)du$ in basic integration tables. Thus
$$g(x)dx = f[\varphi(x)]\varphi'(x)dx = f(u)du \quad (u=\varphi(x)).$$
The key is whether $u=\varphi(x)$'s change of variable exists and how to seek it out. This has strong flexibility that needs to practice more and summarize more. The basic kinds of indefinite integrals that can be calculated through gathering differential together are:

$$\int f(ax+b)dx = \frac{1}{a}\int f(ax+b)d(ax+b);$$

$$\int f(x^2)xdx = \frac{1}{2}\int f(x^2)d(x^2);$$

$$\int f(e^x)e^x dx = \int f(e^x)d(e^x);$$

$$\int \frac{f(\ln x)}{x}dx = \int f(\ln x)d(\ln x);$$

$$\int f(\sin x)\cos xdx = \int f(\sin x)d(\sin x);$$

$$\int f(\cos x)\sin xdx = -\int f(\cos x)d(\cos x);$$

$$\int f(\tan x)\sec^2 xdx = \int f(\tan x)d(\tan x);$$

$$\int f(\arcsin x)\frac{1}{\sqrt{1-x^2}}dx = \int f(\arcsin x)d(\arcsin x);$$

$$\int f(\arctan x)\frac{1}{1+x^2}dx = \int f(\arctan x)d(\arctan x);$$

$$\int f(\sqrt{x^2+a^2})\frac{x}{\sqrt{x^2+a^2}}dx = \int f(\sqrt{x^2+a^2})d(\sqrt{x^2+a^2}).$$

③ The second substitution method.

$\int g(x)dx = \int g[\varphi(t)]\varphi'(t)dt$ is the one that transforms integral $\int g(x)dx$ into the integral $\int g[\varphi(t)]\varphi'(t)dt$ whose integral variable is t through substituting $x=\varphi(t)$. If this integral can be found, it can obtain the answer of indefinite integral by substituting

$x=\varphi(t)$'s inverse function $t=\varphi^{-1}(x)$ back. The method is often used when integrands have surds. For example, it should assume that when the integrand has $\sqrt{a^2-x^2}$; it should assume that $x=a\tan t$ when the integrand has $\sqrt{a^2+x^2}$; it should assume that $x=a\sec t$ when the integrand has $\sqrt{x^2-a^2}$.

When the integrand has other irrational expressions that doesn't belong to typical conditions above, it often regards the whole irrational expression in the integrand as t and then infers the expression of $x=\varphi(t)$.

④ Integration by parts.

Integration by parts is another basic method of finding integrals. The key of using the formula $\int u dv = uv - \int v du$ is to select proper u and dv. The principles of selecting u, v are:

a. v is easy to obtain; b. $\int v du$ is easier than $\int u dv$ to integrate.

2. Basic requirements

(1) Understanding the concepts of primitive function, indefinite integral and the relation between them.

(2) Understanding the characters and geometric meanings of indefinite integral.

(3) Remember basic integral formulas and fundamentums of indefinite integral firmly.

(4) Remember and apply the integration by substitution and integration by parts.

(5) Be able to find simple rational functions and indefinite integrals that can transform into rational functions.

(6) Be able to apply indefinite integral to solve problems in economy.

Quiz

1. Choice question.

(1) If $\int df(x) = \int dg(x)$, the following forms that may not stand up is (   ).

A. $f(x)=g(x)$
B. $f'(x)=g'(x)$
C. $df(x)=dg(x)$
D. $d\int f'(x)dx = d\int g'(x)dx$

(2) If $F(x)$ is one primitive function of $f(x)$ and C is a constant. Which one is the indefinite integral of $f(x)$ in the following forms? (   )

A. $F(x)$
B. $CF(x)$
C. $F\left(\dfrac{x}{C}\right)$
D. $F(x)+C$

(3) If $\int f(x)dx = x^2 + C$, then $f(x)=$ (   ).

A. $2x$
B. $2x+1$
C. $\dfrac{1}{3}x^3$
D. x

(4) If $F'(x)=f(x)$, then the following equation that stands up is (   ).

A. $\int F'(x)dx = f(x) + C$ B. $\int f(x)dx = F(x) + C$

C. $x\int F(x)dx = f(x) + C$ D. $\int f'(x)dx = F(x) + C$

2. Competition.

(1) Assumed that $\frac{1}{x^2}dx = dF(x)$, then $F(x) = $ _____.

(2) $\int x\sqrt[3]{x}dx = $ _____.

(3) $\int 4^x e^x dx = $ _____.

(4) If $\int f(x)dx = F(x) + C$, then $\int f(3x+5)dx = $ _____.

(5) $\int (e^x + x^e)dx = $ _____.

3. Find the following indefinite in tegrals:

(1) $\int (3x-8)^{20} dx$;   (2) $\int \frac{1}{2x^2+9}dx$;

(3) $\int e^x \left(1 - \frac{e^{-x}}{\sqrt{x}}\right)dx$;   (4) $\int \frac{1}{x\sqrt{2x-1}}dx$;

(5) $\int x^2 \ln x dx$;   (6) $\int \frac{x^2}{x-2}dx$.

4. Find the following integrals:

(1) $\int \frac{dx}{x\sqrt{1-x^2}}$;   (2) $\int \frac{\sec^2(\ln x)}{x}dx$;

(3) $\int \sin\sqrt{x+1}dx$;

(4) Assumed that $f''(x)$ is a continuous function, find $\int xf''(x)dx$.

5. Suppose that the curve passes through point $(3,2)$, and the tangent slope in an arbitrary point (x,y) of the curve is $\frac{1}{\sqrt{x+1}}$. Find the curve.

Exercises

1. Find the following indefinite integrals:

(1) $\int x\sqrt[3]{1-3x}dx$;   (2) $\int \frac{dx}{1+e^x}$;

(3) $\int \frac{e^x}{e^x+2+2e^{-x}}dx$;   (4) $\int \frac{x^4+1}{x^6+1}dx$;

(5) $\int \frac{\sin 2x dx}{\sqrt{4-\cos^4 x}}$;   (6) $\int \frac{\ln \tan x}{\cos x \sin x}dx$;

(7) $\int \frac{dx}{\sqrt{1+e^{2x}}}$;   (8) $\int \frac{2^x 3^x}{9^x - 4^x}dx$;

(9) $\int \dfrac{1-\ln x}{(x-\ln x)^2}\mathrm{d}x$;

(10) $\int \dfrac{\mathrm{d}x}{2+\sqrt{1-x^2}}$;

(11) $\int \dfrac{\mathrm{d}x}{x^2\sqrt{a^2+x^2}}\,(a>0)$;

(12) $\int \dfrac{\mathrm{d}x}{x\sqrt{x^2-1}}$;

(13) $\int \dfrac{\mathrm{d}x}{\sqrt{x-x^2}}$;

(14) $\int \dfrac{\ln x}{(x-2)^2}\mathrm{d}x$;

(15) $\int \arctan(1+\sqrt{x})\mathrm{d}x$;

(16) $\int \mathrm{e}^{\sin x}\sin 2x\,\mathrm{d}x$;

(17) $\int \dfrac{x+\sin x}{1+\cos x}\mathrm{d}x$;

(18) $\int \dfrac{x\mathrm{e}^x}{\sqrt{\mathrm{e}^x-2}}\mathrm{d}x$;

(19) $\int \sin(\ln x)\mathrm{d}x$;

(20) $\int x\arctan x\ln(1+x^2)\mathrm{d}x$;

(21) $\int \dfrac{x^9-8}{x^{10}+8x}\mathrm{d}x$;

(22) $\int \dfrac{x^2}{(1-x^2)^3}\mathrm{d}x$;

(23) $\int \dfrac{x^2+1}{x^4+1}\mathrm{d}x$;

(24) $\int \dfrac{\mathrm{d}x}{x^4-2x^2+1}$;

(25) $\int \dfrac{2x+2}{(x-1)(x^2+1)^2}\mathrm{d}x$;

(26) $\int \dfrac{\mathrm{d}x}{1+\cos^2 x}$;

(27) $\int \dfrac{1+\sin x}{\sin 3x+\sin x}\mathrm{d}x$;

(28) $\int \dfrac{\mathrm{d}x}{1+\sin x}$;

(29) $\int \dfrac{3\sin x+4\cos x}{2\sin x+\cos x}\mathrm{d}x$;

(30) $\int \dfrac{\mathrm{d}x}{(2+\sin^2 x)\cos x}$;

(31) $\int \dfrac{5+4\cos x}{(2+\cos x)^2\sin x}\mathrm{d}x$;

(32) $\int \dfrac{\sin x}{\sin^3 x+\cos^3 x}\mathrm{d}x$;

(33) $\int \dfrac{\sin 2x}{\sin^6 x+\cos^6 x}\mathrm{d}x$;

(34) $\int \dfrac{\sin x}{1+\sin x+\cos x}\mathrm{d}x$;

(35) $\int \dfrac{1}{x}\sqrt{\dfrac{x+1}{x-1}}\mathrm{d}x$;

(36) $\int \dfrac{\mathrm{d}x}{x+\sqrt{x^2-1}}$.

2. Solve the following problems.

(1) $f(x)$ is derivable on the interval $[1,+\infty)$. $f(1)=0, f'(\mathrm{e}^x+1)=\mathrm{e}^{3x}+2$, find the $f(x)$.

(2) $f'(\ln x)=\begin{cases}1, & 0<x\leqslant 1,\\ x, & 1<x+\infty,\end{cases}$ find the $f(\ln x)$.

(3) Assumed that $F(x)$ is one antiderivative of $f(x)$ and $F(1)=\dfrac{\sqrt{2}}{4}\pi$. When $x>0$, $f(x)F(x)=\dfrac{\arctan\sqrt{x}}{\sqrt{x}(1+x)}$. Find the $f(x)$.

(4) Assumed that $y=y(x)$ is the implicit function confirmed by $y^2(x-y)=x^2$, then find $\int \dfrac{\mathrm{d}x}{y^2}$.

(5) Assumed that functions $f(x)$ and $g(x)$ are continuous on the interval $[a,b]$ and derivable within (a,b). For any $x\in(a,b)$, there will be $g'(x)\neq 0$. Prove:

① For any two different points x_1 and x_2 on the interval $[a,b]$, there will be $g(x_1) \neq g(x_2)$.

② There will exist at least one point $\xi \in (a,b)$ that makes $\dfrac{f(\xi)-f(a)}{g(b)-g(\xi)} = \dfrac{f'(\xi)}{g'(\xi)}$.

(6) Given that the function $y=f(x)$ has extremum 2 and its figure passes the point $(0,3)$. The figure of its derived function is a straight line which passes the point $(1,0)$ and not parallels to y-axis.

(7) Find the following indefinite integral $\displaystyle\int \left\{ \dfrac{f(x)}{f'(x)} - \dfrac{[f(x)]^2 f''(x)}{[f'(x)]^3} \right\} \mathrm{d}x$.

Chapter 6　Definite integrals

The present chapter is based on a definition of the definite integral, which is another important concept of calculus. We with two typical examples in calculus. Then we give the definition of definite integral and its basic properties. In addition, it will also introduce the fundamental theorem of calculus which communicates the relationship between differential method and integral method. Therefore, it turns the differential calculus and integral calculus into an organic whole.

6.1　The concept of definite integral

6.1.1　Examples of definite integral problems

1) Area of a trapezoid with a curved top

Let function $y=f(x)$ be nonnegative and continuous on an interval $[a,b]$. The graph surrounded by the curve $y=f(x)$, the vertical lines $x=a$, $x=b$ and the x-axis is called a trapezoid with a curved top (Figure 6-1).

We all know that the height of the rectangle is fixed. Its area can be found, according to the formula: rectangle area = bottom × height. Comparing the curved trapezoid with rectangle, you can see the height $f(x)$ of each point on the bottom edge of the curved trapezoid is variable in the domain

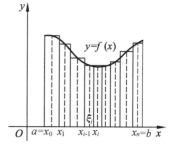

Figure 6-1

$[a,b]$. So it cannot directly be found by the rectangle area formula, but we note that the height $f(x)$ of the curved trapezoid is the continuous function of x. In a small domain, its change is very small, especially when the interval is very small, the height $f(x)$ is approximately unchanged. Therefore, if the interval $[a,b]$ is divided into many small parts, in each interval, use the height of some point to approximately represent the height of the "small curved trapezoid" in the same area, then each small curved trapezoid can be seen approximately as "small rectangle". We consider the sum of the area of all these small rectangles as the approximate value of the area of the curved trapezoid area. Subdividing the domain, we can find out the area of the curved trapezoid by the idea of limit, which is divided into four steps to be discussed concretely.

(1) **Partition** Arbitrarily insert n points of division x_1, x_2, $\cdots$, x_{n-1}, x_n between a and b such that $a = x_0 < x_1 < x_2 < \cdots < x_{n-1} < x_n = b$. Then the interval $[a,b]$ is divided into n subintervals $[x_0, x_1], [x_1, x_2], \cdots, [x_{n-1}, x_n]$. The length of the i th subinterval is
$$\Delta x_i = x_i - x_{i-1}, \quad i = 1, 2, \cdots, n.$$
If we draw a vertical line through each point of division x_i, then the trapezoid with the curved top $f(x)$ is partitioned into n smaller trapezoid with a curved top.

(2) **Homogenization** Although the height $f(x)$ of the graph on the ith subinterval $[x_{i-1}, x_i]$ is still variable, it varies by a very small amount if the length Δx_i is very small. Hence, this graph may be regarded as a rectangle approximately and the height of the rectangle may be chosen as a functional value at any point $\xi_i \in [x_{i-1}, x_i]$, $f(\xi_i)$. Thus, the area of this graph ΔA_i can be computed by multiplication approximately.
$$\Delta A_i \approx f(\xi_i) \Delta x_i, \quad i = 1, 2, \cdots, n.$$

(3) **Summation** Doing the same thing for each subinterval, and combining all the approximate values, we obtain the approximate value of the area for the trapezoid with the curved top:
$$A = \sum_{i=1}^{n} \Delta A_i \approx \sum_{i=1}^{n} f(\xi_i) \Delta x_i.$$

(4) **Precision** It is easy to see that if the interval $[a,b]$ is partitioned more finely, that is, if the length of each subinterval is made smaller, then $\sum_{i=1}^{n} f(\xi_i) \Delta x_i$ is a closer approximation to the total area. Therefore,
$$A = \lim_{\lambda \to 0} \sum_{i=1}^{n} f(\xi_i) \Delta x_i,$$
where $\lambda = \max\{\Delta x_1, \Delta x_2, \cdots \Delta x_n\}$.

2) Solve the problem of total output according to the change rate of it

Assume the change rate of the total output of a product is a function $p = p(t)$ of time t. It is assumed that in the time interval which is from α to β, the production is continuous. Find the total output Q during this period $[\alpha, \beta]$.

When the change rate of the total output rate is a constant, namely, when the number of products produced per unit time remains unchanged, the total output is the product of its rate of change and time. But in the actual work, the change rate of the product is not a constant, but changes with time. If the time interval is very small, then the rate of change also is very small, which can be approximately considered as a constant. Then we can model the way of calculating the area of the curved trapezoid to calculate the total output in $[\alpha, \beta]$.

(1) **Partition** Using the separating points $\alpha < t_0 < t_1 < t_2 < \cdots < t_{n-1} < t_n = \beta$ to divide the interval $[\alpha, \beta]$ into n small intervals, in which the i th small interval is $[t_{i-1}, t_i]$. The length of each small interval is $\Delta t_i = t_i - t_{i-1} (i = 1, 2, \cdots, n)$. Thus, we can divide the total

output into the sum of the outputs in n intervals. We use ΔQ_i to represent the total output in the ith interval $[t_{i-1}, t_i]$, then we have

$$Q = \Delta Q_1 + \Delta Q_2 + \cdots + \Delta Q_n = \sum_{i=1}^{n}\Delta Q_i.$$

(2) **Homogenization** Take any point ξ_i in the ith small interval $[t_{i-1}, t_i]$. Take $p(\xi_i)$ as the change rate of output in the interval $[t_{i-1}, t_i]$. When Δt_i is quite small, $p(\xi_i)$ can be considered as the constant in $[t_{i-1}, t_i]$, and then the total output ΔQ_i on the interval $[t_{i-1}, t_i]$ can be approximately represented by $p(\xi_i)\Delta t_i$. That is

$$\Delta Q_i \approx p(\xi_i)\Delta t_i, \quad i=1,2,\cdots,n.$$

(3) **Summation** Recording Δt as the maximum length of the time interval in each period, namely, $\Delta t = \max_i\{\Delta t_i\}$. Then when Δt is sufficiently small, add up the outputs of each period, then we have the approximate value of the total output:

$$\Delta Q_i \approx \sum_{i=1}^{n} p(\xi_i)\Delta t_i.$$

(4) **Precision** When the number of the separating points n increases infinitely while Δt tends to zero, and the limit of the unity formula $\sum_{i=1}^{n} p(\xi_i)\Delta t_i$ exists, then we can define the limit value of it as the exact value of the total output. That is

$$Q = \lim_{\Delta t \to 0} \sum_{i=1}^{n} p(\xi_i)\Delta t_i.$$

6.1.2 The definition of definite integral

The two examples mentioned above belong to different fields. But from a qualitative point of view, they are both summation problems of a quantity distributed on an interval. It has been seen that the idea and steps of solving these problems are the same, and they can be both viewed as the limit of a sum with the same mathematical structure. In fact, the same type sum of that mathematical structure occurs in a wide variety of situations. Therefore, we define such limit of a sum as the definite integral abstractly.

Definition Let $f(x)$ be a bounded function defined on a closed interval $[a,b]$. Partition the interval $[a,b]$ by inserting arbitrarily $n-1$ points in the colse interval $[a,b]$: $a = x_0 < x_1 < x_2 < \cdots < x_{n-1} < x_n = b$. Divide $[a,b]$ into n small domains $[x_0, x_1], [x_1, x_2], \cdots, [x_{n-1}, x_n]$, and make $\Delta x_i = x_i - x_{i-1}$ the length of the small interval ($i=1,2,\cdots,n$). Take $\lambda = \max_i\{\Delta x_i\}$. Arbitrarily take a point $\xi_i (i=1,2,\cdots,n)$ in the interval $[x_{i-1}, x_i]$. Make the product $f(\xi_i)\Delta x_i$, and make the summation $S_n = \sum_{i=1}^{n} f(\xi_i)\Delta x_i$. No matter how to divide the interval $[a,b]$ and how to take the value of ξ_i, when $\lambda \to 0$, the limit of the summation $\sum_{i=1}^{n} f(\xi_i)\Delta x_i$ always exists. Then we call the function $f(x)$ is integrable in the interval $[a,b]$, and also call the limit value the definite integral of the function $f(x)$

in $[a,b]$, recorded as $\int_a^b f(x)dx$. That is

$$\int_a^b f(x)dx = \lim_{\lambda \to 0} \sum_{i=1}^n f(\xi_i)\Delta x_i,$$

where x is called the variable integral, $f(x)$ is called the integrand, $[a,b]$ is called the interval of integration, a and b are called the lower limit and upper limit of integration respectively, in which $\int$ is called the integral sign and $f(x)dx$ is called the element of integration.

Note According to the concept of definite integral, we can know that the separation of the interval $[a,b]$ is arbitrary in definition. For different separations, we have different summations S_n. Even for the same separation, because of the arbitrariness of the taking method of ξ_i in $[x_{i-1}, x_i]$, we will have infinite summations S_n. According to the request of definition, no matter how to divide the interval and how to take the value of ξ_k, when $\lambda \to 0$, all the summations S_n approach to a same limit, then definite integral exists.

Because the definite integral $\int_a^b f(x)dx$ is the limit of the summation, then the integral is a certain number. But the summation is only related to the integrand $f(x)$ and the interval $[a,b]$, and has nothing to do with the mark of the integral variable x. So we can change the remark of the integral into other letters which will not change the value of the integral. For example,

$$\int_a^b f(x)dx = \int_a^b f(t)dt = \cdots = \int_a^b f(u)du.$$

The geometric meaning of definite integral:

When the curve is $y = f(x) \geq 0$, then the integral $\int_a^b f(x)dx$ represents the area S of the curved trapezoid which is surrounded by the lines $x=a$, $x=b$, $y=0$ and the curve $y=f(x)$. That is

$$S = \int_a^b f(x)dx.$$

When $f(x) \leq 0$, according to the definition, $\int_a^b f(x)dx \leq 0$.

At this time, the curved trapezoid is under the axis. The value of the integral is the negative value of its area. So when $f(x)$ has both negative and positive values in the interval, the geometric meaning of the definite integral $\int_a^b f(x)dx$ is the algebraic sum of the areas of each curved trapezoids in $[a,b]$ (Figure 6-2).

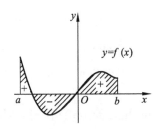

Figure 6-2

The economic significance of definite integral:

In the cited case, solving the total output problem according to the change rate of it is the most typical case of the definite integral applied in economy. Generally speaking, if the change rate of the economic indicator about the independent variable is $f(x)$, then when x changes from a to b, the definite integral $\int_a^b f(x)\,\mathrm{d}x$ represents the total amount of economic indicator y in $[a,b]$. In practical problems, when finding the total amount of some economic indicator according to definite integral, the economic meaning of the independent variable mostly represents time or output. Then we can find the total amount function according to the marginal function and we can find the total amount of the economic variable according to the change rate of some economic variable.

For example, we already know that the change rate of the sales of some product towards time t is $p(t)$, then the total amount of the sales in the interval $[\alpha,\beta]$ is $\int_\alpha^\beta p(t)\,\mathrm{d}t$. And because the change rate of population towards time is $p(t)$, then when on the interval $[t_1,t_2]$, the total amount of population is $\int_{t_1}^{t_2} p(t)\,\mathrm{d}t$.

According to the definition, the existence of the definite integral of the function $f(x)$ in the interval $[a,b]$ depends on the existence of the limit of the summation. It is complex to judge whether a function is integrable by limit. The content is beyond the request of this book. Then under what conditions is the function $f(x)$ integrable in the find? We give the following two sufficient conditions for this question.

Theorem 1 If the function $f(x)$ is continuous on the interval $[a,b]$, then it is integrable on $[a,b]$.

In Theorem 1, the continuity condition of $f(x)$ is very strong on $[a,b]$, we can release it and have

Theorem 2 If the function $f(x)$ is bounded on the interval $[a,b]$, and has only finite numbers of discontinuity points, then $f(x)$ is integrable on $[a,b]$.

Finally, we give an example of finding definite integral by definition.

Example 1 Find the definite integral $\int_0^1 x^2\,\mathrm{d}x$ by definition.

Solution Because the integration function $f(x)=x^2$ is continuous on the integral interval $[0,1]$, by Theorem 1, the continuous function is integrable. So integral has nothing to do with the separation method of the interval $[0,1]$ and the taking method of the point ξ_i. Therefore, in order to work out it easily, we can divide the interval $[0,1]$ equally into n parts, and the separating point is $x_i=\dfrac{i}{n}$ $(i=1,2,\cdots,n-1)$. Thus, the length of each small interval $[x_{i-1},x_i]$ is $\Delta x_i=\dfrac{1}{n}$ $(i=1,2,\cdots,n)$. Take $\xi_i=x_i(i=1,2,\cdots,$

n), then we have the summation

$$\sum_{i=1}^{n} f(\xi_i)\Delta x_i = \sum_{i=1}^{n} \xi_i^2 \Delta x_i = \sum_{i=1}^{n} x_i^2 \Delta x_i$$
$$= \sum_{i=1}^{n} \left(\frac{i}{n}\right)^2 \cdot \frac{1}{n} = \frac{1}{n^3}\sum_{i=1}^{n} i^2$$
$$= \frac{1}{n^3} \cdot \frac{1}{6}n(1+n)(1+2n)$$
$$= \frac{1}{6}\left(1+\frac{1}{n}\right)\left(2+\frac{1}{n}\right).$$

when $\lambda \to 0$, we have $n \to \infty$. Take the limit, we have

$$\int_0^1 x^2 \mathrm{d}x = \lim_{\lambda \to 0} \sum_{i=1}^{n} \xi_i^2 \Delta x_i = \lim_{n \to \infty} \frac{1}{6}\left(1+\frac{1}{n}\right)\left(2+\frac{1}{n}\right) = \frac{1}{3}.$$

According to this example, we can find that although the integration function $f(x) = x^2$ is easy, the processes of finding definite functions by definition is complicated. Therefore, we must find effective methods to find definite integral.

Exercises 6-1

1. What is the difference between definite integral and indefinite integral?

2. What is the geometric meaning of the definite integral $\int_0^1 x^2 \mathrm{d}x$? And what is the geometric meaning of the indefinite integral $\int x^2 \mathrm{d}x$?

3. What are the factors that related to the definite integral $\int_a^b f(x)\mathrm{d}x$? And what are not?

4. Use the integral summation to represent the approximate area of the curved trapezoid which is surrounded by the parabola $y = \frac{x^2}{2}$, the lines $x=3$, $x=6$ and the x-axis. Also, take the limit of the summation and find the value.

5. Use the geometric meaning of definite integral to prove the following equations:

(1) $\int_0^1 2x\mathrm{d}x = 1$; (2) $\int_0^1 \sqrt{1-x^2}\mathrm{d}x = \frac{\pi}{4}$.

6. Use the geometric meaning of definite integral to explain the meaning of the following equations:

(1) $\int_0^a \sqrt{a^2-x^2}\mathrm{d}x = \frac{\pi}{4}a^2$; (2) $\int_{-\frac{\pi}{2}}^{\frac{\pi}{2}} x^2\cos x\mathrm{d}x = 2\int_0^{\frac{\pi}{2}} x^2\cos x\mathrm{d}x$.

6.2 Properties of definite integrals

Considering the convenience of computation and application, we give two rules:

(1) When $a=b$, $\int_a^b f(x)\mathrm{d}x = 0$.

(2) When $a>b$, $\int_a^b f(x)\mathrm{d}x = -\int_b^a f(x)\mathrm{d}x$.

We can deduct the property of definite integral directly according to the definition of definite integral. The values of the upper and bottom limits of integral in the following properties, if not pointed out specially, all have no limitation. And assume that the definite integrals listed in each property all exist.

Property 1 The definite integral of the sum (difference) of the function is equal to the sum (difference) of the definite integrals. That is

$$\int_a^b [f(x) \pm g(x)] dx = \int_a^b f(x) dx \pm \int_a^b g(x) dx.$$

Proof
$$\int_a^b [f(x) \pm g(x)] dx = \lim_{\lambda \to 0} \sum_{i=1}^n [f(\xi_i) \pm g(\xi_i)] \Delta x_i$$
$$= \lim_{\lambda \to 0} \sum_{i=1}^n f(\xi_i) \Delta x_i \pm \lim_{\lambda \to 0} \sum_{i=1}^n g(\xi_i) \Delta x_i$$
$$= \int_a^b f(x) dx \pm \int_a^b g(x) dx.$$

Property 1 is also right for any infinite function.

Property 2 The constant factors of the integration function can be moved outside the bracket. That is

$$\int_a^b k f(x) dx = k \int_a^b f(x) dx \quad (k \text{ is a constant}).$$

Proof $\int_a^b k f(x) dx = \lim_{\lambda \to 0} \sum_{i=1}^n k f(\xi_i) \Delta x_i = k \lim_{\lambda \to 0} \sum_{i=1}^n f(\xi_i) \Delta x_i = k \int_a^b f(x) dx.$

According to Property 1 and 2, we can deduct the linear property of the definite integral. That is

$$\int_a^b [k_1 f_1(x) \pm k_2 f_2(x) \pm \cdots \pm k_n f_n(x)] dx$$
$$= k_1 \int_a^b f_1(x) dx \pm k_2 \int_a^b f_2(x) dx \pm \cdots \pm k_n \int_a^b f_n(x) dx.$$

Property 3 (Additivity for intervals) If the integral interval is divided into two parts, the definite integral in the whole interval is equal to the sum of the definite integral in these two intervals. Assume $a < c < b$, then

$$\int_a^b f(x) dx = \int_a^c f(x) dx + \int_c^b f(x) dx.$$

Proof Because the function $f(x)$ is integrable on $[a, b]$, so no matter how to divide $[a, b]$, the limit of the summation always exists. Therefore, when dividing, we take c always as a separating point, and divide the summation into two parts:

$$\sum_{[a,b]} f(\xi_i) \Delta x_i = \sum_{[a,c]} f(\xi_i) \Delta x_i + \sum_{[c,b]} f(\xi_i) \Delta x_i.$$

Make $\lambda \to 0$, take limits of both sides of the above equation, we have
$$\int_a^b f(x)\,dx = \int_a^c f(x)\,dx + \int_c^b f(x)\,dx.$$

According to the supplementary provision of definite integral, no matter how is the relative position of a, b, c, we always have the right equation
$$\int_a^b f(x)\,dx = \int_a^c f(x)\,dx + \int_c^b f(x)\,dx$$

For example, when $a < b < c$, because $\int_a^c f(x)\,dx = \int_a^b f(x)\,dx + \int_b^c f(x)\,dx$, then
$$\int_a^b f(x)\,dx = \int_a^c f(x)\,dx - \int_b^c f(x)\,dx = \int_a^c f(x)\,dx + \int_c^b f(x)\,dx.$$

Corollary 1 If the function $f(x)$ is integrable in $[\alpha, \beta]$, and a, b, c are any three points in $[\alpha, \beta]$, then $\int_a^b f(x)\,dx = \int_a^c f(x)\,dx + \int_c^b f(x)\,dx$.

Corollary 2 If the function $f(x)$ is integrable among $[c_0, c_1], \cdots, [c_{i-1}, c_i]\,(i=1, 2, \cdots, n)$, then
$$\int_{c_0}^{c_n} f(x)\,dx = \int_{c_0}^{c_1} f(x)\,dx + \int_{c_1}^{c_2} f(x)\,dx + \cdots + \int_{c_{n-1}}^{c_n} f(x)\,dx.$$

Property 4 If on $[a, b]$, $f(x) \equiv 1$, then $\int_a^b dx = b - a$.

We leave the proof of this property to our readers.

Property 5 If on $[a, b]$, we have $f(x) \geqslant g(x)$, then $\int_a^b f(x)\,dx \geqslant \int_a^b g(x)\,dx$.

Proof $\int_a^b f(x)\,dx - \int_a^b g(x)\,dx = \int_a^b [f(x) - g(x)]\,dx = \lim\limits_{\lambda \to 0} \sum\limits_{i=1}^n [f(\xi_i) - g(\xi_i)]\Delta x_i$.
Because for any $\xi_i \in [a, b]$, $f(\xi_i) \geqslant g(\xi_i)$, $\Delta x_i > 0$, the summation is non-negative. Therefore, we have $\int_a^b f(x)\,dx \geqslant \int_a^b g(x)\,dx$.

Corollary 3 If in the interval $[a, b]$, $f(x) \geqslant 0$. Then we have $\int_a^b f(x)\,dx \geqslant 0 \ (a < b)$.

Corollary 4 $\left|\int_a^b f(x)\,dx\right| \leqslant \int_a^b |f(x)|\,dx \ (a < b)$.

Proof Because $-|f(x)| \leqslant f(x) \leqslant |f(x)|$, according to Property 2 and 5, we shall have

$$-\int_a^b |f(x)|\,dx \leqslant \int_a^b f(x)\,dx \leqslant \int_a^b |f(x)|\,dx,$$

That is
$$\left|\int_a^b f(x)\,dx\right| \leqslant \int_a^b |f(x)|\,dx.$$

Note The integrability of $|f(x)|$ on $[a,b]$ can be deducted by the integrability of $f(x)$ on $[a,b]$. Here we ignore the proof.

Property 6 Assume M and m are respectively the maximum and minimum of the function $f(x)$ on the interval $[a,b]$. Then
$$m(b-a) \leqslant \int_a^b f(x)\,dx \leqslant M(b-a).$$

Proof Because for any $x \in [a,b]$, we have $m \leqslant f(x) \leqslant M$. According to Property 5, we have
$$\int_a^b m\,dx \leqslant \int_a^b f(x)\,dx \leqslant \int_a^b M\,dx.$$

And
$$\int_a^b M\,dx = M\int_a^b dx = M(b-a).$$

Similarly, we have
$$\int_a^b m\,dx = m(b-a).$$

Therefore,
$$m(b-a) \leqslant \int_a^b f(x)\,dx \leqslant M(b-a).$$

The property is called the inequality of estimate of definite integral. It indicates that the general range of the integral value can be estimated by the maximum and minimum of the integration on the interval.

Property 7 If the function $f(x)$ is continuous on the closed interval $[a,b]$, then there at least exists a point ξ on $[a,b]$, which can make
$$\int_a^b f(x)\,dx = f(\xi)(b-a) \quad (a \leqslant \xi \leqslant b).$$

This formula is called mean value theorem of definite integrals.

Proof Because $f(x)$ is continuous on $[a,b]$, then it is integrable on $[a,b]$. And it has the maximum M and the minimum m. According to Property 6, we have
$$m(b-a) \leqslant \int_a^b f(x)\,dx \leqslant M(b-a)$$

or
$$m \leqslant \frac{1}{b-a}\int_a^b f(x)\,dx \leqslant M.$$

According to the intermediate value theorem of continuous functions on a closed interval, there at least exists a point ξ on the interval $[a,b]$, $a \leqslant \xi \leqslant b$, which can make
$$f(\xi) = \frac{1}{b-a}\int_a^b f(x)\,dx.$$

That is
$$\int_a^b f(x)\,dx = f(\xi)(b-a).$$

No matter what is the size relationship of a, b, mean value theorem of definite integrals is always right. If $a > b$, then there is

$$\int_a^b f(x)\,dx = f(\xi)(b-a) \quad (b \leqslant \xi \leqslant a).$$

The $f(\xi)$ in the mean value theorem of definite integrals is the average value of the function $f(x)$ on $[a,b]$.

The geometric meaning:

For a curved trapezoid which takes the interval $[a,b]$ as the bottom, its area is equal to the area of a rectangle which has the same bottom and takes the ordinate $f(\xi)$ of a point ξ on the curve as the height (Figure 6-3).

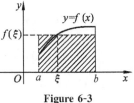

Figure 6-3

Example 1 Compare the size of the integrals $\int_1^2 x^2\,dx$ and $\int_1^2 x^3\,dx$.

Solution Because on $[1,2]$, we have $x^2 \leqslant x^3$. According to Property 5, we know

$$\int_1^2 x^2\,dx \leqslant \int_1^2 x^3\,dx.$$

Example 2 Prove $\dfrac{2}{\sqrt[4]{e}} \leqslant \int_0^2 e^{x^2-x}\,dx \leqslant 2e^2$.

Proof Because the maximum and the minimum of the integration function $f(x) = e^{x^2-x}$ on $[0,2]$ are respectively $\dfrac{1}{\sqrt[4]{e}}$ and e^2. That is

$$\frac{1}{\sqrt[4]{e}} \leqslant f(x) \leqslant e^2, \quad x \in [0,2].$$

According to Property 6, we have

$$\frac{1}{\sqrt[4]{e}}(2-0) \leqslant \int_0^2 e^{x^2-x}\,dx \leqslant e^2(2-0).$$

That is

$$\frac{2}{\sqrt[4]{e}} \leqslant \int_0^2 e^{x^2-x}\,dx \leqslant 2e^2.$$

Example 3 Use the integral properties to estimate the value of the integral $\int_0^3 \sqrt{4+x^2}\,dx$.

Solution Make $f(x) = \sqrt{4+x^2}$, and we have $f'(x) = \dfrac{x}{\sqrt{4+x^2}}$. On the domain $[0,3]$, we have $f'(x) \geqslant 0$. So $f(x)$ increases monotonously on $[0,3]$.

The minimum is $m = f(0) = 2$ and the maximum is $M = f(3) = \sqrt{13}$. Thus, according to Property 6, we have

$$m(3-0) \leqslant \int_0^3 \sqrt{4+x^2}\,dx \leqslant M(3-0).$$

That is

$$6 \leqslant \int_0^3 \sqrt{4+x^2}\,dx \leqslant 3\sqrt{13}.$$

Exercises 6-2

1. Write down the results of the following formulas (assume the functions $f(x)$ in the following questions are continuous in $(-\infty,\infty)$).

   (1) $\int_1^3 f(x)dx + \int_3^1 f(t)dt$;

   (2) $\int_2^3 f(x)dx + \int_3^2 f(u)du + \int_1^2 dx$;

   (3) $\int_1^2 f(x)dx + \int_2^{-1} f(u)du + \int_{-1}^3 f(t)dt$.

2. Compare the size of the following integrals:

   (1) $\int_0^1 x\,dx$ and $\int_0^1 x^2\,dx$;   (2) $\int_0^{\frac{\pi}{2}} x\,dx$ and $\int_0^{\frac{\pi}{2}} \sin x\,dx$;

   (3) $\int_1^e \ln x\,dx$ and $\int_1^e \ln^2 x\,dx$;   (4) $\int_1^2 x\,dx$ and $\int_1^2 x^4\,dx$.

3. Estimate the values of the following integrals:

   (1) $\int_1^4 (x^2+1)dx$;   (2) $\int_{\frac{\pi}{4}}^{\frac{5}{4}\pi} (1+\sin^2 x)dx$;

   (3) $\int_0^{-2} xe^x\,dx$;   (4) $\int_0^1 e^{x^3}\,dx$;

   (5) $\int_{\frac{\pi}{4}}^{\frac{\pi}{3}} \frac{dx}{1+\sin^2 x}$;   (6) $\int_1^2 \frac{x}{1+x^2}dx$.

4. Use definite integral to represent the area of the figure surrounded by the curve $y=f(x)=x(x-1)(2-x)$ and the x-axis.

5. Prove the equation $\dfrac{1}{2} \leqslant \int_{\frac{\pi}{4}}^{\frac{\pi}{2}} \dfrac{\sin x}{x}dx \leqslant \dfrac{\sqrt{2}}{2}$.

6.3 Fundamental formula of calculus

After the establishment of the definite integral concept, how to find the definite integral is an urgent problem to be solved. Because we cannot all use the definition of definite integral of partition, homogenization, summation, precision to find the definite integral.

6.3.1 The relationship between the displacement and the velocity

Suppose that a body moves along a straight line with a varying velocity $v=v(t)$ and the equation of motion is $s=s(t)$. We find the distance traveled by the object in this period from $t=a$ to $t=b$ from two aspects as follows.

On one hand, because the velocity function is the change rate of the distance function towards time t. It is known by the definite integral concept that the total distance traveled by the body in this time period t from a to b is the definite integral $v(t)$ on $[a,b]$. That is

$$\int_a^b v(t)dt.$$

On the other hand, from the motion equation $s=s(t)$ of an object, it can be seen that

the distance traveled by the body during this period from a to b can be expressed
$$s(b)-s(a).$$
Synthesizing these two aspects, we can see that
$$\int_a^b v(t)\,dt = s(b) - s(a).$$

We already know the relationship between $v(t)$ and $s(t)$ is: $s'(t)=v(t)$. That is, $s(t)$ is one of the original functions of $v(t)$. Then from the formula above, we can get some enlightenment: $\int_a^b v(t)\,dt$ is equal to the difference value of a primary function of the integrand $v(t)$ at the places of the upper and the bottom limits. We will see this conclusion is generally correct.

6.3.2 A function of upper limit of integral

Assume the function $f(x)$ is continuous on the interval $[a,b]$. We know that the definite integral $\int_a^b f(x)\,dx$ is a certain number. It is only related to the integrand $f(x)$ and the upper and lower limits of integral a,b, and has nothing to do with the mark of the integral variable x. Assume x is a point on $[a,b]$. Investigate the definite integral $\int_a^x f(x)\,dx$ of $f(x)$ on part of the interval $[a,x]$. In the formula, the integral variable and the upper limit are both expressed by x, but they have different meanings. In order to distinguish them, we often use t to express the integral variable. That is, $\int_a^x f(t)\,dt$. Because the integrand $f(x)$ is continuous on $[a,b]$, then the definite integral exists and changes with the change of $x \in [a,b]$. That is, for any $x \in [a,b]$, the definite integral $\int_a^x f(t)\,dt$ has the only certain corresponding value. According to the definition of function, it is the function of the integral upper limit x which is defined on the interval $[a,b]$, recorded as
$$\Phi(x) = \int_a^x f(t)\,dt, \ a \leqslant x \leqslant b.$$
Since its independent variable x is the upper limit of integral, then $\Phi(x)$ is called integral upper limit function (or integral with variable upper limit).

For example, $\int_0^x (t^2+1)\,dt$, $0 \leqslant x \leqslant 1$ is a integral upper limit function which is defined on the interval $[0,1]$.

The geometric meaning of integral upper limit function $\Phi(x)$ is: if $f(x) \geqslant 0$, for any x on $[a,b]$, there is a corresponding area $\Phi(x)$ of a curved trapezoid. This function $\Phi(x)$ has the following important properties.

Theorem 1 If the function $f(x)$ is continuous on the interval $[a,b]$, then $\Phi(x) = \int_a^x f(t)\,dt$ is differentiable on $[a,b]$, namely

$$\Phi'(x) = \frac{\mathrm{d}}{\mathrm{d}x}\int_a^x f(t)\mathrm{d}t = f(x) \ (a \leqslant x \leqslant b).$$

Proof As shown in Figure 6-4, for any $x \in (a,b)$, assume x has the increment Δx. Make $|\Delta x|$ small enough and $x + \Delta x \in (a,b)$, then we have

$$\Delta\Phi(x) = \Phi(x+\Delta x) - \Phi(x) = \int_a^{x+\Delta x} f(t)\mathrm{d}t - \int_a^x f(t)\mathrm{d}t$$
$$= \int_a^x f(t)\mathrm{d}t + \int_x^{x+\Delta x} f(t)\mathrm{d}t - \int_a^x f(t)\mathrm{d}t = \int_x^{x+\Delta x} f(t)\mathrm{d}t.$$

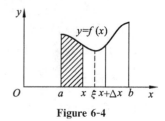

Figure 6-4

According to mean value theorem of integral, we have $\int_x^{x+\Delta x} f(t)\mathrm{d}t = f(\xi)\Delta x$. So $\Delta\Phi(x) = f(\xi)\Delta x$, where ξ lies between x and $x + \Delta x$. Dividing Δx at both sides of the equation above, we have the ratio of the increment of the function and the independent variable: $\frac{\Delta\Phi(x)}{\Delta x} = f(\xi)$. Since $f(x)$ is continuous on $[a, b]$, when $\Delta x \to 0$, we have $\xi \to x$. Therefore,

$$\Phi'(x) = \lim_{\Delta x \to 0}\frac{\Phi(x+\Delta x)-\Phi(x)}{\Delta x} = \lim_{\xi \to x} f(\xi) = f(x).$$

That is to say, $\Phi(x) = \int_a^x f(t)\mathrm{d}t$ is derivable on $[a,b]$ and $\Phi'(x) = f(x)$.

If $x = a$, take $\Delta x > 0$, then similarly we have $\Phi'_+(a) = f(a)$.
If $x = b$, take $\Delta x < 0$, then similarly we have $\Phi'_-(b) = f(b)$.

This theorem gives an important conclusion:

Taking derivation of x, the definite integral of the variable upper bound x of continuous function equals put the variable upper limit into the integration function.

Theorem 1 gives a connection between integral calculus and differential calculus.

Theorem 2 If the function $f(x)$ is continuous on the interval $[a,b]$, then the function $\Phi(x) = \int_a^x f(t)\mathrm{d}t$ is an original function of $f(x)$ on the interval $[a,b]$.

The meaning of this theorem is important. On the one side, it ensures us that there exists an antiderivative for a continuous function. On the other side, it shows us the relationship between definite integral and antiderivative.

6.3.3 Newton–Leibnitz formula

Theorem 3 If $f(x)$ is continuous on the interval $[a,b]$ and $F(x)$ is an antiderivative of $f(x)$ on $[a,b]$, then

$$\int_a^b f(x)\mathrm{d}x = F(b) - F(a).$$

Proof We already know that $F(x)$ is an antiderivative of the continuous function $f(x)$. According to Theorem 2, we know that, the variable upper limit definite integral

$$\Phi(x) = \int_a^x f(t)\,dt$$

is also an antiderivative of $f(x)$. Then the difference between $\Phi(x)$ and $F(x)$ on $[a,b]$ is only a constant C. That is

$$\Phi(x) = F(x) + C,$$

or

$$\int_a^x f(t)\,dt = F(x) + C.$$

Make $x = a$, we have

$$0 = \int_a^a f(x)\,dx = F(a) + C.$$

That is $C = -F(a)$. Therefore,

$$\int_a^x f(t)\,dt = F(x) - F(a).$$

Again make $x = b$, we have

$$\int_a^b f(x)\,dx = F(b) - F(a).$$

In order to express the relationship between the definite integral of $f(x)$ on $[a,b]$ and the antiderivative $F(x)$, we mark $F(b) - F(a)$ as $[F(x)]_a^b$, which is called the increment of the function $F(x)$ on the interval $[a,b]$. Thus, we have

$$\int_a^b f(x)\,dx = [F(x)]_a^b = F(b) - F(a).$$

The formula is called Newton – Leibnitz formula (also called the fundamental formula of calculus). It is the fundamental tool for finding definite integral. According to the formula, in order to find definite integral, we only have to find any antiderivative of the integration function and again find the increment $[F(x)]_a^b$ of the antiderivative $F(x)$ on $[a,b]$. Thus, the problem of finding definite integral can be concluded to the problem of finding antiderivative or indefinite integral.

Newton – Leibnitz formula reveals the relationship between definite integral and indefinite integral: the value of the definite integral of the continuous function $f(x)$ on the interval $[a,b]$ is equal to the increment of one of its antiderivatives $F(x)$ on the interval $[a, b]$. Because of this relationship, we call the whole of antiderivative which has a constant indefinite integral. Therefore, it makes the find of indefinite integral have more practical meaning.

Example 1 Find $\int_{-1}^{1} \dfrac{dx}{1+x^2}$.

Solution Because $\arctan x$ is an antiderivative of $\dfrac{1}{1+x^2}$, then

$$\int_{-1}^{1} \frac{dx}{1+x^2} = [\arctan x]_{-1}^{1} = \arctan 1 - \arctan(-1) = \frac{\pi}{4} - \left(-\frac{\pi}{4}\right) = \frac{\pi}{2}.$$

Example 2 Find $\int_{-2}^{-1} \dfrac{dx}{x}$.

Solution When $x<0$, an antiderivative of $\dfrac{1}{x}$ is $\ln|x|$. Now the integral interval is $[-2,-1]$. So

$$\int_{-2}^{-1} \dfrac{dx}{x} = [\ln|x|]_{-2}^{-1} = \ln 1 - \ln 2 = -\ln 2.$$

Example 3 Find $\int_0^2 f(x)dx$, where

$$f(x) = \begin{cases} x+1, & x \leqslant 1, \\ \dfrac{1}{2}x^2, & x > 1. \end{cases}$$

Solution Because $x=1$ is the segment point of the piecewise function $f(x)$, we divide the interval $[0,2]$ into two parts $[0,1]$ and $[1,2]$. Therefore, we have

$$\int_0^2 f(x)dx = \int_0^1 (x+1)dx + \int_1^2 \dfrac{1}{2}x^2 dx = \left[\dfrac{1}{2}(x+1)^2\right]_0^1 + \left[\dfrac{1}{6}x^3\right]_1^2$$

$$= \dfrac{3}{2} + \dfrac{7}{6} = \dfrac{8}{3}.$$

Note The function in Newton – Leibnitz formula indicates that $F(x)$ must be the antiderivatives of $f(x)$ on the interval $[a,b]$. For example, in Example 2, the antiderivative of $\dfrac{1}{x}$ on $[-2,-1]$ is $\ln|x|$. And in Example 3, respectively find the antiderivatives of $f(x)$ on $[0,1]$ and $[1,2]$.

According to Theorem 1, we already know that

$$\dfrac{d}{dx}\int_a^x f(t)dt = f(x).$$

Use the derivation rule of function composition, we can further have

$$\dfrac{d}{dx}\int_a^{\varphi(x)} f(t)dt = f[\varphi(x)]\varphi'(x).$$

If the upper and below limits of the definite integral $\int_a^b f(x)dx$ are respectively the functions $a(x)$ and $b(x)$. That is $\int_{a(x)}^{b(x)} f(t)dt$. Because

$$\int_{a(x)}^{b(x)} f(t)dt = \int_c^{b(x)} f(t)dt - \int_c^{a(x)} f(t)dt,$$

according to the derivation rule of function composition, we have

$$\dfrac{d}{dx}\int_{a(x)}^{b(x)} f(t)dt = f[b(x)]b'(x) - f[a(x)]a'(x).$$

Example 4 We already know that $\Phi(x) = \int_0^x \dfrac{dt}{1+t^2}$. Find $\Phi'(2)$.

Solution Because $\Phi'(x) = \dfrac{1}{1+x^2}$, then $\Phi'(2) = \dfrac{1}{1+2^2} = \dfrac{1}{5}$.

Example 5 Assume $\Phi(x) = \int_0^{x^2} \sqrt{1+t^3}\, dt$. Find $\Phi'(x)$.

Solution $\Phi'(x) = \left(\int_0^{x^2} \sqrt{1+t^3}\,dt\right)' = \sqrt{1+(x^2)^3} \cdot (x^2)' = 2x\sqrt{1+x^6}$.

Example 6 We already know that $\Phi(x) = \int_{x^2}^{x^3} \sqrt{1+t^3}\,dt$. Find $\Phi'(x)$.

Solution $\Phi'(x) = \left(\int_{x^2}^{x^3} \sqrt{1+t^3}\,dt\right)$
$$= \sqrt{1+x^9} \cdot (x^3)' - \sqrt{1+x^6} \cdot (x^2)' = 3x^2\sqrt{1+x^9} - 2x\sqrt{1+x^6}.$$

Example 7 Find the limit of $\Phi(x) = \int_0^x t e^{-t^2}\,dt$.

Solution Because $\Phi'(x) = \left[\int_0^x t e^{-t^2}\,dt\right]' = x e^{-x^2}$. Make $\Phi'(x) = 0$, we have the stagnation point $x=0$. Then
$$\Phi''(x) = e^{-x^2} + x e^{-x^2} \cdot (-2x) = e^{-x^2}(1-2x^2).$$
But $$\Phi''(0) = 1 > 0.$$
Therefore, when $x=0$, $\Phi(x)$ has the minimum $\Phi(0)=0$.

Example 8 Find $\lim\limits_{x \to 0} \dfrac{\int_{\cos x}^1 e^{-t^2}\,dt}{x^2}$.

Solution This is the indeterminate limit of the form $\dfrac{0}{0}$. Using the L' Hospital's rule. Molecular can be written as $-\int_1^{\cos x} e^{-t^2}\,dt$. Then
$$\frac{d}{dx}\int_{\cos x}^1 e^{-t^2}\,dt = -\frac{d}{dx}\int_1^{\cos x} e^{-t^2}\,dt = -e^{-\cos^2 x} \cdot (-\sin x) = \sin x\, e^{-\cos^2 x}.$$

Therefore, $\lim\limits_{x\to 0} \dfrac{\int_{\cos x}^1 e^{-t^2}\,dt}{x^2} = \lim\limits_{x\to 0} \dfrac{\sin x\, e^{-\cos^2 x}}{2x} = \dfrac{1}{2e}$.

Example 9 Assume $f(x)$ is a continuous function. Prove.
$$\int_0^x f(t)(x-t)\,dt = \int_0^x \left[\int_0^t f(u)\,du\right] dt.$$

Analysis Make $F(x) = \int_0^x f(t)(x-t)\,dt - \int_0^x \left[\int_0^t f(u)\,du\right] dt$. Then we only have to prove that $F(x)=0$. That is to prove $F(x)$ is a constant function. Therefore we have to prove $F'(x)=0$.

Proof Make
$$F(x) = \int_0^x f(t)(x-t)\,dt - \int_0^x \left[\int_0^t f(u)\,du\right] dt$$
$$= x\int_0^x f(t)\,dt - \int_0^x t f(t)\,dt - \int_0^x \left[\int_0^t f(u)\,du\right] dt.$$

Take derivation of $F(x)$, then we have
$$F'(x) = \int_0^x f(t)\,dt + x f(x) - x f(x) - \int_0^x f(u)\,du = 0.$$

Therefore, $F(x)$ is a constant.

Again obviously we have $F(0)=0$, so $F(x) \equiv 0$.

Exercises 6-3

1. Solve the following problems:

    (1) Assume $\Phi(x) = \int_0^x \sin t \, dt$. Find $\Phi'\left(\dfrac{\pi}{4}\right)$.

    (2) Assume $\Phi(x) = \int_0^{x^2} \sqrt{1+t^2} \, dt$. Find $\Phi'(x)$.

    (3) $\lim\limits_{x \to 0} \dfrac{\int_0^x \sin 2t \, dt}{x^2}$.

    (4) $\lim\limits_{x \to +\infty} \dfrac{\int_0^x (\arctan t)^2 \, dt}{\sqrt{x^2+1}}$.

    (5) $\lim\limits_{x \to 0} \dfrac{\left(\int_0^x t^2 \cos t^2 \, dt\right)^2}{\int_0^{x^2} \sin t^2 \, dt}$.

2. Find the following definite integrals:

    (1) $\int_1^2 (3x-1) \, dx$;

    (2) $\int_4^9 \sqrt{x}(1+\sqrt{x}) \, dx$;

    (3) $\int_{\frac{1}{\sqrt{3}}}^{\sqrt{3}} \dfrac{dx}{1+x^2}$;

    (4) $\int_{-\frac{1}{2}}^{\frac{1}{2}} \dfrac{dx}{\sqrt{1-x^2}}$;

    (5) $\int_0^4 (2-\sqrt{x})^2 \, dx$;

    (6) $\int_0^1 10^{2x+1} \, dx$;

    (7) $\int_0^{\pi} (1 - \sin^3 \theta) \, d\theta$;

    (8) $\int_{\frac{\pi}{4}}^{\frac{\pi}{3}} \dfrac{\ln(\tan x)}{\cos x \sin x} \, dx$;

    (9) $\int_0^{\frac{1}{2}} \dfrac{2x+1}{\sqrt{1-x^2}} \, dx$;

    (10) $\int_0^{\pi} \sqrt{1+\cos 2x} \, dx$.

3. (1) Assume $f(x) = \begin{cases} 1+x^2, & 0 \leqslant x \leqslant 1, \\ 2-x, & 1 \leqslant x \leqslant 2. \end{cases}$ Find $\int_0^2 f(x) \, dx$.

    (2) Find $\int_0^{2\pi} |\sin x| \, dx$.

4. Assume $f(x) = \begin{cases} 1+x, & 0 \leqslant x < 1, \\ \dfrac{1}{2}x^2, & 1 \leqslant x \leqslant 2. \end{cases}$ Find out the expression of $\Phi(x) = \int_0^x f(t) \, dt$ on $[0,2]$, and discuss the continuity of $\Phi(x)$ on $[0,2]$.

5. Assume $f(x) = \begin{cases} \dfrac{1}{2} \sin x, & 0 \leqslant x \leqslant \pi, \\ 0, & x < 0 \text{ or } x > \pi. \end{cases}$ Find out the expression of $\Phi(x) = \int_0^x f(t) \, dt$ on $(-\infty, \infty)$.

6.4 Integration by substitution and parts for definite integrals

According to Newton – Leibniz formula, to find definite integral, you only have to find the antiderivative of the integrand. But definite integral is the limit of a "summation" and is a constant, while indefinite integral belongs to antiderivative. They are two completely different concepts. By Newton – Leibniz formula, we give out the relationship between them. So strictly speaking, the find of integral problem has been solved. But this section again puts forward two basic integration methods. Its main purpose is to obtain integral results more easily.

6.4.1 Integration by substitution for definite integrals

Theorem If the function $f(x)$ is continuous on the interval $[a,b]$ and the function $x=\varphi(t)$ satisfies the following conditions:
(1) $\varphi(\alpha)=a, \varphi(\beta)=b$;
(2) $\varphi(t)$ is continuous differentiable on the interval $[\alpha,\beta]$ (or $[\beta,\alpha]$), and its interval is $W(\varphi) \subset [a,b]$.
Then we have
$$\int_a^b f(x) dx = \int_\alpha^\beta f[\varphi(t)] \varphi'(t) dt.$$

Proof Assume $F(x)$ is an antiderivative of $f(x)$. That is $F'(x)=f(x)$. According to the derivation rule of function composition, $F[\varphi(t)]$ is the antiderivative of $f[\varphi(t)]\varphi'(t)$. According to Newton – Leibniz formula, we have
$$\int_a^b f(x) dx = F(b) - F(a).$$
Again
$$\int_\alpha^\beta f[\varphi(t)] \varphi'(t) dt = F[\varphi(\beta)] - F[\varphi(\alpha)] = F(b) - F(a).$$
Then
$$\int_a^b f(x) dx = \int_\alpha^\beta f[\varphi(t)] \varphi'(t) dt.$$

We can see that, the essence of the substitution method of definite integral is to extend the substitution method of indefinite integral to definite integral. The difference between them is that in the substitution method of definite integral, when changing the element of the integrand, the upper and lower limit of integral should also make the corresponding change. So after finding out the antiderivative of t, we do not need to change back into the original variable x. We only have to apply Newton – Leibniz formula to the new variable.

Example 1 Find the definite integral $\int_0^{\ln 2} \sqrt{e^x - 1} \, dx$.

Solution Assume $\sqrt{e^x - 1} = t$. Then we have
$$x = \ln(1+t^2), \quad t>0,$$
$$dx = \frac{2t}{1+t^2} dt.$$

When $x=0$, $t=0$. When $x=\ln 2$, $t=1$.
Therefore,

$$\int_0^{\ln 2} \sqrt{e^x - 1}\,dx = 2\int_0^1 \frac{t^2}{1+t^2}\,dt = 2\int_0^1 \left(1 - \frac{1}{1+t^2}\right)dt$$
$$= 2[t - \arctan t]_0^1 = 2(1 - \arctan 1)$$
$$= 2 - \frac{\pi}{2}.$$

Example 2 Assume $f(x)$ is a continuous function whose period is T. Prove for any real number a, we have
$$\int_a^{a+T} f(x)\,dx = \int_0^T f(x)\,dx.$$

Proof
$$\int_a^{a+T} f(x)\,dx = \int_a^0 f(x)\,dx + \int_0^T f(x)\,dx + \int_T^{a+T} f(x)\,dx.$$

For
$$\int_a^{a+T} f(x)\,dx,$$

make $x = t + T$, we have
$$\int_T^{a+T} f(x)\,dx = \int_0^a f(t+T)\,dt = \int_0^a f(t)\,dt = \int_0^a f(x)\,dx = -\int_a^0 f(x)\,dx.$$

Therefore, we have
$$\int_a^{a+T} f(x)\,dx = \int_0^T f(x)\,dx.$$

Note (1) The substitution method of definite integral: actually, it is an application of the second substitution method of indefinite integral in definite integral.

(2) The upper and lower limit of definite integral a,b is the value range of integrand $f(x)$. When doing the substitution of variables, because of the change of the independent variable, then we must give out the value range of the new variable. This is one of the differences of the substitution methods of definite integral and indefinite integral.

(3) When finding the antiderivative by changing variables in indefinite integral, the variable must be replaced back (because it is the antiderivative of the integrand). But in the find of definite integral, you don't need to replace back to the substitution variable because the definite integral is a "number", whatever method you want to use, just find the value.

Example 3 Find the definite integral $\int_0^\pi \sqrt{\sin^3 x - \sin^5 x}\,dx$.

Solution Because of
$$\sqrt{\sin^3 x - \sin^5 x} = \sqrt{\sin^3 x(1 - \sin^2 x)} = \sin^{\frac{3}{2}} x |\cos x|.$$

When $x \in \left[0, \frac{\pi}{2}\right]$, $|\cos x| = \cos x$. When $x \in \left[\frac{\pi}{2}, \pi\right]$, $|\cos x| = -\cos x$.

Therefore,
$$\int_0^\pi \sqrt{\sin^3 x - \sin^5 x}\,dx = \int_0^{\frac{\pi}{2}} \sin^{\frac{3}{2}} x \cos x\,dx + \int_{\frac{\pi}{2}}^\pi \sin^{\frac{3}{2}} x(-\cos x)\,dx$$
$$= \int_0^{\frac{\pi}{2}} \sin^{\frac{3}{2}} x\,d(\sin x) - \int_{\frac{\pi}{2}}^\pi \sin^{\frac{3}{2}} x\,d(\sin x)$$

$$= \left[\frac{2}{5}\sin^{\frac{5}{2}}x\right]_0^{\frac{\pi}{2}} - \left[\frac{2}{5}\sin^{\frac{5}{2}}x\right]_{\frac{\pi}{2}}^{\pi}$$

$$= \frac{2}{5} - \left(-\frac{2}{5}\right) = \frac{4}{5}.$$

Example 4 Prove:

(1) If $f(x)$ is continuous on $[-a, a]$ and is an even function, then

$$\int_{-a}^{a} f(x)dx = 2\int_0^a f(x)dx.$$

(2) If $f(x)$ is continuous in $[-a, a]$ and is an odd function, then

$$\int_{-a}^{a} f(x)dx = 0.$$

Proof Because

$$\int_{-a}^{a} f(x)dx = \int_{-a}^{0} f(x)dx + \int_{0}^{a} f(x)dx.$$

For the integral $\int_{-a}^{0} f(x)dx$, make a substitution $x=-t$, then we have

$$\int_{-a}^{0} f(x)dx = -\int_{a}^{0} f(-t)dt = \int_{0}^{a} f(-t)dt = \int_{0}^{a} f(-x)dx.$$

Therefore,

$$\int_{-a}^{a} f(x)dx = \int_{0}^{a} f(-x)dx + \int_{0}^{a} f(x)dx = \int_{0}^{a} [f(-x) + f(x)]dx. \quad (*)$$

(1) If $f(x)$ is an even function, then $f(x)+f(-x)=2f(x)$. Therefore,

$$\int_{-a}^{a} f(x)dx = 2\int_{0}^{a} f(x)dx.$$

(2) If $f(x)$ is an odd function, then $f(x)+f(-x)=0$. Therefore,

$$\int_{-a}^{a} f(x)dx = 0.$$

The conclusion is obvious from the perspective of geometry. Because the figure of the odd function is symmetrical about the origin, so the two curved trapezoid on the intervals $[-a,0]$ and $[0,a]$ are respectively located at both sides of the x-axis and their areas are the same. Therefore, the algebra sum of the two areas is zero. And the figure of the even function is symmetrical about the y-axis, so the area of the curved trapezoid on $[-a,a]$ is naturally two times of that on $[0, a]$.

Using this conclusion, it is often possible to simplify the find of the definite integral of the even function and the odd function on the interval which is symmetric about the origin. In particular, the integral of the odd function in the symmetric interval is equal to 0, which is a very useful conclusion.

Example 5 Assume $f(x)$ is continuous on the integral interval. Find the definite integral

$$\int_{-a}^{a} [f(x) - f(-x)]\cos x dx.$$

Solution Becauses $\cos x$ is an even function and $f(x)-f(-x)$ is an odd function, then $\cos x[f(x)-f(-x)]$ is also an odd function. And the integral interval $[-a, a]$ is a symmetrical interval which is centered in the origin. Thus,

$$\int_{-a}^{a} [f(x) - f(-x)]\cos x \, dx = 0.$$

Example 6 Find the definite integral

$$\int_{-1}^{1} (|x| + \sin x)x^2 \, dx.$$

Solution Because $|x|x^2$ is an even function and $x^2 \sin x$ is an odd function. The integral interval $[-1, 1]$ is a symmetrical interval which is centered in the origin. Therefore,

$$\int_{-1}^{1} (|x| + \sin x)x^2 \, dx = \int_{-1}^{1} |x|x^2 \, dx = 2\int_{0}^{1} x^3 \, dx = \frac{1}{2}.$$

Example 7 Find the definite integral

$$I = \int_{-2}^{2} \min\left\{x^2, \frac{1}{|x|}\right\} dx.$$

Solution We can see that the integrand is an even function. Therefore,

$$I = 2\int_{0}^{2} \min\left\{x^2, \frac{1}{x}\right\} dx = 2\left[\int_{0}^{1} x^2 \, dx + \int_{1}^{2} \frac{1}{x} dx\right] = 2\left(\frac{1}{3} + \ln 2\right).$$

Example 8 Prove $\int_{0}^{a} x^5 f(x^3) \, dx = \frac{1}{3}\int_{0}^{a^3} x f(x) \, dx$, where $f(x)$ is a continuous function.

Proof Make $x = \sqrt[3]{t}$, and we have $x^3 = t$, $dt = 3x^2 dx$. When $x=0$, $t=0$. When $x=a$, $t=a^3$. Therefore,

$$\int_{0}^{a} x^5 f(x^3) \, dx = \int_{0}^{a^3} t^{\frac{5}{3}} f(t) \left(\frac{1}{3} t^{-\frac{2}{3}}\right) dt = \frac{1}{3}\int_{0}^{a^3} t f(t) \, dt = \frac{1}{3}\int_{0}^{a^3} x f(x) \, dx.$$

Note All the methods above include proper substitution of variables, according to the property that the value of definite integral is related to the integrand and the integral interval, but has nothing to do with the mark of the integral variable.

Example 9 Assume the function

$$f(x) = \begin{cases} x e^{-x^2}, & x \geq 0, \\ \dfrac{1}{1+\cos x}, & -1 < x < 0. \end{cases}$$

Find $\int_{1}^{4} f(x-2) \, dx$.

Solution Assume $x - 2 = t$. Then $dx = dt$. And when $x=1$, $t=-1$. When $x=4$, $t=2$.

Therefore,

$$\int_{1}^{4} f(x-2) \, dx = \int_{-1}^{2} f(t) \, dt = \int_{-1}^{0} \frac{dt}{1+\cos t} + \int_{0}^{2} t e^{-t^2} \, dt$$

$$= \left[\tan \frac{t}{2}\right]_{-1}^{0} - \left[\frac{1}{2} e^{-t^2}\right]_{0}^{2} = \tan \frac{1}{2} - \frac{1}{2} e^{-4} + \frac{1}{2}.$$

Example 10 Assume $f(x)$ is continuous on $[a, b]$, and $f(x) > 0$.

$$F(x) = \int_a^x f(t)\,\mathrm{d}t + \int_b^x \frac{\mathrm{d}t}{f(t)}, x \in [a,b],$$

Prove:

(1) $F'(x) \geqslant 2$;

(2) The equation $F(x)=0$ has only one real root on (a,b).

Proof

(1) $F'(x) = \dfrac{\mathrm{d}}{\mathrm{d}x}\left[\int_a^x f(t)\,\mathrm{d}t + \int_b^x \dfrac{1}{f(t)}\mathrm{d}t\right] = f(x) + \dfrac{1}{f(x)}.$

And because $f(x) > 0$,

then

$$f(x) + \frac{1}{f(x)} \geqslant 2\sqrt{f(x) \cdot \frac{1}{f(x)}} = 2.$$

That is $F'(x) \geqslant 2$.

(2) Because $F(x)$ is continuous on $[a,b]$, and $f(x)>0$, then

$$F(a) = \int_a^a f(x)\,\mathrm{d}x + \int_b^a \frac{1}{f(x)}\mathrm{d}x = \int_b^a \frac{1}{f(x)}\mathrm{d}x < 0,$$

$$F(b) = \int_a^b f(x)\,\mathrm{d}x + \int_b^b \frac{1}{f(x)}\mathrm{d}x = \int_a^b \frac{1}{f(x)}\mathrm{d}x > 0.$$

According to the zero point theorem, there exists $\xi \in (a,b)$, which can make $F(\xi)=0$. Thus $F(x)=0$ at least has a real root on (a,b).

Again because $F'(x) \geqslant 2 > 0$, then $F(x)$ is monotonously increase on $[a,b]$. Therefore, $F(x)=0$ only has one real root on (a,b).

6.4.2 Integration by parts for definite integrals

Integration by parts for definite integrals and integration by parts for indefinite integrals are completely the same in finding methods. Namely, assume the functions $u(x)$, $v(x)$ have derivatives $u'(x)$, $v'(x)$ on the interval $[a,b]$, then we have

$$(uv)' = u'v + uv'.$$

Respectively finding the definite integral of both sides of the equation on $[a,b]$, we have

$$\int_a^b (uv)'\,\mathrm{d}x = \int_a^b u'v\,\mathrm{d}x + \int_a^b uv'\,\mathrm{d}x.$$

Move the items

$$\int_a^b uv'\,\mathrm{d}x = [uv]_a^b - \int_a^b u'v\,\mathrm{d}x,$$

or simplify it as

$$\int_a^b u\,\mathrm{d}v = [uv]_a^b - \int_a^b v\,\mathrm{d}u.$$

It is called the integration by parts formula in definite integrals.

Note Every item of this formula has a integral limit. When concretely using the integration by parts for definite integrals, the parts that have already been found in the antiderivative should be replaced by the upper and lower limit of integral immediately.

Thus, we can simplify the calculation process.

Example 11 Find $\int_0^1 \arcsin x \, dx$.

Solution
$$\int_0^1 \arcsin x \, dx = [x\arcsin x]_0^1 - \int_0^1 x \, d(\arcsin x)$$
$$= \arcsin 1 - \int_0^1 \frac{x}{\sqrt{1-x^2}} dx$$
$$= \frac{\pi}{2} + [\sqrt{1-x^2}]_0^1 = \frac{\pi}{2} - 1.$$

Example 12 Find $\int_0^{\sqrt{3}} \ln(x+\sqrt{1+x^2}) \, dx$.

Solution
$$\int_0^{\sqrt{3}} \ln(x+\sqrt{1+x^2}) \, dx = [x\ln(x+\sqrt{1+x^2})]_0^{\sqrt{3}} - \int_0^{\sqrt{3}} \frac{x}{\sqrt{1+x^2}} dx$$
$$= \sqrt{3}\ln(\sqrt{3}+2) - [\sqrt{1+x^2}]_0^{\sqrt{3}}$$
$$= \sqrt{3}\ln(\sqrt{3}+2) - 2 + 1$$
$$= \sqrt{3}\ln(\sqrt{3}+2) - 1.$$

Example 13 Find $\int_0^1 e^{\sqrt{x}} \, dx$.

Solution Make $\sqrt{x}=t$, then $x=t^2$, $dx=2t\,dt$. And when $x=0$, $t=0$. When $x=1$, $t=1$.

Thus,
$$\int_0^1 e^{\sqrt{x}} \, dx = 2\int_0^1 t e^t \, dt = 2[te^t]_0^1 - 2\int_0^1 e^t \, dt$$
$$= 2(e - [e^t]_0^1) = 2[e-(e-1)] = 2.$$

Example 14 Assume $I_n = \int_0^{\frac{\pi}{2}} \sin^n x \, dx$, $J_n = \int_0^{\frac{\pi}{2}} \cos^n x \, dx$, $n \in \mathbf{N}$. Prove $I_n = J_n$, and Find I_n.

Proof Assume $x = \frac{\pi}{2} - y$. Then $dx = -dy$. Thus,
$$J_n = \int_0^{\frac{\pi}{2}} \cos^n x \, dx = \int_{\frac{\pi}{2}}^0 \cos^n\left(\frac{\pi}{2}-y\right)(-dy) = \int_0^{\frac{\pi}{2}} \sin^n y \, dy = I_n.$$

Then find I_n,
$$I_n = \int_0^{\frac{\pi}{2}} \sin^n x \, dx = \int_0^{\frac{\pi}{2}} \sin^{n-1} x \, d(-\cos x) = [-\sin^{n-1} x \cos x]_0^{\frac{\pi}{2}} + \int_0^{\frac{\pi}{2}} \cos x \, d(\sin^{n-1} x)$$
$$= (n-1)\int_0^{\frac{\pi}{2}} \sin^{n-2} x \cos^2 x \, dx = (n-1)\int_0^{\frac{\pi}{2}} \sin^{n-2} x (1-\sin^2 x) \, dx$$
$$= (n-1)\int_0^{\frac{\pi}{2}} \sin^{n-2} x \, dx - (n-1)\int_0^{\frac{\pi}{2}} \sin^n x \, dx = (n-1)I_{n-2} - (n-1)I_n.$$

Thus we have a recursive formula
$$I_n = \frac{n-1}{n} I_{n-2}.$$

When n is an even number, assume $n = 2k (k \in \mathbf{N}^+)$, and we have

$$I_{2k} = \frac{2k-1}{2k} \cdot \frac{2k-3}{2k-2} \cdot \frac{2k-5}{2k-4} \cdot \cdots \cdot \frac{5}{6} \cdot \frac{3}{4} \cdot \frac{1}{2} I_0.$$

When n is an odd number, assume $n = 2k+1 (k \in \mathbf{N})$, and we have

$$I_{2k+1} = \frac{2k}{2k+1} \cdot \frac{2k-2}{2k-1} \cdot \frac{2k-4}{2k-3} \cdot \cdots \cdot \frac{6}{7} \cdot \frac{4}{5} \cdot \frac{2}{3} I_1.$$

While, $\quad I_0 = \int_0^{\frac{\pi}{2}} dx = \frac{\pi}{2}, I_1 = \int_0^{\frac{\pi}{2}} \sin x \, dx = 1.$

We have

$$\int_0^{\frac{\pi}{2}} \sin^n x \, dx = \int_0^{\frac{\pi}{2}} \cos^n x \, dx = \begin{cases} \dfrac{n-1}{n} \cdot \dfrac{n-3}{n-2} \cdot \cdots \cdot \dfrac{1}{2} \cdot \dfrac{\pi}{2}, & \text{if } n \text{ is even,} \\ \dfrac{n-1}{n} \cdot \dfrac{n-3}{n-2} \cdot \cdots \cdot \dfrac{2}{3} \cdot 1, & \text{if } n \text{ is odd.} \end{cases}$$

This formula is convenient for computing certain definite integrals.

Example 15 Find $\int_{-\frac{\pi}{2}}^{\frac{\pi}{2}} \frac{e^x}{1 + e^x} \sin^6 x \, dx.$

Solution Using the parity in integrand and the ($*$) equation in Example 4, we have

$$\int_{-\frac{\pi}{2}}^{\frac{\pi}{2}} \frac{e^x}{1+e^x} \sin^6 x \, dx = \int_0^{\frac{\pi}{2}} \left(\frac{e^x}{1+e^x} + \frac{e^{-x}}{1+e^{-x}} \right) \sin^6 x \, dx$$

$$= \int_0^{\frac{\pi}{2}} \sin^6 x \, dx = \frac{5}{6} \cdot \frac{3}{4} \cdot \frac{1}{2} \cdot \frac{\pi}{2} = \frac{15}{96} \pi.$$

Exercises 6-4

1. Find the following definite integrals:

(1) $\int_3^8 \dfrac{x}{\sqrt{1+x}} dx;$

(2) $\int_1^2 \dfrac{\sqrt{x^2-1}}{x} dx;$

(3) $\int_0^1 (x-1)^{10} x^2 \, dx;$

(4) $\int_0^1 x^2 \sqrt{1-x^2} \, dx;$

(5) $\int_0^1 \dfrac{1}{1+e^x} dx;$

(6) $\int_1^{e^2} \dfrac{1}{x\sqrt{1+\ln x}} dx;$

(7) $\int_1^4 \dfrac{1}{1+\sqrt{x}} dx;$

(8) $\int_0^1 (1+x^2)^{-\frac{3}{2}} dx;$

(9) $\int_0^1 \dfrac{x^{\frac{3}{2}}}{1+x} dx;$

(10) $\int_1^e \dfrac{1+\ln x}{x} dx;$

(11) $\int_0^1 \dfrac{x}{1+\sqrt{x}} dx;$

(12) $\int_0^2 \sqrt{4-x^2} \, dx;$

(13) $\int_0^{10\pi} \sqrt{1-\cos 2x} \, dx.$

2. Using the parity of function to find the following definite integrals:

(1) $\int_{-\pi}^{\pi} x^4 \sin x \, dx;$

(2) $\int_{-\frac{\pi}{2}}^{\frac{\pi}{2}} x^4 \cos^4 \theta \, d\theta;$

(3) $\int_{-\frac{1}{2}}^{\frac{1}{2}} \dfrac{(\arcsin x)^2}{\sqrt{1-x^2}} dx;$

(4) $\int_{-5}^{5} \dfrac{x^3 \sin^2 x}{x^4 + 2x^2 + 1} dx.$

3. Prove the following equations:

(1) $\int_{-a}^{a} \varphi(x^2)\,dx = 2\int_{0}^{a} \varphi(x^2)\,dx$.

(2) Assume $f(x)$ is continuous on $[a,b]$. Try to Prove $\int_{a}^{b} f(x)\,dx = \int_{a}^{b} f(a+b-x)\,dx$.

(3) $\int_{x}^{1} \dfrac{dx}{1+x^2} = \int_{1}^{\frac{1}{x}} \dfrac{dx}{1+x^2}$ ($x > 0$).

(4) $\int_{-\pi}^{\pi} \sin^n x\,dx = 2\int_{0}^{\frac{\pi}{2}} \sin^n x\,dx$.

4. Find the following definite integrals:

(1) $\int_{0}^{1} \arctan x\,dx$;

(2) $\int_{-\pi}^{\pi} \sin^3 \dfrac{\theta}{2}\,d\theta$;

(3) $\int_{0}^{\frac{\pi}{2}} e^{2t} \cos t\,dt$;

(4) $\int_{0}^{\pi} x^2 \cos 2x\,dx$;

(5) $\int_{\frac{1}{e}}^{e} |\ln x|\,dx$;

(6) $\int_{-\frac{1}{2}}^{\frac{1}{2}} \dfrac{x \arcsin x}{\sqrt{1-x^2}}\,dx$;

(7) $\int_{0}^{1} x e^{-x}\,dx$;

(8) $\int_{1}^{4} \dfrac{1}{\sqrt{x}} \ln x\,dx$;

(9) $\int_{1}^{e} \sin(\ln x)\,dx$;

(10) $\int_{0}^{\frac{1}{e}} \ln(x+1)\,dx$.

5. Assume $f(x)$ has a second order continuous derivative on $[0,1]$, and $f(0)=1$, $f(1)=3$, $f'(1)=5$. Find $\int_{0}^{1} x f''(x)\,dx$.

6. Find the definite integral $I = \int_{0}^{n\pi} x|\sin x|\,dx$ (n is a positive integer).

7. Assume $f(x), g(x)$ are derivable and the derivatives are continuous on $[0,1]$, $f(0)=0$, $f'(x) \geq 0$, $g'(x) \geq 0$. Prove: for any $a \in [0,1]$, we have

$$\int_{0}^{a} g(x)f'(x)\,dx + \int_{0}^{1} f(x)g'(x)\,dx \geq f(a)g(1).$$

8. Assume $f''(x)$ is continuous on $[0,\pi]$, and $f(0)=2$, $f(\pi)=1$. Prove:

$$\int_{0}^{\pi} [f(x) + f''(x)]\sin x\,dx = 3.$$

6.5 Improper integrals

When discussing definite integral ahead, there are two basic conditions, namely, the limitation of integral interval and the boundedness of integrand on the integral interval. But in some practical problems, we often need to break through these constraints, considering the integral on infinite interval or finding integral of unbounded function. This kind of integral is called the improper integral, and accordingly the integral discussed earlier is called normal integral or proper integral. For anomalous integral, actually, we are used to define it by proper integration.

6.5.1 Improper integrals on an infinite interval

Definition 1 Assume the function $f(x)$ is continuous on the infinite interval $[a, +\infty)$. For any $b > a$, the $\lim\limits_{b \to +\infty} \int_a^b f(x)\,dx$ exists. Then $\lim\limits_{b \to +\infty} \int_a^{+\infty} f(x)\,dx$ is called the improper integral of the function $f(x)$ on the infinite interval $[a, \infty)$, denoted by $\int_a^{+\infty} f(x)\,dx$.

That is
$$\int_a^{+\infty} f(x)\,dx = \lim_{b \to +\infty} \int_a^b f(x)\,dx.$$

At this time, it can also say that the improper integral $\int_a^{+\infty} f(x)\,dx$ converges. If the above limit exists, then we say that the improper integral $\int_a^{+\infty} f(x)\,dx$ of the function does not exist or diverge on the infinite interval $[a, +\infty)$.

Similarly, assume the function $f(x)$ is continuous on the interval $(-\infty, b]$. Take any $a < b$, if $\lim\limits_{a \to -\infty} \int_a^b f(x)\,dx$ exists, then we say that this limit ($\lim\limits_{a \to -\infty} \int_a^b f(x)\,dx$) is the improper integral of the function $f(x)$ on the infinite interval $[-\infty, b)$, denoted by $\int_{-\infty}^b f(x)\,dx$. That is

$$\int_{-\infty}^b f(x)\,dx = \lim_{a \to -\infty} \int_a^b f(x)\,dx.$$

At this time, we also say that the improper integral $\int_{-\infty}^b f(x)\,dx$ converges. If the above limit does not exist, then we say that the improper integral $\int_{-\infty}^b f(x)\,dx$ does not exist or diverge.

Also, if the function $f(x)$ is continuous on the interval $(-\infty, +\infty)$, and the improper integrals $\int_{-\infty}^0 f(x)\,dx$ and $\int_0^{+\infty} f(x)\,dx$ both converge, then we call the sum of the two improper integrals above the improper integral of the function $f(x)$ on the infinite interval $(-\infty, +\infty)$, denoted by $\int_{-\infty}^{+\infty} f(x)\,dx$. That is

$$\int_{-\infty}^{+\infty} f(x)\,dx = \int_{-\infty}^0 f(x)\,dx + \int_0^{+\infty} f(x)\,dx$$
$$= \lim_{a \to -\infty} \int_a^0 f(x)\,dx + \lim_{b \to +\infty} \int_0^b f(x)\,dx.$$

At this time, we also say that the improper integral $\int_{-\infty}^{+\infty} f(x)\,dx$ converges, otherwise we say the improper integral $\int_{-\infty}^{+\infty} f(x)\,dx$ diverges.

If the improper integrals $\int_a^{+\infty} f(x)dx, \int_{-\infty}^b f(x)dx, \int_{-\infty}^{+\infty} f(x)dx$ constrain, and $F'(x) = f(x)$, then we record that $\int_a^{+\infty} f(x)dx = [F(x)]_a^{+\infty}$, $\int_{-\infty}^b f(x)dx = [F(x)]_{-\infty}^b$, $\int_{-\infty}^{+\infty} f(x)dx = [F(x)]_{-\infty}^{+\infty}$.

Example 1 Judge the convergence and divergence of the improper integral $\int_a^{+\infty} \dfrac{1}{x^p} dx$ which has infinite limit, where $a > 0$, and p is a constant.

Solution According to definition:

When $p \neq 1$, for any $b > a$, we have

$$\int_a^b \frac{1}{x^p} dx = \left[\frac{1}{1-p} x^{1-p}\right]_a^b = \frac{1}{1-p}(b^{1-p} - a^{1-p}).$$

Therefore, we have

$$\int_a^{+\infty} \frac{1}{x^p} dx = \lim_{b \to +\infty} \int_a^b \frac{1}{x^p} dx = \frac{1}{1-p} \lim_{b \to +\infty}(b^{1-p} - a^{1-p})$$

$$= \begin{cases} \dfrac{1}{p-1} a^{1-p}, & p > 1, \\ +\infty, & p < 1. \end{cases}$$

When $p = 1$, we have

$$\int_a^{+\infty} \frac{1}{x^p} dx = \lim_{b \to +\infty} \int_a^b \frac{1}{x} dx = [\ln x]_a^{+\infty} = +\infty.$$

Therefore, when $p > 1$, the improper integral converges. Its value is $\dfrac{a^{1-p}}{1-p}$. When $p \leqslant 1$, the improper integral diverges.

Example 2 Find $\int_0^{+\infty} e^{-x} \sin x \, dx$.

Solution $\int_0^{+\infty} e^{-x} \sin x \, dx = -\int_0^{+\infty} \sin x \, d(e^{-x}) = [-e^{-x} \sin x]_0^{+\infty} + \int_0^{+\infty} e^{-x} \cos x \, dx$

$$= -\int_0^{+\infty} \cos x \, d(e^{-x}) = [-e^{-x} \cos x]_0^{+\infty} - \int_0^{+\infty} e^{-x} \sin x \, dx.$$

Therefore, $\int_0^{+\infty} e^{-x} \sin x \, dx = \left[-\dfrac{1}{2} e^{-x} \cos x\right]_0^{+\infty} = \dfrac{1}{2}$.

Example 3 Find the improper integral $\int_0^{+\infty} x e^{-px} dx$ (p is a constant, and $p > 0$).

Solution $\int_0^{+\infty} x e^{-px} dx = -\dfrac{1}{p} \int_0^{+\infty} x \, d(e^{-px}) = \left[-\dfrac{1}{p} x e^{-px}\right]_0^{+\infty} + \dfrac{1}{p} \int_0^{+\infty} e^{-px} dx$

$$= -\frac{1}{p} \lim_{x \to +\infty} x e^{-px} + \frac{1}{p^2}$$

$$= \frac{1}{p^2}.$$

Example 4 Find the improper integral $\int_{-\infty}^{+\infty} \dfrac{dx}{1+x^2}$.

Solution According to definition

$$\int_{-\infty}^{+\infty} \frac{\mathrm{d}x}{1+x^2} = \int_{-\infty}^{0} \frac{\mathrm{d}x}{1+x^2} + \int_{0}^{+\infty} \frac{\mathrm{d}x}{1+x^2}$$

$$= \lim_{a \to -\infty} \int_{a}^{0} \frac{1}{1+x^2} \mathrm{d}x + \lim_{b \to +\infty} \int_{0}^{b} \frac{1}{1+x^2} \mathrm{d}x$$

$$= \lim_{a \to -\infty} [\arctan x]_{a}^{0} + \lim_{b \to +\infty} [\arctan x]_{0}^{b}$$

$$= -\lim_{a \to -\infty} \arctan a + \lim_{b \to +\infty} \arctan b$$

$$= \frac{\pi}{2} + \frac{\pi}{2} = \pi.$$

6.5.2 Improper integral of unbounded functions

With respect to the concept of improper integral on infinite interval, we can generalize to the case of unbounded function.

Definition 2 Assume the function $f(x)$ is continuous in $(a,b]$, and the point a is unbounded in the right neighborhood. Take $\varepsilon > 0$. If

$$\lim_{\varepsilon \to 0^+} \int_{a+\varepsilon}^{b} f(x) \mathrm{d}x$$

exists, then the limit is called the improper integral of the function $f(x)$ on $(a,b]$, still denoted by $\int_{a}^{b} f(x) \mathrm{d}x$. That is

$$\int_{a}^{b} f(x) \mathrm{d}x = \lim_{\varepsilon \to 0^+} \int_{a+\varepsilon}^{b} f(x) \mathrm{d}x.$$

We also say that the improper integral $\int_{a}^{b} f(x) \mathrm{d}x$ converges. If the above limit does not exist, we say that the improper integral $\int_{a}^{b} f(x) \mathrm{d}x$ diverges.

Similarly, assume the function $f(x)$ is continuous on $[a,b)$, and the point b is unbounded in the left neighborhood. Take $\varepsilon > 0$. If the limit $\lim_{\varepsilon \to 0^+} \int_{a}^{b-\varepsilon} f(x) \mathrm{d}x$ exists, then define

$$\int_{a}^{b} f(x) \mathrm{d}x = \lim_{\varepsilon \to 0^+} \int_{a}^{b-\varepsilon} f(x) \mathrm{d}x.$$

Otherwise, we say that the improper integral $\int_{a}^{b} f(x) \mathrm{d}x$ diverges.

Assume the function $f(x)$ is continuous on $[a,b]$ except the point $c(a<c<b)$, and is unbounded in the neighborhood of the point c. At this time, if the two improper integrals $\int_{a}^{c} f(x) \mathrm{d}x$ and $\int_{c}^{b} f(x) \mathrm{d}x$ both converge, then we define

$$\int_{a}^{b} f(x) \mathrm{d}x = \int_{a}^{c} f(x) \mathrm{d}x + \int_{c}^{b} f(x) \mathrm{d}x$$

$$= \lim_{\varepsilon \to 0^+} \int_{a}^{c-\varepsilon} f(x) \mathrm{d}x + \lim_{\varepsilon \to 0^+} \int_{c+\varepsilon}^{b} f(x) \mathrm{d}x.$$

Otherwise, we say that the improper integral $\int_a^b f(x)\mathrm{d}x$ diverges.

Example 5 Prove the improper integral $\int_a^b \dfrac{1}{(x-a)^p}\mathrm{d}x$ converges when $p<1$ and diverges when $p\geq 1$.

Proof When $p=1$,
$$\int_a^b \frac{1}{(x-a)^p}\mathrm{d}x = \int_a^b \frac{1}{x-a}\mathrm{d}x = \lim_{\varepsilon\to 0^+}\int_{a+\varepsilon}^b \frac{1}{x-a}\mathrm{d}x$$
$$= \lim_{\varepsilon\to 0^+}[\ln(x-a)]_{a+\varepsilon}^b = \lim_{\varepsilon\to 0^+}[\ln(b-a)-\ln\varepsilon] = +\infty.$$

When $p\neq 1$,
$$\int_a^b \frac{1}{(x-a)^p}\mathrm{d}x = \begin{cases}\dfrac{(b-a)^{1-p}}{1-p}, & p<1,\\ +\infty, & p>1.\end{cases}$$

Therefore, when $p<1$, the improper integral converges and its value is $\dfrac{(b-a)^{1-p}}{1-p}$. When $p\geq 1$, the improper integral diverges.

Example 6 Find the improper integral $\int_0^1 \ln x\,\mathrm{d}x$.

Solution The function $f(x)=\ln x$ is continuous on $(0,1]$, and is unbounded at the point 0. Take $\varepsilon>0$, we have
$$\int_0^1 \ln x\,\mathrm{d}x = \lim_{\varepsilon\to 0^+}\int_\varepsilon^1 \ln x\,\mathrm{d}x = \lim_{\varepsilon\to 0^+}[x\ln x - x]_\varepsilon^1$$
$$= \lim_{\varepsilon\to 0^+}(-1-\varepsilon\ln\varepsilon+\varepsilon) = -1.$$

Example 7 Find the improper integral $\int_{-1}^1 \dfrac{1}{\sqrt{1-x^2}}\mathrm{d}x$.

Solution The function $f(x)$ is continuous on $(-1,1)$ and is unbounded in the neighborhood of the two endpoints. According to definition,
$$\int_{-1}^1 \frac{1}{\sqrt{1-x^2}}\mathrm{d}x = \int_{-1}^0 \frac{1}{\sqrt{1-x^2}}\mathrm{d}x + \int_0^1 \frac{1}{\sqrt{1-x^2}}\mathrm{d}x,$$
$$\int_{-1}^0 \frac{1}{\sqrt{1-x^2}}\mathrm{d}x = \lim_{\varepsilon\to 0^+}\int_{-1+\varepsilon}^0 \frac{1}{\sqrt{1-x^2}}\mathrm{d}x$$
$$= \lim_{\varepsilon\to 0^+}[-\arcsin(-1+\varepsilon)] = \frac{\pi}{2},$$
$$\int_0^1 \frac{1}{\sqrt{1-x^2}}\mathrm{d}x = \lim_{\varepsilon\to 0^+}\int_0^{1-\varepsilon} \frac{1}{\sqrt{1-x^2}}\mathrm{d}x$$
$$= \lim_{\varepsilon\to 0^+}\arcsin(1-\varepsilon) = \frac{\pi}{2}.$$

Therefore, $\int_{-1}^1 \dfrac{1}{\sqrt{1-x^2}}\mathrm{d}x = \dfrac{\pi}{2}+\dfrac{\pi}{2} = \pi$.

Example 8 Judge the convergence and divergence of the integral $\int_{-1}^1 \dfrac{1}{x}\mathrm{d}x$.

Solution Because $x=0$ is the unbounded point of $\dfrac{1}{x}$.

And because
$$\int_{-1}^{1}\frac{1}{x}dx = \int_{-1}^{0}\frac{1}{x}dx + \int_{0}^{1}\frac{1}{x}dx,$$

and
$$\int_{-1}^{0}\frac{1}{x}dx = \lim_{\varepsilon\to 0^{-}}\int_{-1}^{\varepsilon}\frac{1}{x}dx = \lim_{\varepsilon\to 0^{-}}[\ln|x|]_{-1}^{\varepsilon} = -\infty,$$

also
$$\int_{0}^{1}\frac{1}{x}dx = \lim_{\varepsilon\to 0^{+}}\int_{\varepsilon}^{1}\frac{1}{x}dx = \lim_{\varepsilon\to 0^{+}}[\ln x]_{\varepsilon}^{1} = -\infty,$$

then the improper integral diverges.

Exercises 6-5

1. Judge the convergence of the following improper integrals. If it converges, find the value of the improper integral.

(1) $\displaystyle\int_{1}^{+\infty}\frac{1}{x^{6}}dx$;

(2) $\displaystyle\int_{-1}^{+\infty}\frac{1}{\sqrt[3]{x}}dx$;

(3) $\displaystyle\int_{0}^{+\infty}e^{-ax}dx\ (a>0)$;

(4) $\displaystyle\int_{-\infty}^{+\infty}\frac{x}{1+x^{2}}dx$;

(5) $\displaystyle\int_{0}^{1}\frac{xdx}{\sqrt{1-x^{2}}}$;

(6) $\displaystyle\int_{0}^{2}\frac{dx}{1-x^{2}}$;

(7) $\displaystyle\int_{1}^{e}\frac{dx}{x\sqrt{1-\ln^{2}x}}$;

(8) $\displaystyle\int_{e}^{+\infty}\frac{dx}{x\ln^{2}x}$;

(9) $\displaystyle\int_{0}^{+\infty}\frac{dx}{1+e^{x}}$;

(10) $\displaystyle\int_{0}^{+\infty}\frac{x^{2}}{x^{6}+x^{3}+1}dx$;

(11) $\displaystyle\int_{2}^{+\infty}\frac{1+\ln x}{x\ln^{3}x}dx$.

2. Find $\displaystyle\int_{1}^{3}\frac{xdx}{\sqrt{|x^{2}-4|}}$.

3. Find the maximum and the minimum of the function $f(x) = \displaystyle\int_{0}^{x^{2}}(2-t)e^{-t}dt$.

4. Judge the convergence of the following improper integrals:

(1) $\displaystyle\int_{\frac{1}{e}}^{e}\frac{\ln|x-1|}{x-1}dx$;

(2) $\displaystyle\int_{0}^{+\infty}\frac{1}{e^{x}\sqrt{x}}dx$.

Summary

1. Main contents

(1) The concept of definite integral.

Definite integral $\displaystyle\int_{a}^{b}f(x)dx$ is a bounded function which is defined on the interval [a,b]. After segmentation, production, summation and limitation, we have a summation limit $\lim\limits_{\lambda \to 0} \sum\limits_{k=1}^{n} f(\xi_k) \Delta x_k$. This kind of operation reveals the same characteristic of many different questions. For example, finding the sum area of the curved trapezoid, the change rate of the total output and the total output, which all reveals this kind of operation.

① The value of definite integral is a constant. It is only related to the integrand $f(x)$ and the integral interval $[a,b]$, while has nothing to do with the mark of the integral variable. Please deeply understand this point from the definition of definite integral.

② The continuous function $f(x)$ must be integrable on the interval $[a,b]$.

(2) The property of definite integral.

The definite integral of the function $f(x)$ on the interval $[a,b]$ has linearity, inequality, interval additivity, valuation and the mean value theorem of definite integrals. Those properties have important significance for the calculation and valuation of definite integral.

The mean value theorem of definite integrals is corresponding with the Lagrange theorem of differential calculus. In the theorem, ξ is some value in $[a,b]$, but it does not know the exact location of it. According to the theorem, we know that there must be a mean value $\dfrac{1}{b-a}\int_a^b f(x)\mathrm{d}x$ of the continuous function $f(x)$ on its interval $[a,b]$.

(3) Integral upper limit function.

Upper limit function $\Phi(x) = \int_a^x f(t)\mathrm{d}t$ has the following conclusions: $\Phi'(x) = f(x)$, which explains that $\Phi(x) = \int_a^x f(t)\mathrm{d}t$ is an antiderivative of $f(x)$. Also, if $f(x)$ is a continuous function on the interval $[a,b]$, then the antiderivative of $f(x)$ must exist. That is $\Phi(x) = \int_a^x f(t)\mathrm{d}t$.

This conclusion reveals the reciprocal relation between differential calculus and integral calculus.

Note $\Phi'(x) = \left[\int_a^x f(t)\mathrm{d}t\right]_x' = f(x)$ is the derivation of the integral upper limit x.

If assume $\alpha(x)$, $\beta(x)$ are derivable functions, and $\Phi(x) = \int_{\alpha(x)}^{\beta(x)} f(t)\mathrm{d}t$, then

$$\Phi'(x) = \frac{\mathrm{d}}{\mathrm{d}x}\left[\int_{\alpha(x)}^{\beta(x)} f(t)\mathrm{d}t\right] = f[\beta(x)]\beta'(x) - f[\alpha(x)]\alpha'(x).$$

Note The structure form of the integral upper limit function is different from the general function. Its variable is the upper or lower limit of the integral. There are two basic theorems about this function, which are not only basic but also important in the calculus.

(4) Find of definite integral.

Newton – Leibniz formula not only provides a simple and effective method to find definite integral, but also lays the foundation for the wide application of definite integral. If $F(x)$ is any antiderivative of the continuous function $f(x)$ on $[a, b]$, then

$$\int_a^b f(x)\mathrm{d}x = [F(x)]_a^b = F(b) - F(a).$$

The importance of this theorem is that it changes the integral calculation problem into a problem for antiderivative, so we can use the method of indefinite integral to find definite integral.

At the same time, according to the characteristics of the definite integral, we introduce the element method and partial method of definite integral. Thus, a method is proposed which is simple and unified for the calculation of integral.

Newton - Leibniz formula is the effective tool for the calculation of definite integral. When applying we should pay attention to the condition that the integrand is continuous on the integral interval. For the condition that there is finite first type of point of discontinuity on integral interval, we must pay attention to the use of Newton - Leibniz formula which should be after using discontinuity point to divide to the integral interval $[a,b]$.

The characteristic of substitution method is to transform the integral variables at the same time to transform the integral upper limit and lower limit. So, after finding the original function of the integrand of the new variable, we do not have to change back the original variable, but directly use Newton - Leibniz formula, which is convenient. But we must pay attention to the condition of the theorem in application. Also, we should pay attention to the use of the parity of the integrand, which will bring convenience to the calculation.

① In the calculation of definite integral, by using the parity of the integrand $f(x)$ in the symmetric domain, we can simplify the calculation of definite integral. But the premise of this simplification is the integral of odd and even function on the symmetric interval $[-a, a]$.

If $f(x)$ is an odd function on $[-a,a]$, then $\int_{-a}^{a} f(x)\mathrm{d}x = 0$;

If $f(x)$ is an even function on $[-a,a]$, then $\int_{-a}^{a} f(x)\mathrm{d}x = 2\int_{0}^{a} f(x)\mathrm{d}x$.

② In the substitution method of definite integral, "substitution must change the limit". At the same time, when using indefinite integral substitution method, the accumulated experience and skills can be generally used in the definite integral substitution method.

2. Basic requirements

(1) Correctly understand the concepts of definite integral and the ideas, methods and steps generalized in the definition of definite integral for solving practical problems.

(2) Understand the meaning and importance of definite integral properties, the mean value theorem of definite integrals, derivative of integral upper limit function and Newton - Leibniz formula.

(3) With the help of antiderivative, skillfully use the natures of definite integral and Newton – Leibniz formula to find the integral value.

(4) Skillfully master the substitution method and the integration by parts for definite integrals.

Quiz

1. Choice question.

(1) The value of the definite integral $\int_a^b dx$ is (   ).

A. 0   B. $b-a$ C. $a-b$ D. any function

(2) Assume $f(x) = x^3 + x$, then the value of the definite integral $\int_{-2}^{2} f(x)dx$ is (   ).

A. 0   B. 8   C. $\int_0^2 f(x)dx$ D. $2\int_0^2 f(x)dx$

(3) $\int_0^a (\arcsin x)' dx = ($ $)$.

A. $\dfrac{1}{\sqrt{1-x^2}}$ B. $-\dfrac{1}{\sqrt{1-x^2}}$ C. $\arcsin x - \dfrac{\pi}{2}$ D. $\arcsin a$

(4) Which of the following equations is right. (   )

A. $\int_{-2}^{2} x^3 \sin x\,dx = 0$ B. $\int_{-1}^{1} 2e^x dx = 0$

C. $\left[\int_3^5 \ln x\,dx\right]' = \ln 5 - \ln 3$ D. $\int_{-1}^{1} x\cos x\,dx = 0$

2. Competition.

(1) Assume the function $\Phi(x) = \int_1^{x^2} e^t dt$. Then $\Phi'(x) = $ _____.

(2) $\lim\limits_{x \to 0} \dfrac{\int_0^x \sin^2 t\,dt}{x^2} = $ _____.

(3) $\int_{-2}^{2} x^3 e^{x^2} \cos x\,dx = $ _____.

(4) $\int_0^1 \sqrt{x\sqrt{x}}\,dx = $ _____.

(5) $\int_1^e \dfrac{\ln^2 x}{x}\,dx = $ _____.

3. Find:

(1) $\int_0^\pi x\cos x\,dx$;

(2) $\int_0^4 \dfrac{1}{1+\sqrt{x}}\,dx$;

(3) $\int_0^2 |1-x|\,dx$;

(4) $\int_1^{e^2} \dfrac{1}{x\sqrt{1+\ln x}}\,dx$;

(5) $\int_0^1 e^{\sqrt{x}}\,dx$.

4. Comprehensive question.

(1) Assume $f(x)=\begin{cases} x+1, & x<0, \\ 0, & x=0, \\ x^2, & x>0. \end{cases}$ Find $\int_{-1}^{1} f(x)dx$.

(2) Prove: if the function $f(x)$ is continuous, then $\int_{0}^{\frac{\pi}{2}} f(\sin x)dx = \int_{0}^{\frac{\pi}{2}} f(\cos x)dx$.

(3) Assume $f(x)=\begin{cases} -1, & -1\leq x<0, \\ 0, & x=0, \\ 1, & 0<x\leq 1. \end{cases}$ Find $F(x) = \int_{-1}^{x} f(x)dx$ and $F'(x)$.

Exercises

1. Find the following limits:

(1) $\lim\limits_{n\to\infty}\dfrac{1}{n^2}(\sqrt{n}+\sqrt{2n}+\cdots+\sqrt{n^2})$.

(2) $\lim\limits_{n\to\infty}\left(\dfrac{1}{4n^2-2^2}+\dfrac{2}{4n^2-3^2}+\cdots+\dfrac{n-1}{4n^2-n^2}\right)$.

(3) $\lim\limits_{x\to 0}\dfrac{\int_{0}^{\sin^2 x} \ln(1+t)dt}{\sqrt{1+x^4}-1}$.

(4) Assume $f(x)$ is continuous. Find $\lim\limits_{x\to a}\dfrac{x}{x-a}\int_{a}^{x} f(t)dt$.

2. Find the following definite integrals:

(1) $\int_{\frac{1}{2}}^{2}\left(1+x-\dfrac{1}{x}\right)e^{x+\frac{1}{x}}dx$;

(2) $\int_{0}^{\pi}\dfrac{|\cos x|}{\cos^2 x+2\sin^2 x}dx$;

(3) $\int_{0}^{\pi}\sqrt{1-\sin x}\,dx$;

(4) $\int_{a}^{b}\dfrac{dx}{\sqrt{(x-a)(b-x)}}$ (a, b are constant numbers).

3. Prove the following questions:

(1) Try to prove $\int_{0}^{\frac{\pi}{2}}\dfrac{\sin x}{1+x^2}dx \leq \int_{0}^{\frac{\pi}{2}}\dfrac{\cos x}{1+x^2}dx$.

(2) If $f(x)$ and $g(x)$ are both continuous on the interval $[a,b]$. Prove:

① $\left[\int_{a}^{b} f(x)g(x)dx\right]^2 \leq \int_{a}^{b} f^2(x)dx \int_{a}^{b} g^2(x)dx$ (Cauchy - Schwartz inequality);

② $\left\{\int_{a}^{b} [f(x)+g(x)]^2 dx\right\}^{\frac{1}{2}} \leq \left[\int_{a}^{b} f^2(x)dx\right]^{\frac{1}{2}} + \left[\int_{a}^{b} g^2(x)dx\right]^{\frac{1}{2}}$ (Minkowski inequality).

(3) Assume $f(x)$ is continuous and monotonously increase on $[a,b]$. Try to prove

$$\int_{a}^{b} xf(x)dx \geq \dfrac{a+b}{2}\int_{a}^{b} f(x)dx.$$

(4) Assume $f(x)$ is defined on $[a,b]$, and for any two points x, y on $[a,b]$ and $0<\lambda<1$, we have

$$f[\lambda x+(1-\lambda)y] \leq \lambda f(x)+(1-\lambda)f(y).$$

Try to prove $f\left(\dfrac{a+b}{2}\right) \leqslant \dfrac{\int_a^b f(x)\,dx}{b-a} \leqslant \dfrac{f(a)+f(b)}{2}$.

(5) Assume $f(x)$ is derivable on $[0,1]$, and meets the condition that $f(1) - 2\int_0^{\frac{1}{2}} xf(x)\,dx = 0$. Try to prove there exists $\xi \in (0,1)$, which can make $f'(\xi) = -\dfrac{f(\xi)}{\xi}$.

4. Try to solve the following questions:

(1) Assume $f(x) = x - \int_0^{\pi} f(x)\cos x\,dx$. Find $f(x)$.

(2) Assume $f(x) = \begin{cases} 0, & |x| \leqslant 1, \\ \dfrac{1}{x^2}, & |x| > 1. \end{cases}$ Find $\int_0^3 xf(x-1)\,dx$.

(3) Assume $f(x) = \int_0^1 t|t-x|\,dt$. Find $f'(x)$.

(4) Assume $y = y(x)$ is determined by $x - \int_1^{y+x} e^{-u^2}\,du = 0$. Find $\left.\dfrac{d^2 y}{dx^2}\right|_{x=0}$.

(5) Assume $f(x)$ is continuous in $[0, +\infty)$, and for any positive numbers a, b, the integral $\int_a^{ab} f(x)\,dx$ has nothing to do with a and $f(1) = 1$. Find $f(x)$.

(6) Assume $f(x)$ is continuous and $\int_0^x tf(2x - t)\,dt = \dfrac{1}{2}\arctan x^2$, $f(1) = 1$. Find $\int_1^2 f(x)\,dx$.

Chapter 7 Applications of definite integral

> Definite integral arises from the need of the practical problems. Therefore, it is widely used to many disciplines such as geometry, physics and economics etc.. This chapter will chiefly discuss the application of definite integral to geometry and economics on the basis of the theories and computational methods of the previous chapter, which aims to establish formulas and the analytic methods of solving practical problems by definite integral, which is called element method.

7.1 The element method of definite integral

The quantity calculated by definite integral must has a feature of "additivity", that is, the quantity could be divided into small quantities to obtain the approximate values of each smally quantity, and then sum these approximate values up to get the approximate value of total amount, and calculate the exact total amount by finding the limit of the qpproximate vaule. So the method of solving practical problems by definite integral is called "element method" or "differential element method".

Definite integral usually deals with the non-uniform distribution of integral quantity. The basic principle of solving such kind of problem is "segmentation—approximate—sum up—limit". First, segmentation means to divide the whole into several parts; then substitute straight lines for curve lines, substitute uniformity for non-uniformity to find the approximate values of the quantities in these parts; then add these approximate values up to obtain the approximate value of the total amount; finally find the limit of the approximate value and get the exact value of the total amount.

In order to further illustrate the basic idea of the element method, we first review the problems of the area of trapezoid with curved edge discussed in the sixth chapter in which the concept of definite integral is introduced.

To be specific, assume that $f(x)$ is continuous and nonnegative on $[a,b]$. Suppose that A is the area of the trapezoid with the curved edge $y=f(x)$ and the bottom $[a,b]$, we have

$$A = \int_a^b f(x) \mathrm{d}x.$$

The steps are as follows:

(1) **Segmentation**   Divide the interval $[a,b]$ into n subintervals and denote the length of the ith subinterval by $\Delta x_i (i=1,2,\cdots,n)$,
$$[x_0,x_1],[x_1,x_2],\cdots,[x_{n-1},x_n], \ \Delta x_i=x_i-x_{i-1}.$$

(2) **Approximate**   Substitute the area of a small rectangular approximately for the ith small trapezoid with curved edge with the area ΔA_i,
$$\Delta A_i \approx f(\xi_i)\Delta x_i \ (x_{i-1} \leqslant \xi_i \leqslant x_i).$$

(3) **Sum**   Add all these approximate areas up to obtain the approximate value of the trapezoid with curved edge with area A,
$$A = \sum_{i=1}^{n} \Delta A_i \approx \sum_{i=1}^{n} f(\xi_i)\Delta x_i.$$

(4) **Limit**   Let $\lambda = \max\{\Delta x_i\}$, the limit of the above-mentioned formula as $\lambda \to 0$ is the exact area of A of the trapezoid with curved edge,
$$A = \lim_{\lambda \to 0} \sum_{i=1}^{n} f(\xi_i)\Delta x_i = \int_a^b f(x)\mathrm{d}x.$$

The above four steps are generally simplified to the following two steps:

(1) Find the "area element" $\mathrm{d}A$ on an infinitely thin segment interval $[x,x+\mathrm{d}x]$,
$$\Delta A \approx f(x)\mathrm{d}x \quad \text{or} \quad \mathrm{d}A = f(x)\mathrm{d}x.$$

(2) Integrate the area element on the interval $[a,b]$,
$$A = \int_a^b \mathrm{d}A = \int_a^b f(x)\mathrm{d}x.$$

In general, if a quantity U has the feature of "additivity" in a practical problem and is relevant to a continuous function $f(x)$ in the interval $[a,b]$, we can use the element method to calculate the quantity of U by the following steps.

(1) According to the specific problem, we will select a variable (such as x) to be the integral variable and determine the interval of x, $[a, b]$. And then, we divide $[a, b]$ into n subintervals. For any infinitely thin subinterval $[x, x + \mathrm{d}x]$ (Figure 7-1), we calculate the infinitesimal of U on this subinterval which is also called the element of U,
$$\mathrm{d}U = f(x)\mathrm{d}x.$$

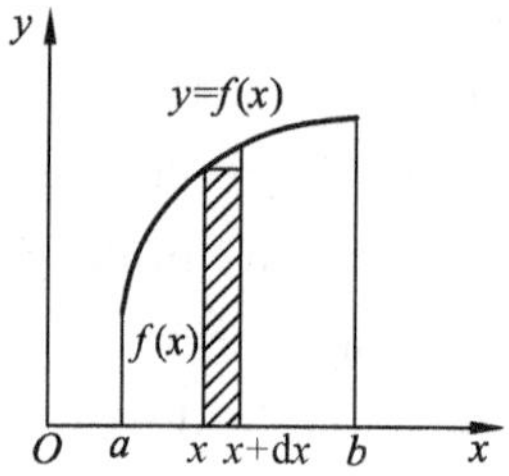

Figure 7-1

(2) Take integral of the element $\mathrm{d}U = f(x)\mathrm{d}x$ over the interval $[a,b]$,
$$U = \int_a^b f(x)\mathrm{d}x,$$
and then, evaluate the definite integral.

This technique is usually called "element method" or "differential element method" of definite integral. The most important step is the first one, in which we find the element of U, $\mathrm{d}U = f(x)\mathrm{d}x$, based on $\mathrm{d}x$. Then the method will be discussed in the application of definite integral to some problems in geometry and physics.

7.2 Application of definite integral in geometry

7.2.1 The area of a region

1) The area of a region in a rectangular coordinate system

By the geometric explanation of definite integral, we know that the area of a trapezoid with curved edge which is bounded by the curve $y=f(x)(f(x)\geq 0)$ and the straight lines $x=a$, $x=b(a<b)$, and $y=0$ is

$$A = \int_a^b f(x)\mathrm{d}x.$$

In which $f(x)\mathrm{d}x$ is the area element in a rectangular coordinate system and is denoted by $\mathrm{d}A$, that is, $\mathrm{d}A=f(x)\mathrm{d}x$.

Generally, the area of trapezoid with curved edge surrounded by curve $y=f(x)$ and straight lines $x=a$, $x=b(a<b)$ and $y=0$ (Figure 7-2) is

$$A = \int_a^b |f(x)| \, \mathrm{d}x. \tag{1}$$

If the region is surrounded by curves $y=f_1(x)$, $y=f_2(x)(f_2(x)>f_1(x))$ and the straight lines $x=a$, $x=b$ (Figure 7-3), we can evaluate the area A by the element method.

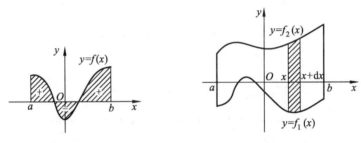

Figure 7-2   **Figure 7-3**

By the element method, we take x to be the integral variable, divide the interval $[a,b]$ into n subintervals and take a small subinterval $[x, x+\mathrm{d}x]$, then the area of the rectangle with height $f_2(x)-f_1(x)$ can replace approximately the area of the small strip in this subinterval. Since the width of the rectangle is $\mathrm{d}x$, the area element is

$$\mathrm{d}A=[f_2(x)-f_1(x)]\mathrm{d}x,$$

therefore

$$A = \int_a^b [f_2(x) - f_1(x)]\mathrm{d}x. \tag{2}$$

Similarly, the area of the region surrounded by curves $x=\varphi_1(y)$, $x=\varphi_2(y)(\varphi_2(y)>\varphi_1(y))$ and the straight lines $y=c$, $y=d(c<d)$ is

$$A = \int_c^d [\varphi_2(y) - \varphi_1(y)]\mathrm{d}y, \tag{3}$$

where y is the variable of integration and

$$\mathrm{d}A=[\varphi_2(y)-\varphi_1(y)]\mathrm{d}y$$

is the area element (Figure 7-4).

In this way, we can apply the element method of definite integral to calculate the area of trapezoid with curved edges, and even the area of some more complex regions.

Example 1 Find the area of the region bounded by the parabola $y=x^2$ and the straight line $y=2x+3$.

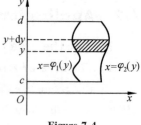

Figure 7-4

Solution As shown in Figure 7-5, we first find the points of intersection of the parabola and the straight line by solving their equations

$$\begin{cases} y=x^2, \\ y=2x+3, \end{cases}$$

thus, the points of intersection are $(-1,1)$ and $(3,9)$.

We see from Figure 7-5 that the top boundary is $y=2x+3$ and the bottom is $y=x^2$. Take x to be the integral variable, the area element is

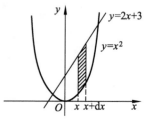

Figure 7-5

$$dA=(2x+3-x^2)dx.$$

Since the region lies between $x=-1$ and $x=3$, the total area is

$$A = \int_{-1}^{3} (2x+3-x^2)dx = \left[x^2+3x-\frac{x^3}{3}\right]_{-1}^{3} = \frac{32}{3}.$$

Example 2 Find the area of the region bounded by the curve $y=\ln x$, x-axis and the straight lines $x=\frac{1}{2}$ and $x=2$.

Solution As shown in Figure 7-6, the region lies between $x=\frac{1}{2}$ and $x=2$. Take x to be the integral variable and notice that $\ln x \leqslant 0$ if $x\in\left[\frac{1}{2},2\right]$. So the area element on the interval $\left[\frac{1}{2},1\right]$ is

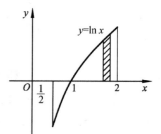

Figure 7-6

$$dA=-\ln x dx.$$

Since $\ln x \geqslant 0$ if $x\in[1,2]$, the area element on the interval $[1,2]$ is

$$dA=\ln x dx.$$

Therefore, the area of the region is

$$A = \int_{\frac{1}{2}}^{2} |\ln x| \, dx = \int_{\frac{1}{2}}^{1} (-\ln x)dx + \int_{1}^{2} \ln x dx$$

$$= [x-x\ln x]_{\frac{1}{2}}^{1} + [x\ln x - x]_{1}^{2} = \frac{3}{2}\ln 2 - \frac{1}{2}.$$

Example 3 Find the area of the region enclosed by the parabola $y^2=2x$ and the straight line $y=x-4$.

Solution As shown in Figure 7-7, we first find the points of intersection of the parabola and the straight line by solving their equations

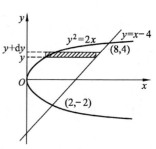

Figure 7-7

$$\begin{cases} y^2 = 2x, \\ y = x - 4, \end{cases}$$

so the points of intersection are $(2,-2)$ and $(8,4)$.

Take x to be the integral variable, and it is on the interval $[0,8]$, but the area element on the interval $[0,2]$ is

$$dA = 2\sqrt{2x}\,dx,$$

while the area element on the interval $[2,8]$ is

$$dA = (\sqrt{2x} - x + 4)\,dx.$$

So the area of the region is

$$A = \int_0^2 2\sqrt{2x}\,dx + \int_2^8 (\sqrt{2x} - x + 4)\,dx$$

$$= \left[\frac{4\sqrt{2}}{3} x^{\frac{3}{2}}\right]_0^2 + \left[\frac{2\sqrt{2}}{3} x^{\frac{3}{2}} - \frac{x^2}{2} + 4x\right]_2^8 = 18.$$

On the other hand, we take y to be the integral variable, and it is on the interval $[-2,4]$. Then, the area element on a small interval $[y, y+dy]$ is

$$dA = \left(y + 4 - \frac{1}{2} y^2\right) dy,$$

so

$$A = \int_{-2}^4 \left(y + 4 - \frac{1}{2} y^2\right) dy = \left[\frac{y^2}{2} + 4y - \frac{1}{6} y^3\right]_{-2}^4 = 18.$$

In this example, we see that if the integral variables are chosen properly, the area can be easily calculated.

2) The area of a region in polar coordinates

For some regions, it is easy to calculate their areas by polar coordinates.

In polar coordinates, suppose that the region is surrounded by the polar curve $r = r(\theta)$ ($\alpha \leqslant \theta \leqslant \beta$) and the rays $\theta = \alpha$ and $\theta = \beta$, which is called curved sector (Figure 7-8). We will find the area of the curved sector by the element method.

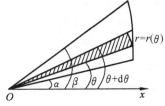

Figure 7-8

Take the polar angle θ to be the integral variable, and it is on the interval $[\alpha, \beta]$. On any small interval $[\theta, \theta + d\theta]$, the area of the small curved sector can be approximated by the area of the sector of a circular with radius $r = r(\theta)$ and central angle $d\theta$.

Therefore, the area element of curved sector is

$$dA = \frac{1}{2}[r(\theta)^2]\,d\theta,$$

and the area of the curved sector is

$$A = \frac{1}{2}\int_\alpha^\beta r^2(\theta)\,d\theta. \tag{4}$$

Example 4 Find the area of the region bounded by cardioid $r = a(1 + \cos\theta)$ ($a > 0$).

Solution The region (Figure 7-9) is surrounded by the heart-shaped line which is symmetric about the polar axis, so the area of the region denoted by A is double of the area of the region lying above the polar axis denoted by A_1.

Take the angle θ to be the integral variable, and it is on the interval $[0,\pi]$. We have

$$A = 2A_1 = 2 \times \frac{1}{2} \int_0^\pi a^2 (1+\cos\theta)^2 \,d\theta$$

$$= a^2 \int_0^\pi (1 + 2\cos\theta + \cos^2\theta)\,d\theta$$

$$= a^2 \left[\frac{3}{2}\theta + 2\sin\theta + \frac{1}{4}\sin 2\theta \right]_0^\pi$$

$$= \frac{3}{2}\pi a^2.$$

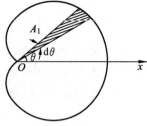

Figure 7-9

Example 5 Evaluate the area of the region bounded by the lemniscate $r^2 = a^2 \cos 2\theta$ $(a>0)$.

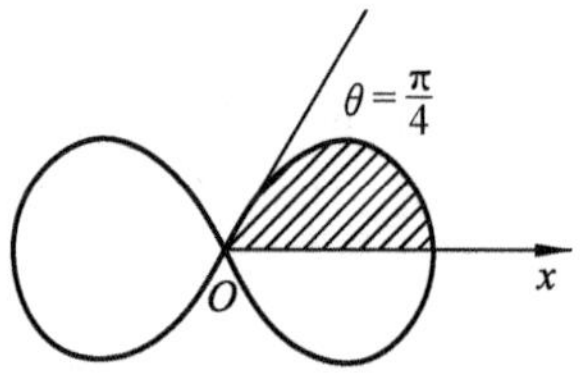

Solution According to the symmetry of the lemniscate (Figure 7-10), the area of the region A is four times of the area of the shadow area A_1

$$A = 4A_1 = 4\int_0^{\frac{\pi}{4}} \frac{1}{2} a^2 \cos 2\theta \,d\theta$$

$$= [a^2 \cdot \sin 2\theta]_0^{\frac{\pi}{4}} = a^2.$$

Figure 7-10

7.2.2 Volume

1) The volume of a solid with the known cross-sectional area

For a solid (Figure 7-11), suppose the solid lies between $x=a$ and $x=b$, and the area of the cross-section (a plane region obtained by intersecting the solid by a plane perpendicular to the x-axis and passing through the point x) is a continuous function of x, which is denoted by $A(x)$ where $a \leqslant x \leqslant b$. We can also use the element method to find the volume of the solid.

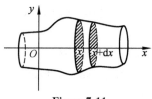

Figure 7-11

Take x to be the integral variable which is on the interval $[a,b]$. We can divide the solid into some thin "slices" by parallel planes which are perpendicular to the x-axis. The volume of the slice on the thin interval $[x, x+\mathrm{d}x]$ can be approximated by a flat cylinder with base area $A(x)$ and height $\mathrm{d}x$. So the volume element is

$$\mathrm{d}V = A(x)\mathrm{d}x.$$

Therefore, the volume of the solid lies between $x=a$ and $x=b$ with the cross-sectional area $A(x)$ is

$$V = \int_a^b A(x)\,\mathrm{d}x. \tag{5}$$

Example 6 Find the volume of solid taking the circle whose radius is R and all sections of a fixed diameter perpendicular to the base are equilateral triangles.

Solution As shown in Figure 7-12, the coordinate system is established with the fixed diameter as the x-axis and the center of the circle as the origin. Over the point x on the x-axis, we make a cross-section and get a equilateral triangle with the length of side $a=2\sqrt{R^2-x^2}$, so the area of the cross-section is

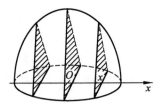

Figure 7-12

$$A(x)=\frac{1}{2}a \cdot \frac{\sqrt{3}}{2}a=\sqrt{3}(R^2-x^2).$$

Since the solid lies between $x=-R$ and $x=R$ and is symmetric about the plane $x=0$, the volume of the solid is

$$V = 2\int_0^R A(x)\mathrm{d}x = 2\sqrt{3}\int_0^R (R^2-x^2)\mathrm{d}x = \frac{4\sqrt{3}}{3}R^3.$$

Formula (5) is very important. It is not only the formula for calculating the volume of a solid with the known cross-sectional area $A(x)$, but also the key formula for expressing a double integral to an iterated integral we will discuss later in this book.

2) The volume of solid of rotation

The solid of rotation is a solid generated by rotating a plane region around a straight line in the same plane and the straight line is called the axis of rotation.

To be specific, we suppose that the solid of rotation is generated by rotating the plane region bounded by the curve $y=f(x)$ and straight lines $x=a$, $x=b$ and $y=0$. Find the volume of the solid.

Over the interval $[a,b]$, the cross-section perpendicular to the x-axis is a circle with radius $y=f(x)$ (Figure 7-13), thus, the area of cross-section is

$$A(x)=\pi y^2 =\pi [f(x)]^2.$$

So the volume of the solid of rotation is

$$V = \int_a^b \pi [f(x)]^2 \mathrm{d}x. \qquad (6)$$

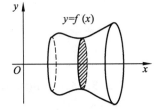

Figure 7-13

Example 7 Find the volume of the solid of revolution generated by rotating the region surrounded by $xy=2$, $x=2$, $x=4$ around the x-axis.

Solution From $xy=2$ we have $y=\dfrac{2}{x}$. The volume element is

$$\mathrm{d}V=\pi y^2 \mathrm{d}x=\pi \left(\frac{2}{x}\right)^2 \mathrm{d}x,$$

Since the solid lies between $x=2$ and $x=4$, the volume of the solid of revolution is,

$$V = \int_2^4 \pi \left(\frac{2}{x}\right)^2 \mathrm{d}x = 4\pi \left[-\frac{1}{x}\right]_2^4 = \pi.$$

Similarly (Figure 7-14), the solid of revolution is generated by rotating the trapezoid with curved edge bounded by curve $x = \varphi(y)$ and the straight lines $y = c$, $y = d (c < d)$ and y-axis about the y-axis, the volume is

$$V = \pi \int_c^d [\varphi(y)]^2 \, dy. \qquad (7)$$

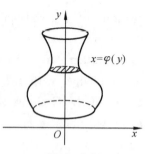

Figure 7-14

Example 8 Find the volume of the solid of revolution generated by rotating the plane region bounded by the curve $y = \sqrt{2x-4}$ and straight lines $x = 2$ and $x = 4$ about the x-axis and y-axis respectively.

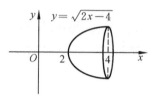

Solution (1) Since the axis of rotation is the x-axis (Figure 7-15), we take x to be the integral variable, the volume element is

$$dV = \pi y^2 \, dx = \pi (2x - 4) \, dx,$$

and then

Figure 7-15

$$V = \int_2^4 \pi (2x - 4) \, dx = \pi [(x^2 - 4x)]_2^4 = 4\pi.$$

(2) Since the axis of rotation is the y-axis, we take y to be the integral variable, and it is on the interval $[0, 2]$ (Figure 7-16). We let

$$x_1 = \frac{y^2 + 4}{2}, \quad x_2 = 4,$$

then the volume is

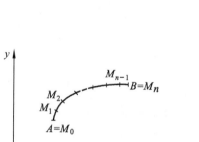

Figure 7-16

$$V = \pi \int_0^2 x_2^2 \, dy - \pi \int_0^2 x_1^2 \, dy = \pi \int_0^2 4^2 \, dy - \pi \int_0^2 \left(\frac{y^2+4}{2}\right)^2 dy$$

$$= 32\pi - \frac{224}{15}\pi = \frac{256}{15}\pi.$$

7.2.3 Arc length of plane curve

Suppose there is a curve from A to B in the plane, we put points $A = M_0, M_1, \cdots, M_n = B$ along the curve $\overset{\frown}{AB}$ (Figure 7-17) and joint adjacent pairs of these points with straight line segment to form a polygonal line. Let $|\overline{M_{k-1}M_k}|$ denote the distance between M_{k-1} and M_k. L is $L = \sum_{k=1}^n |\overline{M_{k-1}M_k}|$, and as

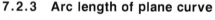

Figure 7-17

the number of points increasing infinitely each segment $|\overline{M_{k-1}M_k}|$ will shrink to a single point. If the limit L as n approaches to infinity exists, this limit is called the arc length of $\overset{\frown}{AB}$ and the curve is called rectifiable.

For smooth curves, we have the following conclusions.

Theorem Let the function of curve $\overset{\frown}{AB}$ be $y=f(x)$ that have continuous first-order derivative in the interval $[a,b]$, then the curve is rectifiable (this curve is called smooth curve).

Since the smooth curve is rectifiable, we will not prove the theorem. We will discuss the calculation of the arc length by using the element method of definite integral.

(1) Let the function of curve $\overset{\frown}{AB}$ be $y=f(x)$, $x\in [a,b]$, and $f(x)$ have continuous first-order derivative.

Take x to be the integral variable. The length of a section of curve $y=f(x)$ on any small interval $[x, x+\mathrm{d}x]$ can be approximated by a straight line segment of the tangent line at point $(x,f(x))$ (Figure 7-18). And the arc length element (it is also called the differential of arc length, see formula in 4.5.1) is

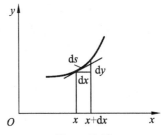

Figure 7-18

$$\mathrm{d}s=\sqrt{(\mathrm{d}x)^2+(\mathrm{d}y)^2}=\sqrt{1+y'^2}\,\mathrm{d}x,$$

so the arc length is

$$s=\int_a^b \sqrt{1+y'^2}\,\mathrm{d}x. \tag{8}$$

(2) The function of curve $\overset{\frown}{AB}$ is defined by the parameter equation $\begin{cases} x=\varphi(t), \\ y=\psi(t), \end{cases}$ $t\in [\alpha,\beta]$ and $\varphi(t),\psi(t)$ have continuous derivatives in the interval $[\alpha,\beta]$. Take t to be the integral variable, and the arc length element is

$$\mathrm{d}s=\sqrt{(\mathrm{d}x)^2+(\mathrm{d}y)^2}=\sqrt{\varphi'^2(t)+\psi'^2(t)}\,\mathrm{d}t,$$

so the arc length is

$$s=\int_\alpha^\beta \sqrt{\varphi'^2(t)+\psi'^2(t)}\,\mathrm{d}t. \tag{9}$$

(3) The function of curve AB is defined by the polar equation $r=r(\theta)$, $\theta\in [\theta_1, \theta_2]$ and $r(\theta)$ has continuous derivative in the interval $[\theta_1, \theta_2]$.

From the relationship between Cartesian coordinates and polar coordinates, we have

$$\begin{cases} x=r(\theta)\cos\theta, \\ y=r(\theta)\sin\theta, \end{cases} \theta\in [\theta_1,\theta_2],$$

and the arc length element is

$$\mathrm{d}s=\sqrt{x'^2(\theta)+y'^2(\theta)}\,\mathrm{d}\theta=\sqrt{r^2(\theta)+r'^2(\theta)}\,\mathrm{d}\theta,$$

so the arc length is

$$s=\int_{\theta_1}^{\theta_2} \sqrt{r^2(\theta)+r'^2(\theta)}\,\mathrm{d}\theta. \tag{10}$$

Example 9 Find the arc length of the curve $y=\ln(1-x^2)$ from $x=0$ to $x=\dfrac{1}{2}$.

Solution Since $y'=\dfrac{-2x}{1-x^2}$, $1+y'^2=\left(\dfrac{1+x^2}{1-x^2}\right)^2$, we have the arc length element

$$ds = \sqrt{1+y'^2}\,dx = \frac{1+x^2}{1-x^2}dx,$$

so the arc length is

$$s = \int_0^{\frac{1}{2}} \frac{1+x^2}{1-x^2}dx = \int_0^{\frac{1}{2}} \left(-1 + \frac{2}{1-x^2}\right)dx$$

$$= \left[-x + \ln\left|\frac{1+x}{1-x}\right|\right]_0^{\frac{1}{2}} = -\frac{1}{2} + \ln 3.$$

Example 10 The position function of an object is $x = 3\cos 2t$, $y = 3\sin 2t$, find the distance that the object moves from $t=0$ to $t=\pi$.

Solution Since the distance is an arc length and

$$x'(t) = -6\sin 2t,\quad y'(t) = 6\cos 2t,$$

the distance is given by

$$s = \int_0^\pi \sqrt{x'^2(t) + y'^2(t)}\,dt = \int_0^\pi \sqrt{(-6\sin 2t)^2 + (6\cos 2t)^2}\,dt$$

$$= \int_0^\pi 6\,dt = 6\pi.$$

Example 11 Find the arc length of the heart-shaped line $r = a(1+\cos\theta)$, where $a > 0$.

Solution As shown in Figure 7-9, the heart-shaped line is symmetric about the polar axis and has the equal arc length within the interval $[0,\pi]$ and $[\pi,2\pi]$. Since

$$r = a(1+\cos\theta),\quad r' = -a\sin\theta,$$
$$r^2 = a^2(1+\cos\theta)^2,\quad r'^2 = a^2\sin^2\theta.$$

By Formula (10), the arc length of the heart-shaped line is

$$s = 2\int_0^\pi \sqrt{a^2(1+\cos\theta)^2 + a^2\sin^2\theta}\,d\theta = 2a\int_0^\pi \sqrt{2(1+\cos\theta)}\,d\theta$$

$$= 4a\int_0^\pi \cos\frac{\theta}{2}\,d\theta = 8a.$$

Exercises 7-2

1. Find the area of the region bounded by the following plane curves:
   (1) $y = x^3, y = 1$;
   (2) $y = x^2, y^2 = x$;
   (3) $y = \frac{1}{x}, y = x, x = 2$;
   (4) $y = e^x, y = e, x = 0$;
   (5) $y = \sqrt{x}, y = x$;
   (6) $y = 2x, y = 3 - x^2$;
   (7) $y = x^2, y = x, y = 2x$;
   (8) $y = \sin x, y = \cos x, x = 0, x = \frac{\pi}{2}$.

2. The parabola curve $y^2 = 2x$ cuts the circle $x^2 + y^2 = 8$ into two parts and what's the ratio of the area of the two parts?

3. Find the area of the region bounded by the parabola curve $y = -x^2 + 4x - 3$ and the tangents at points $(0, -3)$ and $(3, 0)$.

4. Find the area of the region bounded by two parabola curves $y^2 = 1 + x$, $y^2 = 1 - x$.

5. Find the area of the region bounded by the following plane curves:

(1) $r=2a\cos\theta$;   (2) $r=a\sin 3\theta$.

6. Find the volumes of the following solids of revolutions:

(1) The plane region bounded by curves $y=x^2$, $x=y^2$, rotating about the x-axis.

(2) The plane region bounded by curve $y=\sin x$, $y=0 (0\leqslant x \leqslant \pi)$, rotating about the x-axis and y-axis respectively.

(3) The plane region bounded by curve $y=x^3$, $x=2$, $y=0$, rotating about the x-axis and y-axis respectively.

7. The plane region is bounded by the curves $y=e^x$, $y=e^{-x}$ and the straight $x=1$, find:

(1) The area of the plane region.

(2) The volume of the solid of revolution generated by the plane region rotating about the x-axis.

8. Find the volume of the right vertebral taking a circle with radius R to be the base, and the top is the line segment parallel to the base and equal to the diameter of the base circle, and the height is h (Figure 7-19).

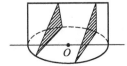

Figure 7-19

9. Find the arc length of the following curves:

(1) $y=\ln x, \sqrt{3}\leqslant x\leqslant\sqrt{8}$;

(2) $y=x^{\frac{3}{2}}, 0\leqslant x\leqslant 4$;

(3) $x=t^2, y=t^3, 1\leqslant t\leqslant 2$;

(4) $x=a(t-\sin t), y=a(1-\cos t), 0\leqslant t\leqslant 2\pi$;

(5) $r=e^{a\theta}, 0\leqslant\theta\leqslant\varphi_0$.

10. Find the circumference of astroid $x=a\cos^3 t, y=a\sin^3 t$.

7.3 Application of definite integral in economy

Known from the discussion of Chapter 5, we can calculate the economic function by indefinite integral if the rate of change of the economic function is given. Then according to the relationship of indefinite integral and definite integral, we can translate the problem of finding total amount of economic function on the interval $[a,b]$ into finding the value of the definite integral of the rate of change on the interval $[a,b]$. This chapter introduces three examples about definite integrals applied in economy: the problem for calculating total output if its rate of change is given, the problem for calculating amount function if marginal function is given and the discount problem.

7.3.1 Calculate total output if its rate of change is given

It is known that the rate of change of one commodity's total output Q is a continuous function of time t, suppose the function is $f(t)$, that is, $Q'(t)=f(t)$. Thus the total output function of this commodity is

$$Q(t) = Q(t_0) + \int_{t_0}^{t} f(z)dz, \quad t \geqslant t_0,$$

where t_0 is a given initial moment and $t_0 \geq 0$. We usually believe $Q(0)=0$ which means the total output at the beginning is zero. According to the definition of definite integral, the increment of total output during the time interval $[t_0, t_1]$ is

$$\Delta Q = Q(t_1) - Q(t_0) = \int_{t_0}^{t_1} f(t)\,dt.$$

Example 1 Provided that the rate of change of the commodity's total output is

$$f(t) = 100 + 10t - 0.45t^2\ (t/h),$$

find:

(1) Total output function $Q(t)$.

(2) Total output within the time interval $[4,8]$.

Solution (1) The total output function is

$$Q(t) = \int_0^t f(z)\,dz = \int_0^t (100 + 10z - 0.45z^2)\,dz = 100t + 5t^2 - 0.15t^3\ (t).$$

(2) The total output within the time interval $[4,8]$ is

$$Q(8) - Q(4) = \int_4^8 (100 + 10t - 0.45t^2)\,dt = 572.8\,(t).$$

7.3.2 Calculate amount function when marginal function is given

According to the discussion of marginal analysis in Chapter 2, we know that for a given economic function $F(x)$ (such as demand function $Q(p)$, total cost function $C(x)$, total revenue function $R(x)$ and profit function $L(x)$), the corresponding marginal function is $F'(x)$. According to the discussion in Chapter 5, if $F'(x)$ is a given marginal function, the indefinite integral $\int F'(x)\,dx$ will be the original economic function, and the constant C in this function will be determined by the specific condition $F(x_0) = F_0$ and usually $x_0 = 0$.

We know that by Newton-Leibniz formula, we have

$$\int_0^x F'(x)\,dx = F(x) - F(0).$$

Thus, with variable upper limit, we will obtain the economic function $F(x)$,

$$F(x) = \int_0^x F'(x)\,dx + F(0). \tag{1}$$

In addition, the increment of economic function from a to b is

$$\Delta F = F(b) - F(a) = \int_a^b F'(x)\,dx. \tag{2}$$

Example 2 The demand function of a commodity Q is a function of price p and the marginal demand is $Q'(p) = -4$. The maximal demand of this commodity is 80, find the function of demand and price.

Solution Generally the demand reach the maximum value at the price $p=0$, so

$$Q(0) = 80.$$

By the Formula (1), the expression of demand function is

$$Q(p) = \int_0^p Q'(p)\,dp + Q(0) = \int_0^p (-4)\,dp + 80 = -4p + 80\ (p > 0).$$

Example 3 The marginal revenue of a commodity is $R'(x)=100-2x$ (Yuan/unit). Find the total revenue and average revenue after producing 40 units and the revenue increased if producing ten more units.

Solution For revenue functions, we usually assume that the total revenue is zero, if the production or sales volume is zero, that is, $R(0)=0$. By the Formula (1), the total revenue at $x=40$ is

$$R(40) = \int_0^{40} (100-2x)\,dx = [100x-x^2]_0^{40} = 2\,400(\text{Yuan}).$$

The average revenue is

$$\overline{R}(40) = \frac{R(40)}{40} = \frac{2\,400}{40} = 60(\text{Yuan}).$$

By the Formula (2), if producing ten more units, the total revenue increased, which is denoted by ΔR, so

$$\Delta R = R(50) - R(40) = \int_{40}^{50} R'(x)\,dx = \int_{40}^{50} (100-2x)\,dx$$
$$= [100x-x^2]_{40}^{50} = 100(\text{Yuan}).$$

Example 4 Suppose that the marginal revenue function of a commodity is $R'(x)=9-x$ (ten-thousand Yuan/ten thousand Pcs) and the marginal cost function is $C'(x)=4+\dfrac{x}{4}$ (ten-thousand Yuan/ten thousand Pcs), where the unit of the production x is ten-thousand Pcs.

(1) Find the variation of profit when the production increases from forty thousand Pcs to fifty thousand Pcs.

(2) What is the production that maximizes the profit?

(3) Find the total cost function and gross profit function if the fixed cost is ten thousand Yuan.

Solution (1) Firstly, calculate the marginal profit function

$$L'(x) = R'(x) - C'(x) = (9-x) - \left(4+\frac{x}{4}\right) = 5 - \frac{5}{4}x.$$

By the Formula (2), when the production increases from forty thousand to fifty thousand, the variation of profit is

$$\Delta L = L(5) - L(4) = \int_4^5 L'(x)\,dx = \int_4^5 \left(5 - \frac{5}{4}x\right)dx$$
$$= \left[5x - \frac{5}{8}x^2\right]_4^5 = -\frac{5}{8}(\text{ten-thousand Yuan}).$$

Thus we see that producing ten thousand more units on the basis of forty thousand, the profit decreases instead.

(2) According to the method of calculating maximum and minimum, we let $L'(x)=0$, and then, we can obtain the unique stagnation point $x=4$ (ten-thousand Pcs). It means that when the production is forty thousand Pcs, the profit gets the maximum value. This answer also explains the decrease profit in question (1).

(3) Fixed cost being ten thousand Yuan means $C(0)=1$ (ten-thousand Yuan). By the Formula (1), the total cost function is

$$C(x) = \int_0^x C'(t)\,dt + C(0) = \int_0^x \left(4 + \frac{t}{4}\right)dt + 1 = 4x + \frac{1}{8}x^2 + 1.$$

The profit function is

$$L(x) = \int_0^x L'(t)\,dt + L(0) = \int_0^x \left(5 - \frac{5}{4}t\right)dt - C(0)$$

$$= 5x - \frac{5}{8}x^2 - 1.$$

7.3.3 Discount problems

If the revenue (or expenditure) within a period is not a fixed amount but having revenue (or expenditure) in each unit time, it is called cash flows. Suppose R_i is the revenue flow at the end of the i th unit time ($i=1,2,\cdots,n$) and r is the discount rate, then the present value of R_i is

$$R_i(1+r)^{-i} = \frac{R_i}{(1+r)^i}, \quad i=1,2,\cdots,n.$$

The present value of all revenue flows in n periods is a summation

$$M = \sum_{i=1}^{n} \frac{R_i}{(1+r)^i}.$$

The formula above is for the discrete revenue flow. If the revenue flow is continuous, it will be a function $R(t)$. Thus, within a quite short time interval $[t, t+\Delta t]$, the approximate value of revenue flow is

$$R(t)\Delta t.$$

If the discount rate is r, by continuous compounding, the corresponding present value is

$$\Delta M \approx R(t)e^{-rt}\Delta t.$$

Therefore, on the time interval from 0 to n, the present value of total revenue flow is

$$M = \int_0^n R(t)e^{-rt}\,dt.$$

In particular, if $R(t)$ is a constant A (the revenue in each period is A which is called uniform flow), we have

$$M = A\int_0^n e^{-rt}\,dt = \frac{A}{r}(1 - e^{-rn}).$$

Example 5 Continuous revenue flow goes on for two years at a constant ratio 60 000 Yuan a year. If discounting at annual interest rate 6%, what is the present value?

Solution Due to the revenue flow $R(t)=60\ 000$, the present value of total revenue flow is

$$M = A\int_0^n e^{-rt}\,dt = 60\ 000\int_0^2 e^{-0.06t}\,dt = \frac{60\ 000}{0.06}(1 - e^{-0.12}) \approx 113\ 100\ (\text{Yuan}).$$

Thus the present value of total revenue flow is 113 100 Yuan.

Example 6 Suppose we spend twenty million Yuan to build a colliery. The additional

cost and additional revenue after t years are
$$C'(t)=t+2t^{\frac{2}{3}} \text{ (million Yuan/ year)},$$
$$R'(t)=17-t^{\frac{2}{3}} \text{ (million Yuan/ year)}.$$
How long does the colliery exploit that can obtain the maximum profit? And calculate the maximum profit.

Solution This additional cost is the rate of change of the total cost with respect to time and addition revenue is the rate of change of the total revenue with respect to time. Thus, according to the necessary condition that extremum exists, optimal downtime which obtains the maximum profit will satisfy $R'(t)=C'(t)$, that is,
$$17-t^{\frac{2}{3}}=t+2t^{\frac{2}{3}}.$$
Solve the equation for t, we get $t=8$.

Since there exists the maximum value in practice and the unique stagnation point, $t=8$ is the answer. Therefore after exploiting coals for 8 years, the colliery will obtain the maximum profit
$$\pi(8) = \int_0^8 [R'(t)-C'(t)]dt - C_0$$
$$= \int_0^8 [17-t^{\frac{2}{3}}-(t+2t^{\frac{2}{3}})]dt - 20$$
$$= \left[17t-\frac{1}{2}t^2-\frac{9}{5}t^{\frac{5}{3}}\right]_0^8 - 20 = 26.4 \quad \text{(million Yuan)}.$$

The maximum profit is 26.4 million Yuan.

Exercises 7-3

1. The fixed cost of producing some commodity is 36 (ten-thousand Yuan) and marginal cost is $C'(x)=2x+40$ (ten-thousand Yuan/ hundred units). Find the increment of total cost when the production increases from four hundred units to six hundred units. What is the production that minimizes the average cost?

2. Given that marginal revenue function of some commodity is
$$R'(Q)=10(10-Q)e^{-\frac{Q}{10}}.$$
The sales volume is Q and total revenue is $R=R(Q)$. Find the total revenue function of the commodity.

3. Given that one commodity's marginal cost function and marginal revenue function of some commodity are
$$C'(Q)=Q^2-4Q+6, \ R'(Q)=105-2Q.$$
The fixed cost is 100 and the sales volume is Q. $C(Q)$ is the total cost and $R(Q)$ is the total revenue. Find the maximum profit.

4. Suppose the rate of change of total revenue R (marginal revenue) for producing x units of some commodity is
$$R'(x)=200-\frac{x}{100} \quad (x \geqslant 0).$$

(1) Find the total revenue after producing 50 units.

(2) If 100 units have been produced, calculate the total revenue if producing 100 more units.

Summary

The concept of definite integral is derived from practical problems in geometry and physics, therefore it is applied widely in practice. This chapter uses definite integral's "element method" to discuss geometric application of definite integral including calculate areas of plane figures, volumes of solids that the area of revolution or the cross-section are given and arc length of plane curves and physical application of definite integral about working by variable force, static pressure and gravitation.

1. Main contents

(1) The element method.

The element method is that under certain conditions, make partial quantity ΔU corresponding to arbitrary interval $[x, x+dx]$ on $[a,b]$ approximately represent the element dU of U by thoughts that substitute straight lines for curve and substitute uniformity for non-uniformity:

$$\Delta U \approx dU = f(x)dx.$$

Then integrate the element dU on $[a,b]$, we can obtain that

$$U = \int_a^b dU = \int_a^b f(x)dx.$$

(2) Application of definite integral in geometry.

The area of the region bounded by curves $y=f_1(x), y=f_2(x)$ $(f_2(x)>f_1(x))$ and straight lines $x=a$, $x=b$ is

$$A = \int_a^b [f_2(x) - f_1(x)]dx.$$

In polar coordinates, the area of curve sector is

$$A = \frac{1}{2}\int_{\theta_1}^{\theta_2} r^2(\theta)d\theta.$$

The volume of revolution which is generated by rotating the plane region surrounded by the curve $y=f(x)$ and straight lines $x=a$, $x=b$, $y=0$ about the x-axis is

$$V = \int_a^b \pi[f(x)]^2 dx.$$

And the volume of solid given the area of cross-section is $V = \int_a^b A(x)dx$.

By the arc differential formula

$$ds = \sqrt{(dx)^2 + (dy)^2},$$

we can obtain arc length of curves defined by rectangular coordinate function $y=f(x)$, parametric equation $x=\varphi(t)$, $y=\varphi(t)$ and polar coordinate function $r=r(\theta)$ respectively are

$$s = \int_a^b \sqrt{1 + f'^2(x)}\,dx,$$

$$s = \int_\alpha^\beta \sqrt{\varphi'^2 + \psi'^2}\,dt,$$

$$s = \int_{\theta_1}^{\theta_2} \sqrt{r^2 + r'^2}\,d\theta.$$

(3) Application of definite integral in economy.

① Calculate total output if the rate of change is given.

② Calculate the amount function if marginal function is given.

③ Discount problems.

2. Basic requirements

(1) Understand the definite integral's element method, and know the basic idea and specific steps to solve practical problems by the element method.

(2) Understand how to find the area of plane region in rectangular coordinates, know how to find the area of plane region in parametric equations and polar coordinates.

(3) Understand how to find volume of revolution, know how to find volume of solid that the area of cross-section is given.

(4) Understand how to find the arc length of plane curves.

(5) Understand how to find total cost functions, revenue functions and profit functions by indefinite integral and definite integral.

The key points in this chapter are the element method and the application in geometry, especially area of plane region and volume of revolution.

Quiz

1. Find the area of the region bounded by curves $y = x+1, y = x^2 (x \geq 0), y = 1, y = 0$.

2. Find the area of the region bounded by parametric equation $x = a\cos t, y = b\sin t, 0 \leq t \leq 2\pi$.

3. Find the area of the region surrounded by curves $r = 3\cos\theta$ and $r = 1 + \cos\theta$.

4. Find the area of the region surrounded by straight lines $y = 0, x = e$ and curve $y = \ln x$ and the volume of revolution generated by rotating the region about the x-axis.

5. Find the volume of the ellipsoid $\dfrac{x^2}{a^2} + \dfrac{b^2}{b^2} + \dfrac{z^2}{c^2} \leq 1$.

6. Find the arc length of the astroid $x = a\cos^3 t, y = a\sin^3 t, 0 \leq t \leq 2\pi$.

7. Given that the fixed cost of producing some commodity is one hundred thousand Yuan, marginal cost and marginal revenue(Unit: ten-thousand Yuan/t) are $C'(x) = x^2 - 5x + 40$, $R'(x) = 50 - 2x$ respectively. Find:

(1) total cost function.

(2) total revenue function.

(3) gross profit function.

(4) variation of gross profit if produce one more ton on the basis of production that maximizes the profit.

8. Given that marginal cost of some commodity is $C'(x)=2$(Yuan/unit), fixed cost is 0 and marginal revenue is $R'(x)=12-0.02x$. What is the production that maximizes the profit? What is the change in profit if producing 50 more units on the basis of this production?

Exercises

1. Find the area of plane regions surrounded by the following curves:
(1) $y^2=2x, y^2=-(x-1)$ and $x=0$;
(2) $y=x^2, y=x$ and $y=2x$;
(3) $xy=1, y=x, y=2$ and $x=0$;
(4) $y=x^2, y=2+x$.

2. Find the area of the mutual part bounded by curves
$$r=a\sin\theta, r=a(\cos\theta+\sin\theta) \ (a>0).$$

3. Find the volume of revolution generated by rotating the region bounded by $y=x^2$, $y=0$ and $x=1$ about the x-axis.

4. Find the volume of revolution created by rotating the region bounded by curves $y=x^2$ and $x=y^2$ about the y-axis.

5. Suppose that the parabola $y=ax^2+bx+c$ passing through the point $(0,0)$ and $y \geqslant 0$ if $x\in[0,1]$. Find a,b,c such that the area of the region bounded by parabola $y=ax^2+bx+c$ and straight lines $x=1, y=0$ is $\dfrac{4}{9}$ and the volume of revolution generated by rotating the region about the x-axis is the minimum.

6. Find the arc length of parabola $y=\dfrac{1}{2}x^2$ which is cut by $x^2+y^2=3$.

7. By the annual continuous compounding rate $r=0.05$, find the present value after 20 years of the revenue flow, 10 000 Yuan/year.

Answers

Exercises 1-1

2. (1) $(-\infty,-1)\cup(-1,2)\cup(2,\infty)$;   (2) $(-\infty,0)\cup(0,4)$;
   (3) $[-1,0)\cup(0,1]$;   (4) $(-\infty,0]\cup(1,\infty)$;
   (5) $x>-2, x\neq-1,0,1,2,\cdots$;   (6) $|x-k\pi|<\dfrac{\pi}{6}, k\in\mathbf{Z}$;
   (7) $(-\infty,0)\cup(0,3]$;   (8) $(0,1)$.

3. (1) different (the domains are different);   (2) the same;
   (3) different (the domains are different).

4. $(x^2+1)^2, (x+1)^4$.

5. (1) $[-1,0]$;   (2) $[2k\pi,(2k+1)\pi], k\in\mathbf{Z}$;
   (3) $[1,e]$;   (4) If $a\in\left(0,\dfrac{1}{2}\right]$, then $D=[a,1-a]$, if $a>\dfrac{1}{2}$, then $D=\varnothing$.

7. (1) even function;   (2) odd function;   (3) odd function;
   (4) neither even nor odd;   (5) neither even nor odd;   (6) odd function.

8. (1) monotone decreasing;   (2) monotone increasing.

9. (1) bounded;   (2) unbounded;   (3) unbounded;   (4) bounded.

10. (1) $y=\dfrac{1-x}{1+x}, x\neq-1$;   (2) $y=x^3-1, x\in(-\infty,+\infty)$;
    (3) $y=2+e^{x-1}, x\in(-\infty,+\infty)$;   (4) $y=\begin{cases}\sqrt{x}, & x\geq 1,\\ \dfrac{x+1}{2}, & x<1.\end{cases}$

12. (1) $y=u^5, u=\sin v, v=x^2$;   (2) $y=\ln u, u=v^2, v=\tan w, w=x^3$;
    (3) $y=\sqrt[3]{u}, u=\arccos v, v=\dfrac{1}{w}, w=x^3$;   (4) $y=\cos u, u=\sin v, v=e^w, w=\sqrt{x}$.

14. x^2-x+1.

15. $\dfrac{R^3}{24\pi^2}(2\pi-a)^2\sqrt{4\pi a-a^2}$ $(0<x<2\pi)$.   16. 144 dollars.

17. $y=\begin{cases}6x, & 0\leq x\leq 200,\\ 4x+400, & 200<x<500,\\ 3x+900, & 500<x.\end{cases}$ 18. $A=4x+\dfrac{2}{x}+\dfrac{1}{2}, 0<x<+\infty$.

Exercises 1-2

1. (1) convergent, 0;   (2) convergent, 0;   (3) divergent;
   (4) convergent, $\dfrac{2}{3}$;   (5) convergent, 0;   (6) convergent, 0;
   (7) convergent, 1;   (8) divergent.

4. (1) 0;   (2) $\dfrac{1}{2}$;   (3) 1;   (4) Doesn't exist.

8. $f(0-0)=-1, f(0+0)=1, \lim\limits_{x\to 0}f(x)$ doesn't exist;
   $g(0-0)=g(0+0)=0, \lim\limits_{x\to 0}g(x)=0$.

Exercises 1-3

1. (1) 3; (2) 0; (3) 0; (4) $\frac{6}{5}$; (5) 2; (6) $\frac{1}{2}$.

2. (1) 12; (2) $-\frac{1}{3}$; (3) $\frac{4}{3}$; (4) 2; (5) $-\frac{1}{2}$; (6) 1;
   (7) -1; (8) $\frac{n(n-1)}{2}$.

3. (1) 0; (2) 3; (3) $\frac{1}{2}$; (4) $\frac{1}{2}$.

4. $a=1, b=-1$.   5. $a=\frac{3}{2}$.

6. (1) True; (2) False; (3) False; (4) False.

Exercises 1-4

1. (1) 3; (2) k; (3) $\frac{2}{5}$; (4) $\frac{1}{2}$; (5) 2; (6) $2x$;
   (7) $\frac{1}{3}$; (8) $\frac{2}{\pi}$.

2. (1) e^{-2}; (2) $e^{-\frac{1}{2}}$; (3) e^{-3}; (4) e^{-2}.

3. $a=e^2$.   4. 2.

Exercises 1-5

1. (1),(2) are infinitesimals of higher order than $\varphi(x)$;
   (4) is an infinitesimal of lower order than $\varphi(x)$;
   (3),(5),(6) and $\varphi(x)$ are infinitesimals of the same order;
   (3) and $\varphi(x)$ are equivalent infinitesimals.

3. unbounded; not infinity.

4. horizontal asymptote $y=-1$; vertical asymptote $x=2$.

5. (1) $\frac{3}{2}$; (2) $0(m>n)$, $1(m=n)$, $\infty(m<n)$; (3) 2; (4) $\frac{1}{4}$;
   (5) $\frac{1}{3}$; (6) $\frac{1}{2}$;   (7) 2; (8) e^a.

Exercises 1-6

1. (1) not continuous at $x=-1$, continuous at $x=1$; (2) continuous.

2. (1) $x=2$ is a discontinuity of the first kind (a removable discontinuity), $x=-2$ is a discontinuity of the second kind (an infinite discontinuity);
   (2) $x=1$ is a discontinuity of the first kind (a removable discontinuity), $x=2$ is a discontinuity of the second kind (an infinite discontinuity);
   (3) $x=0$ is a discontinuity of the second kind (an oscillating discontinuity);
   (4) $x=0$ is a discontinuity of the first kind (a jump discontinuity);
   (5) $x=1$ is a discontinuity of the second kind (an infinite discontinuity);
   (6) $x=0$ is a discontinuity of the first kind (a removable discontinuity), $x=k\pi (k \in \mathbf{Z}^*)$ is a discontinuity of the second kind (an infinite discontinuity).

3. (1) $a=1$; (2) $a=b=e^2$; (3) $a=2, b=-\frac{3}{2}$.

4. $f(x)=\begin{cases} -x, & |x|>1, \\ 0, & |x|=1, \\ x, & |x|<1. \end{cases}$ $x=-1, 1$ are discontinuities of the first kind (jump discontinuities).

Answers

5. (1) cos 2;　(2) $\sqrt{2}$;　(3) 2;　(4) $\dfrac{1}{2}$;

   (5) $\dfrac{1}{a}$;　(6) 3;　(7) $e^{-\frac{3}{2}}$;　(8) e^3.

Quiz

1. (1) sufficient;　(2) sufficient, necessary, necessary and sufficient;　(3) sufficient, necessary;
   (4) sufficient;　(5) necessary and sufficient, necessary and sufficient.
2. (1) B, D, A, C;　　(2) B;　　(3) B.
3. (1) $(-\infty, -2] \cup [5, 8) \cup (8, +\infty)$;　(2) $(2, 3)$;
   (3) $(-\infty, +\infty)$;　　　(4) $[-4, -\pi] \cup [0, \pi]$.
4. (1) $\dfrac{1}{2}$;　(2) $\dfrac{3}{2}$;　(3) $\dfrac{1}{2}$;　(4) e;　(5) 2ln 2;　(6) 1.
5. (1) $x=0$ is a discontinuity of the first kind (a removable discontinuity), $x=k\pi (k \in \mathbf{Z}^*)$ is a discontinuity of the second kind (an infinite discontinuity);
   (2) $x=0$ is a discontinuity of the first kind (a jump discontinuity);
   (3) $x=0$ is a discontinuity of the first kind (a removable discontinuity).
6. (1) $a=2b$;　(2) $a=2b=2$.
7. $y=\dfrac{(l-4x)^2}{4\pi}+x^2, 0<x<\dfrac{l}{4}$.

Exercises

2. (1) $y=1+\log_3(2+\sin x)$;　　(2) $y=\begin{cases} \ln x, & x \geqslant 1, \\ x-1, & x<1. \end{cases}$
3. $2, -4, -6, -12$.　　　　　4. $f(x)=x(1+x)$.
5. (1) $\dfrac{1}{2}$;　(2) $\dfrac{3}{2}$;　(3) $e^{-\frac{\pi}{2}}$;　(4) $\dfrac{7}{4}$;　(5) e^4;　(6) -3.
6. (1) $\dfrac{\sin x}{x}$;　(2) 1;　(3) $\dfrac{1}{3}$;　(4) 3.
7. (1) $x=0$ is a discontinuity of the first kind (a jump discontinuity), $x=1$ is a discontinuity of the first kind (a removable discontinuity), $x=-1$ is a discontinuity of the second kind (an infinite discontinuity).
   (2) $x=0$ is a discontinuity of the second kind (an infinite discontinuity), $x=1$ is a discontinuity of the first kind (a removable discontinuity).
   (3) $x=-2, 2$ are discontinuities of the first kind (removable discontinuities).
   (4) $x=\dfrac{3\pi}{4}, \dfrac{7\pi}{4}$ are discontinuities of the first kind (removable discontinuities), $x=\dfrac{\pi}{4}, \dfrac{5\pi}{4}$ are discontinuities of the second kind (infinite discontinuities).

Exercises 2-1

1. (1) -64 m/s;　　　(2) -128 m/s.
2. (1) $k=3, y=3x-1$;　(2) $k=\dfrac{1}{2}, y=\dfrac{1}{2}x+\dfrac{1}{2}$.
3. $\omega'(t)=\lim\limits_{\Delta t \to 0}\dfrac{\omega(t+\Delta t)-\omega(t)}{\Delta t}$.
4. (1) $-\dfrac{1}{4}$;　　(2) $-\dfrac{\sqrt{2}}{8}$;　　(3) 0.
5. (1) 1, 0, not derivatie;　(2) 1, 1, derivative;　(3) 1, 1, derivative.

6. $a=b=2$.

7. (1) When $a>0$, it is continuous; (2) When $a>1$, it is derivative; (3) When $a>2$, it is continuous and derivative.

8. (1) $-f'(x_0)$; (2) $2f'(x_0)$; (3) $5f'(x_0)$.

9. $(2,8)$, $(-2,-8)$; $\left(\dfrac{1}{6}, \dfrac{1}{216}\right)$, $\left(-\dfrac{1}{6}, -\dfrac{1}{216}\right)$.

10. (1) $y=x, y=-x$;

    (2) $y=1$, $y=-1$, $y=1$ and the coordinates of the points of intersection is $\left(\dfrac{\pi}{2}+2k\pi, 1\right)$; $y=-1$ and the coordinates of the points of intersection is $\left(-\dfrac{\pi}{2}+2k\pi, -1\right)$ $(k=0,\pm1,\pm2,\cdots)$.

11. $-8\,000$ L/min; $-10\,000$ L/min.

Exercises 2-2

1. (1) $6x^2+\dfrac{2}{x^3}+5$; (2) $2x\cos x - x^2\sin x$; (3) $-\dfrac{1}{3}x^{-\frac{4}{3}}$;

    (4) $3^x\ln 3+4x^3$; (5) $-\dfrac{1+\cos x}{(1+\sin x)^2}$; (6) $1+\ln x+\dfrac{1}{x^2}(1-\ln x)$;

    (7) $3\left(\dfrac{1}{x^4}-x^2\right)$; (8) $4(4-x^2)^{-\frac{3}{2}}$; (9) $-\dfrac{1+t}{\sqrt{t}(1-t)^2}$;

    (10) $2e^x\sin x$; (11) $\dfrac{(1-\ln 2)\cos x-(1+\ln 2)\sin x}{2^x}$;

    (12) $2\cos x-\sin x+\cos x\cot^2 x$.

2. (1) $-\ln a$; (2) $-\dfrac{1}{9}$; (3) $\dfrac{1}{1-\pi}$.

3. $a=d=1$, $b=c=0$.   4. $-\dfrac{64}{5}$.

5. (1) $\dfrac{2x}{\sqrt{1-x^4}}-(1+2x^2)e^{x^2}$; (2) $\arccos\dfrac{x}{2}-\dfrac{x}{\sqrt{4-x^2}}$;

    (3) $-\dfrac{3x^2-2}{2(x^3-2x+1)\sqrt{x^3-2x}}$; (4) $\dfrac{1}{1-x^4}$.

6. (1) $\operatorname{sh}(\operatorname{sh} x)\cdot\operatorname{ch} x$; (2) $e^{\operatorname{ch} x}(\operatorname{ch} x+\operatorname{sh}^2 x)$; (3) $-\dfrac{2x}{\operatorname{ch}^2(1-x^2)}$;

    (4) $\dfrac{1}{\operatorname{ch}^2 x+\operatorname{sh}^2 x}$; (5) $\operatorname{th}^3 x$.

7. (1) $2x\varphi'(x^2)$; (2) $[e^x\varphi'(e^x)+\varphi(e^x)\varphi'(x)]e^{\varphi(x)}$;

    (3) $\varphi\left(\dfrac{1}{x}\right)-\dfrac{1}{x}\varphi'\left(\dfrac{1}{x}\right)$; (4) $\varphi'(x)\varphi'[\varphi(x)]$.

Exercises 2-3

1. (1) $2x(2x^2-3)e^{-x^2}$; (2) $3+2\ln x$; (3) $2\arctan x+\dfrac{2x}{1+x^2}$;

    (4) $\left(\dfrac{1}{x}-\dfrac{2}{x^2}+\dfrac{2}{x^3}\right)e^x$; (5) $-x(1+x^2)^{-\frac{3}{2}}$; (6) $8e^{2x}\cos(2x+1)$.

2. (1) $12\cos 2x-24x\sin 2x-8x^2\cos 2x$; (2) $\dfrac{4a^3}{(a^2+x^2)^2}$.

3. (1) $(-1)^n n!\left[\dfrac{1}{(x+3)^{n+1}}+\dfrac{1}{(x-1)^{n+1}}\right]$; (2) $(n-1)!b^n\left[\dfrac{(-1)^{n-1}}{(a+bx)^n}+\dfrac{1}{(a-bx)^n}\right]$.

4. (1) $2f'(\sin^2 x)\cos 2x+f''(\sin^2 x)\sin^2 2x$;

(2) $\dfrac{2f(x^2)f'(x^2)-4x^2[f'(x^2)]^2+4x^2 f(x^2)f''(x^2)}{f^2(x^2)}$.

6. (1) $3^9 e^{3x}(3x^2+20x+30)$;   (2) $(-4)^n e^x \cos x$;   (3) $x\operatorname{sh} x+100\operatorname{ch} x$.

Exercises 2-4

1. (1) $\dfrac{\cos(x+y)}{e^y-\cos(x+y)}$;   (2) $-\sqrt{\dfrac{y}{x}}$;

   (3) $\dfrac{e^y-y\sin xy}{1-xe^y+x\sin xy}$;   (4) $\dfrac{x+y}{x-y}$.

2. (1) $\dfrac{y}{2}\left(\dfrac{3}{3x-2}+\dfrac{2}{5-2x}-\dfrac{1}{x-1}\right)$;   (2) $y\left[\dfrac{8}{2x-3}+\dfrac{1}{2(x-6)}-\dfrac{1}{3(x+1)}\right]$;

   (3) $y\left[\ln(x+\sqrt{1+x^2})+\dfrac{x}{\sqrt{1+x^2}}\right]$;   (4) $\dfrac{y}{2}\left[\dfrac{1}{x}+\cot x-\dfrac{e^x}{2(1-e^x)}\right]$.

3. (1) $-\sqrt{\dfrac{1+\theta}{1-\theta}}$;   (2) $\dfrac{e^{2t}}{1-t}$;   (3) -1.

4. 3.

5. (1) $x=0$, $y=0$;   (2) $4x+3y-12=0$, $3x-4y+6=0$.

6. (1) $\dfrac{d\rho}{d\theta}=\dfrac{v}{\omega}$;   (2) $\dfrac{d\rho}{d\theta}=R\csc\alpha$.

7. (1) $\dfrac{e^{2y}(3-y)}{(2-y)^3}$;   (2) $-\dfrac{\cos(x+y)}{[1+\sin(x+y)]^3}$;

   (3) $-2\cot^3(x+y)\csc^2(x+y)$ or $-\dfrac{2(1+y^2)}{y^5}$;

   (4) $\dfrac{2(y-e^{x+y})(e^{x+y}-x)-e^{x+y}(y-x)^2}{(e^{x+y}-x)^3}$ or $\dfrac{y[(y-1)^2+(x-1)^2]}{x^2(1-y)^3}$.

8. (1) 8;   (2) $-\dfrac{2}{y^3}$;   (3) $-\dfrac{1+t^2}{4t^3}$;   (4) $\dfrac{1}{f''(t)}$.

9. (1) $-\dfrac{3(t^2+1)}{8t^5}$;   (2) $-6e^{-4\theta}-2e^{-3\theta}$.

Exercises 2-5

1. $\Delta y = 0.61, 0.0601, -1.16$; $dy = 0.6, 0.06, -1.2$.

2. (1) $(1+\ln x)dx$;   (2) $\left(-\dfrac{a}{x^2}-\dfrac{a}{x^2+a^2}\right)dx$;   (3) $[\sin(3-x)-\cos(3-x)]e^{-x}dx$.

   (4) $-\dfrac{2x}{1+x^4}dx$;   (5) $2xe^{\sin x^2}\cos x^2 dx$;   (6) $\dfrac{2\ln 5}{\sin 2x}5^{\ln\tan x}dx$;

   (7) $dy = \begin{cases} \left(\arccos\sqrt{1-x^2}+\dfrac{x}{\sqrt{1-x^2}}\right)dx,\ 0<x<1, \\ \left(\arccos\sqrt{1-x^2}-\dfrac{x}{\sqrt{1-x^2}}\right)dx,\ -1<x<0; \end{cases}$

   (8) $-f'\left(\arctan\dfrac{1}{x}\right)\dfrac{1}{1+x^2}dx$.

3. (1) $-\dfrac{y}{x+2y+\sin y}dx$;   (2) $\dfrac{1}{2}dx$.

4. (1) $\dfrac{du}{v^2}-2\dfrac{u}{v^3}dv$;   (2) $\dfrac{vdu-udv}{u^2+v^2}$;

   (3) $2\cot(u^2+v^2)\cdot(udu+vdv)$;   (4) $f'(uv)(vdu+udv)$.

5. (1) 0.002;   (2) 9.99;   (3) 0.4849.

7. (1) 0.02;   (2) 0.01;   (3) 1.05;   (4) 0.03.

8. -43.63 cm^2, 104.72 cm^2.   9. 100 W.   10. 0.33%.

Exercises 2-6

1. $C'(Q)=0.02Q+10$, $L(Q)=-0.01Q^2+20Q-1\,000$, when the marginal profit is 0, the output is 1 000 unit.

2. $L'(Q)=-0.02Q+5$, $L'(200)=1$(Yuan), $L'(250)=0$(Yuan), $L'(300)=-1$(Yuan). Its economic meaning: when output is 200 kg on that day, the output increase 1 kg, the profit is increase 1 Yuan; when output is 250 kg on that day, the output increase 1 kg, the profit is not increase; when output is 300 kg on that day, the output increase 1 kg, the profit is decrease 1 Yuan.

3. (1) $C'(Q)=3+Q$, $R'(Q)=50Q^{-\frac{1}{2}}$, $L'(Q)=50Q^{-\frac{1}{2}}-Q-3$;

   (2) $\dfrac{ER}{EP}=-1$.

4. (1) $Q'(P)=-200\mathrm{e}^{-0.02P}$, $Q'(100)=-200\mathrm{e}^{-2}$;

   (2) $\dfrac{EQ}{EP}=\dfrac{P}{50}$;

   (3) $\dfrac{EQ}{EP}\bigg|_{P=100}=2$. The economic meaning: when the price is 100 unit, the price is increase 1%, the demand is decrease 2%.

5. (1) $0<P<\dfrac{16}{9}$ is the value of the price P when the demand is low elastic, $\dfrac{16}{9}<P<4$ is the value of the price P when the demand is high elastic.

   (2) $0<P<\sqrt{\dfrac{a}{3}}$ is the value of the price P when the demand is low elastic, $\sqrt{\dfrac{a}{3}}<P<\sqrt{a}$ is the value of the price when the demand is high elastic.

6. (1) a;   (2) kx;   (3) $\dfrac{\sqrt{x}}{2(\sqrt{x}-4)}$;   (4) $\dfrac{x}{2(x-9)}$.

Quiz

1. (1) $\mathrm{e}^x(x^2+x-1)$;

   (2) $\dfrac{1}{3}(x+\sqrt[3]{x+\sqrt[3]{x}})^{-\frac{2}{3}}\left[1+\dfrac{1}{3}(x+\sqrt[3]{x})^{-\frac{2}{3}}\left(1+\dfrac{1}{3}x^{-\frac{2}{3}}\right)\right]$;

   (3) $\dfrac{2}{\cos x\sqrt{\sin x}}$;   (4) $\dfrac{1}{2}\left(\dfrac{\arccos\sqrt{t}}{\sqrt{t}}-\dfrac{1}{\sqrt{1-t}}\right)$;

   (5) $x^{\sin x}\left(\cos x\ln x+\dfrac{\sin x}{x}\right)$;   (6) $\left(1+\dfrac{1}{x}\right)^x\left[\ln\left(1+\dfrac{1}{x}\right)-\dfrac{1}{1+x}\right]$.

2. (1) $\dfrac{3x^2-y^4}{5y^4+4xy^3+2}$;   (2) $\dfrac{y}{x+1}\cdot\dfrac{1-(x+1)y\cos(xy)}{1+xy\cos(xy)}$.

3. The tangent line is $2\sqrt{2}x-3y-1=0$, $2\sqrt{2}x+3y+1=0$;
   The normal line is $3x+2\sqrt{2}y-5\sqrt{2}=0$, $3x-2\sqrt{2}y+5\sqrt{2}=0$.

4. (1) $\dfrac{t+1}{t(t^2+1)}$;   (2) $-\dfrac{\sin\theta+\sin 2\theta}{\cos\theta+\cos 2\theta}$.

5. -2.   6. $\dfrac{1}{3a}\sec^4\theta\csc\theta$.

7. $n!\left[\dfrac{2^n}{(1-2x)^{n+1}}+\dfrac{(-1)^n}{(1+x)^{n+1}}\right]$.

8. $\mathrm{d}y=\dfrac{x+y-1}{x+y+1}\mathrm{d}x$, $\mathrm{d}x=\dfrac{x+y+1}{x+y-1}\mathrm{d}y$, $\dfrac{\mathrm{d}y}{\mathrm{d}x}=\dfrac{x+y-1}{x+y+1}$, $\dfrac{\mathrm{d}x}{\mathrm{d}y}=\dfrac{x+y+1}{x+y-1}$.

9. 4%.

10. (1) $(2x-x^2)e^{-x}, 2-x$;   (2) $(a-bx)x^{a-1}e^{-(bx+c)}, a-bx$.

11. (1) 9.5 Yuan;   (2) 22 Yuan.   12. $L(Q)=-Q^2+28Q-100, Q=14$ (every 100 produces).

Exercises

1. (1) $\dfrac{\cos x(1-\ln 3)-\sin x(1+\ln 3)}{3^x}$;   (2) $\dfrac{1}{2}\left(\dfrac{\sin x}{1+\cos x}-\dfrac{\cos x}{1-\sin x}\right)$;

    (3) $a^{a^x} \cdot a^x \ln^2 a$;   (4) $x^{x^x}[x^{x-1}+x^x(1+\ln x)\ln x]$;

    (5) $-\dfrac{1}{x\ln^2 x}+x^{\frac{1}{x}-2}(1-\ln x)$;   (6) $\begin{cases}(1-2x^2)e^{-x^2}, & x<0, \\ \cos x, & x\geqslant 0.\end{cases}$

2. When $\varphi(a)\neq 0$, $f(x)$ is not derivable at $x=a$;

    When $\varphi(a)=0$, $f(x)$ is derivable at $x=a$, and $f'(a)=0$.

3. $a=2, b=0$.

4. $f(x)=\begin{cases}ax+b, & x<1, \\ \dfrac{a+b+1}{2}, & x=1, \\ x^2, & x>1,\end{cases}$ $a=2, b=-1$; $f'(x)=\begin{cases}2x, & x>1, \\ 2, & x\leqslant 1.\end{cases}$

6. $-\cot\dfrac{3\theta}{2}$.   8. $\dfrac{f''(x+y)}{[1-f'(x+y)]^3}$.

9. $y'(0)=0$, $y''(0)=2, \cdots$, $y^{(n)}=\begin{cases}0, & n=2k-1, \\ (n-1)^2 y^{(n-2)}(0), & n=2k,\end{cases}$ where $k=2, 3, \cdots$.

10. $\dfrac{15}{34-16\cos x}dx$.   11. (1) $y=\pm\dfrac{2}{3\sqrt{3}}$;   (2) $y=\pm x$.

12. $\dfrac{26}{21}$ m/s.   13. 7.6 s.   14. 26 s.

15. (1) $104-0.8Q$;   (2) 64;   (3) $\dfrac{3}{8}$.   16. (1) -0.5;   (2) $\dfrac{40}{21}$.

Exercises 3-2

1. $\tan x = x + \dfrac{2\sin^2(\theta x)+1}{3\cos^4(\theta x)}x^3$ $(0<\theta<1)$.

2. $\sqrt{x}=2+\dfrac{1}{4}(x-4)-\dfrac{1}{64}(x-4)^2+\dfrac{1}{512}(x-4)^3-\dfrac{1}{4!}\cdot\dfrac{15(x-4)^4}{16(\theta x)^{\frac{7}{2}}}$ $(0<\theta<1)$.

3. $\sin^2 x = \dfrac{2}{2!}x^2 - \dfrac{2^3}{4!}x^4 + \cdots + \dfrac{(-1)^{n-1}2^{2n-1}}{(2n)!}x^{2n}+o(x^{2n})$.

4. $\dfrac{e^x+e^{-x}}{2}=1+\dfrac{1}{2!}x^2+\cdots+\dfrac{1}{(2n)!}x^{2n}+o(x^{2n})$.

5. (1) $\dfrac{9}{2}$;   (2) $\dfrac{1}{6}$;   (3) $\dfrac{1}{2}$;   (4) $\dfrac{1}{2^7}$.   6. $P(x)=(\ln 2)^2 x^2+(\ln 2)x+1$.

Quiz

1. $\xi=\dfrac{\sqrt{3}}{3}$.   4. $\arcsin x = x+\dfrac{x^3}{3}+o(x^3)$.   5. Hint: construct the function $\varphi(x)=xf(x)$.

6. Hint: for $f(x)$ and $\ln x$ on $[a,b]$ use Cauchy's theorem.   7. (1) $-\dfrac{1}{12}$;   (2) -1.

Exercises

3. Hint: use the intermediate value theorem to prove the existence of the root. And then use proof by contradiction and Rolle's Theorem to prove the uniqueness.

6. Hint: use taylor's Theorem.

Exercises 4-1

1. (1) 2; (2) 2; (3) $\dfrac{\alpha}{\beta}$; (4) $-\dfrac{1}{2}$; (5) 2; (6) $\dfrac{1}{2}$; (7) 1; (8) 3; (9) -1;

   (10) $\dfrac{1}{2}$; (11) 1; (12) 1; (13) 1; (14) $-\dfrac{1}{3}$; (15) $e^{-\frac{2}{\pi}}$; (16) 0; (17) 0;

   (18) $e^{-\frac{1}{2}}$; (19) $\dfrac{1}{2}$; (20) 0.

2. (1) 0; (2) 0; (3) 1; (4) The limit doesn't exist.

3. (1) low-order; (2) low-order; (3) low-order; (4) equivalent.

Exercises 4-2

1. (1) Increase strictly on $(-\infty, -1)$, $[3,+\infty)$ and decrease strictly on $[-1,3]$.
   (2) Increase strictly on $[-1,1]$ and decrease strictly on $(-\infty,-1]$, $[1,+\infty)$.
   (3) Increase strictly on $[0,1]$ and decrease strictly on $[1,2]$.
   (4) Increase strictly on $\left[\dfrac{\pi}{3},\dfrac{5\pi}{3}\right]$ and decrease strictly on $\left[0,\dfrac{\pi}{3}\right]$, $\left[\dfrac{5\pi}{3},2\pi\right]$.
   (5) Increase strictly on $[0,+\infty)$ and decrease strictly on $(-1,0)$.
   (6) Increase strictly on $(-\infty,0]$ and decrease strictly on $[0,+\infty)$.

3. Hint: use monotonicity of function to demonstrate.

4. (1) Local minimum is $y\left(\dfrac{12}{5}\right)=-\dfrac{1}{24}$;
   (2) Local maximum is $y(0)=a^{\frac{4}{3}}$, local minimum is $y(\pm a)=0$;
   (3) Local maximum is $y\left(\dfrac{3}{4}\right)=\dfrac{5}{4}$;   (4) Local maximum is $y\left(\dfrac{\pi}{4}\right)=\sqrt{2}$;
   (5) Local maximum is $y(\pm 1)=e^{-1}$, local minimum is $y(0)=0$;
   (6) No extreme values;   (7) Local maximum is $y(e)=e^{\frac{1}{e}}$.

7. (1) When $a>\dfrac{1}{e}$, there is no real root;   (2) When $0<a<\dfrac{1}{e}$, there are two real roots;
   (3) When $a=\dfrac{1}{e}$, there is only one real root $x=e$.

10. Hint: demonstrate $g'(x)>0$.

11. $a=2$, maximum point, local maximum is $\sqrt{3}$.

12. If $\varphi(x_0)>0$, n is even number, the minimum value of $f(x)$ is $f(x_0)=0$,
    n is odd number, $f(x)$ has no extreme value.
    If $\varphi(x_0)<0$, n is even number, the maximum value of $f(x)$ is $f(x_0)=0$,
    n is odd number, $f(x)$ has no extreme value.

13. 64.

15. Demonstrate by the definition of extreme value.

16. (1) Maximum value is 1, minimum value is 0;
    (2) Maximum value is 132, minimum value is 0;
    (3) Maximum value is 1;

(4) Maximum value is $\dfrac{1}{\sqrt{2}}e^{-\frac{1}{2}}$, minimum value is $-\dfrac{1}{\sqrt{2}}e^{-\frac{1}{2}}$.

17. $\sqrt[3]{3}$.

19. The length is 6 cm, the width is 3 cm and the height is 4 cm.

20. $\left(2-\dfrac{2\sqrt{6}}{3}\right)\pi$.

21. When $AD=15$ km, the freight is the minimum.

22. The distance away from D is $\dfrac{1\,600}{7}$ m.

23. 70 N, 14 m.

Exercises 4-3

1. (1) It is convex on $\left(-\infty,\dfrac{1}{2}\right]$ and concave on $\left[\dfrac{1}{2},+\infty\right)$; $\left(\dfrac{1}{2},\dfrac{13}{2}\right)$.

   (2) It is convex on $(-\infty,0]$ and concave on $[0,+\infty)$; no inflection point.

   (3) It is convex on $(-\infty,-1]$, $[1,+\infty)$ and concave on $[-1,1]$; $(-1,-2),(1,2)$.

   (4) It is convex on $[-1,0)$ and concave on $(-\infty,-1],[0,+\infty]$;$(-1,0)$.

2. $a=6$.
3. $a=-\dfrac{3}{2}, b=\dfrac{9}{2}$.

Exercises 4-5

1. (1) $0,\infty$; (2) $1,1$; (3) $2,\dfrac{1}{2}$; (4) $5^{-\frac{3}{2}},5^{\frac{3}{2}}$.

2. $K=\dfrac{3ax^2}{(1+a^2x^6)^{\frac{3}{2}}}$.
3. $\rho=\dfrac{(x^2+a^2)^2}{2a(x^2-a^2)}$.
4. About 1 246 N.

Exercises 4-7

1. Produce 50 000 units and the maximum profit is 30 000 Yuan.
2. The demand is 5/4 and the maximum profit is 25/4.
3. 300 sets.
4. 6. 5.
5. (1) 3; (2) 6.
6. 5 batches.
7. 100 sets.
8. (1) 263.01 tons; (2) 19.66 batches/year;
   (3) One cycle has 18.31 days; (4) 22 408.74 Yuan.
9. Order goods for 5 times a year and the batch is 20.
10. (1) When $x=10-2.5t$ (ton), the enterprise achieves the maximum profit.
    (2) When tax of one ton is twenty thousand Yuan, the government obtains maximal total revenue which is one hundred thousand Yuan.

Quiz

1. (1) 1; (2) $\dfrac{1}{2}$; (3) inexistence.

2. (1) increase strictly; (2) $x=1$;

   (3) $\left(-\dfrac{\sqrt{2}}{2},e^{-\frac{1}{2}}\right),\left(\dfrac{\sqrt{2}}{2},e^{-\frac{1}{2}}\right)$; (4) exist, $y=-3$;

(5) no maximum value.

3. (1) 0; (2) increase strictly on $(-\infty, +\infty)$.

4. non-derivable, exist. The minimal value is $f(0)=1$.

5. 7 km.

6. Hint: use monotonicity of function to demonstrate.

7. $k=\dfrac{2x}{(1+x^4)^{\frac{3}{2}}}$, $\rho=\dfrac{(1+x^4)^{\frac{3}{2}}}{2x}$, point of minimal value is $x=\dfrac{1}{\sqrt[4]{5}}$.

8. $x=100$.

9. $x=3$, $P=15e^{-1}$ and maximum revenue is $45e^{-1}$.

10. $P=101$ and the maximum profit is 167 080.

11. (1) $R(x)=800x-x^2$; (2) $L(x)=-x^2+790x-2\,000$; (3) 395 sets;
    (4) 154 025 Yuan; (5) 405 Yuan.

Exercises

1. (1) $\dfrac{2a\ln^2 a}{1-a\ln a}$; (2) $\dfrac{1}{6}$; (3) 0; (4) ∞; (5) $\dfrac{1}{3}$; (6) $\dfrac{\pi}{4}$.

2. $\dfrac{1}{16}$.

3. (1) Increase strictly on $(-\infty, -1]$, $[1, +\infty)$ and decrease strictly on $[-1, 1]$;
   (2) When $a>0$, increase strictly on $(-\infty, a]$, $\left[\dfrac{3}{4}a, +\infty\right)$ and decrease strictly on $\left[a, \dfrac{3}{4}a\right]$.

4. Minimal value is $-\pi$, maximal value is 2π, minimum value is $-\pi$ and maximum value is 2π.

6. $\dfrac{R}{\sqrt{3}}$. 7. $\dfrac{3}{2}R$. 8. $a=-3, b=0, c=1$.

12. $\overline{C}_{\min}=\overline{C}(1\,612)\approx 5.22$.

13. The daily output which minimizes the average cost is 50 tons and the minimal average cost is 300 (Yuan/t).

14. (1) $R(x)=280x-0.4x^2\ (0<x<700)$; (2) $L(x)=-x^2+280x-5\,000$;
    (3) 140 sets; (4) 14 600 Yuan; (5) 224 Yuan.

15. 14.1 Yuan.

16. Order goods for 25 times a year and the batch is 100.

17. It should reorder 35 times and the batch is 71.

Exercises 5-1

1. (1) $\dfrac{2^x}{\ln 2}+\dfrac{x^3}{3}+C$; (2) $\dfrac{1}{1+\ln 2}(2e)^x+C$;

   (3) $-\dfrac{1}{x}+\arctan x+C$; (4) $\dfrac{1}{2}x-\dfrac{1}{2}\sin x+C$;

   (5) $-\cot x+C$; (6) $-\dfrac{1}{x}-\arctan x+C$;

   (7) $\dfrac{2}{5}x^{\frac{5}{2}}+2\sqrt{x}+C$; (8) $x-\arctan x+C$;

   (9) $\tan x-x+C$; (10) $2x-\dfrac{5\cdot\left(\dfrac{2}{3}\right)^x}{\ln 2-\ln 3}+C$;

   (11) $\dfrac{1}{3}x^3-\dfrac{2}{3}x^{\frac{3}{2}}+\dfrac{2}{5}x^{\frac{5}{2}}-x+C$; (12) $\dfrac{1}{2}\tan x+\dfrac{1}{2}x+C$;

(13) $\dfrac{x^5}{5} - \dfrac{x^3}{3} + x - \arctan x + C$; (14) $\dfrac{1}{2}e^{2x} - e^x + x + C$;

(15) $\dfrac{x^5}{5} + \dfrac{x^4}{2} + \dfrac{x^3}{3} + C$; (16) $\dfrac{x^3}{3} + \dfrac{3}{2}x^2 + 9x + C$;

(17) $3x - 3\arctan x + C$; (18) $-\dfrac{1}{2}\cot x + C$;

(19) $\sin x - \cos x + C$; (20) $\dfrac{4}{7}x^{\frac{7}{4}} + \dfrac{4}{9}x^{\frac{9}{4}} + C$;

(21) $-\cot x - \tan x + C$; (22) $\tan x - \cot x + C$.

2. $y = \ln|x| + 1$. 3. $y = x^3 + x^2 - 2x$. 4. $-\dfrac{1}{x\sqrt{1-x^2}}$.

Exercises 5-2

1. (1) $\dfrac{1}{2}\ln|2x+1| + C$; (2) $-\dfrac{1}{3}(1-x^2)^{\frac{3}{2}} + C$;

(3) $\dfrac{1}{3}\arctan 3x + C$; (4) $\dfrac{2}{3}\ln^{\frac{3}{2}} x + C$;

(5) $\dfrac{1}{3}\sin^3 x + C$; (6) $-\dfrac{1}{\arcsin x} + C$;

(7) $-\dfrac{1}{3}e^{-x^3} + C$; (8) $-\dfrac{1}{4}\ln|3-2x^2| + C$;

(9) $\ln(x^2 - 3x + 8) + C$; (10) $\arcsin x + \ln(x + \sqrt{x^2+1}) + C$;

(11) $\dfrac{\sqrt{2}}{2}\arctan \dfrac{\cos x}{\sqrt{2}} + C$; (12) $\ln|1 + \tan x| + C$;

(13) $\dfrac{1}{2}\arctan \dfrac{e^x}{2} + C$; (14) $-2\cot \sqrt{x} + C$;

(15) $\sqrt{2}\arctan \sqrt{2x} + C$; (16) $\arctan e^x + C$;

(17) $\arcsin x - \sqrt{1-x^2} + C$; (18) $\dfrac{1}{4}\ln\left|\dfrac{x+1}{x-3}\right| + C$;

(19) $\ln|\ln \ln x| + C$; (20) $\dfrac{1}{2}\sin x^2 + C$;

(21) $\dfrac{1}{2}\ln|\sin x| + C$.

2. (1) $\dfrac{x}{\sqrt{1-x^2}} + C$; (2) $\dfrac{1}{a}\arccos\left|\dfrac{a}{x}\right| + C$;

(3) $\dfrac{\sqrt{9-x^2}}{27} + C$; (4) $-\arcsin \dfrac{1+\cos 2x}{2\sqrt{2}} + C$;

(5) $\left|\arccos \dfrac{a}{x}\right| + C$; (6) $\dfrac{|x|}{\sqrt{x^2+1}} + C$;

(7) $\dfrac{3}{4}\arcsin\sqrt{x} - \sqrt{x(1-x)} + \dfrac{1}{4}\sqrt{x(1-x)}(1-2x) + C$;

(8) $-\dfrac{1}{47}(x+1)^{-47} + \dfrac{1}{48}(x+1)^{-48} - \dfrac{1}{49}(x+1)^{-49} + C$;

(9) $\dfrac{1}{3}\tan^3 x - \tan x + x + C$; (10) $\dfrac{1}{5}(1+x^2)^{\frac{5}{2}} - \dfrac{1}{3}(1+x^2)^{\frac{3}{2}} + C$;

(11) $\sqrt{1+x^2} - \ln(1 + \sqrt{1+x^2}) + C$.

Exercises 5-3

1. (1) $-\dfrac{1}{2}x\cos 2x+\dfrac{1}{4}\sin 2x+C$; (2) $x^2\sin x+2x\cos x-2\sin x+C$;

   (3) $-\dfrac{e^{-x}}{2}(\sin x+\cos x)+C$; (4) $-\dfrac{1}{x}\ln x-\dfrac{1}{x}+C$;

   (5) $x\arcsin x+\sqrt{1-x^2}+C$; (6) $x\ln(x+\sqrt{1+x^2})-\sqrt{1+x^2}+C$;

   (7) $\dfrac{x^{n+1}\ln x}{n+1}-\dfrac{x^{n+1}}{(n+1)^2}+C$; (8) $\dfrac{x^3}{3}\arctan x-\dfrac{1}{6}x^2+\dfrac{1}{6}\ln(x^2+1)+C$;

   (9) $\dfrac{x^3}{3}\ln(1+x)-\dfrac{x^3}{9}+\dfrac{x^2}{6}-\dfrac{x}{3}+\dfrac{1}{3}\ln|1+x|+C$;

   (10) $-\dfrac{\ln^3 x}{x}-\dfrac{3\ln^2 x}{x}-\dfrac{6\ln x}{x}-\dfrac{6}{x}+C$;

   (11) $x(\arcsin x)^2+2\arcsin x\sqrt{1-x^2}-2x+C$;

   (12) $\dfrac{x^2}{\omega}\sin\omega x+\dfrac{2}{\omega^2}x\cos\omega x-\dfrac{2}{\omega^3}\sin\omega x+C$;

   (13) $2\sqrt{x}\ln(1+x)-4\sqrt{x}+4\arctan\sqrt{x}+C$.

2. (1) $\dfrac{x}{4\cos^4 x}-\dfrac{1}{4}\left(\tan x+\dfrac{1}{3}\tan^3 x\right)+C$; (2) $2\sqrt{1+x^2}+3\ln(x+\sqrt{1+x^2})+C$;

   (3) $\tan x-x+\dfrac{1}{2}\cos(x^2+1)+C$; (4) $x\arctan\sqrt{x}-\sqrt{x}+\arctan\sqrt{x}+C$;

   (5) $\dfrac{1}{8}e^{2x}(2-\cos 2x-\sin 2x)+C$; (6) $\dfrac{1}{2}[x\cos(\ln x)+x\sin(\ln x)]+C$.

3. $-2x^2 e^{-x^2}-e^{-x^2}+C$.

Exercises 5-4

1. (1) $-\dfrac{1}{2}\ln|1+x|+2\ln|2+x|-\dfrac{3}{2}|3+x|+C$;

   (2) $\dfrac{1}{x-2}-\dfrac{1}{3}\ln|x-2|-\dfrac{2}{3}\ln|x+1|+C$;

   (3) $\dfrac{1}{4}\ln\left|\dfrac{x-1}{x+1}\right|+\dfrac{1}{2}\dfrac{1}{1+x}+C$;

   (4) $-3\ln|x-2|+5\ln|x-3|+C$;

   (5) $\dfrac{x^3}{3}+\dfrac{x^2}{2}+x+8\ln|x|-3\ln|x-1|-4\ln|x+1|+C$;

   (6) $\ln|x|-\dfrac{1}{2}\ln(x^2+1)+C$;

   (7) $x+2\ln|x|-4\ln|x+1|+\dfrac{4}{x+1}+C$;

   (8) $\dfrac{1}{x^2}-\dfrac{1}{x^2+1}+\dfrac{1}{2}\ln(x^2+1)+C$;

   (9) $\dfrac{5}{2}\ln|x^2+2x+3|-4\dfrac{1}{\sqrt{2}}\arctan\dfrac{x+1}{\sqrt{2}}+C$;

   (10) $\dfrac{1}{8}\ln\left|\dfrac{x^2-1}{x^2+1}\right|-\dfrac{1}{4}\arctan x^2+C$.

2. (1) $-\dfrac{1}{4}\ln\left|\dfrac{\tan\frac{x}{2}-1}{\tan\frac{x}{2}+1}\right|+C$; (2) $2\tan\dfrac{x}{2}-x+C$;

   (3) $x-\tan\dfrac{x}{2}+C$; (4) $\dfrac{2}{3}\tan^3 x+\tan x+C$;

(5) $\frac{1}{2}\ln\left|\tan\frac{x}{2}\right|-\frac{1}{2}\tan\frac{x}{2}+C$;　　(6) $x+\dfrac{2}{1+\tan\frac{x}{2}}+C$.

3. (1) $\frac{2}{9}(3-x)\sqrt{3-x}-\frac{10}{3}\sqrt{3-x}+C$;

(2) $\frac{1}{8}\left[\dfrac{1}{\left(1-\sqrt{\frac{x-1}{x+1}}\right)^2}-\dfrac{3}{1-\sqrt{\frac{x-1}{x+1}}}-\ln\left|1-\sqrt{\frac{x-1}{x+1}}\right|-\dfrac{1}{\left(1+\sqrt{\frac{x-1}{x+1}}\right)^2}+\dfrac{3}{1+\sqrt{\frac{x-1}{x+1}}}+\ln\left|1+\sqrt{\frac{x-1}{x+1}}\right|\right]+C$;

(3) $-\frac{4}{15}(3x+4\sqrt{x}+8)\sqrt{1-\sqrt{x}}+C$;　　(4) $x+4\ln|\sqrt{x+1}+1|-4\sqrt{x+1}+C$;

(5) $\frac{1}{\sqrt{3}}\ln\left|\dfrac{\sqrt{x+3}-\sqrt{3(x+1)}}{\sqrt{x+3}+\sqrt{3(x+1)}}\right|+C$;　　(6) $3\ln|\sqrt{1+\sqrt[3]{x^2}}+\sqrt[3]{x^2}|+C$.

Exercises 5-5

1. Total cost function is $C(x)=5x+3\sqrt{x}+10$ (ten-thousand Yuan); Average cost function is $\dfrac{C(x)}{x}=5+\dfrac{3}{\sqrt{x}}+\dfrac{10}{x}$ (ten-thousand Yuan).

2. Savings function is $S(x)=0.3x-0.2x^{\frac{1}{2}}-22.5$.

3. Demand function is $Q(P)=1\,000\left(\dfrac{1}{3}\right)^P$.

Quiz

1. (1) A;　(2) D;　(3) A;　(4) B.

2. (1) $-\frac{1}{x}+C$;　　(2) $\frac{3}{7}x^{\frac{7}{3}}+C$;　　(3) $\dfrac{(4e)^x}{1+2\ln 2}+C$;

(4) $\frac{1}{3}F(3x+5)+C$;　　(5) $e^x+\dfrac{x^{e+1}}{1+e}+C$.

3. (1) $\frac{1}{63}(3x-8)^{21}+C$;　　(2) $\frac{\sqrt{2}}{6}\arctan\frac{\sqrt{2}}{3}x+C$;　　(3) $e^x-2\sqrt{x}+C$;

(4) $2\arctan\sqrt{2x-1}+C$;　　(5) $\frac{1}{3}x^3\ln x-\frac{1}{9}x^3+C$;　　(6) $\frac{x^2}{2}+2x+4\ln|x-2|+C$.

4. (1) $\ln\left|\tan\frac{\arcsin x}{2}\right|+C$;　　(2) $\tan(\ln x)+C$;

(3) $-2\sqrt{x+1}\cos\sqrt{x+1}+2\sin\sqrt{x+1}+C$;　　(4) $xf'(x)-f(x)+C$.

5. $y=2\sqrt{x+1}$.

Exercises

1. (1) $\frac{1}{21}(1-3x)^{\frac{7}{3}}-\frac{1}{12}(1-3x)^{\frac{4}{3}}+C$;　　(2) $x-\ln(1+e^x)+C$;

(3) $\frac{1}{2}\ln(e^{2x}+2e^x+2)-\arctan(e^x+1)+C$;　　(4) $\arctan x+\frac{1}{3}\arctan x^3+C$;

(5) $-\arcsin\left(\dfrac{\cos^2 x}{2}\right)+C$;　　(6) $\frac{1}{2}(\ln\tan x)^2+C$;

(7) $-\ln[e^{-x}+\sqrt{1+e^{-2x}}]+C$;　　(8) $\dfrac{1}{2\ln\frac{3}{2}}\ln\left|\dfrac{3^x-2^x}{3^x+2^x}\right|+C$;

(9) $\dfrac{x}{x-\ln x}+C$;

(10) $\dfrac{1}{5}(\arcsin x+2\ln 2x+\sqrt{1-x^2})+C$;

(11) $-\dfrac{\sqrt{a^2+x^2}}{a^2 x}+C$; (12) $\left|\arccos\dfrac{1}{x}\right|+C$;

(13) $2\arcsin\sqrt{x}+C$; (14) $\dfrac{1}{2-x}\ln x+\dfrac{1}{2}\ln\left|1-\dfrac{2}{x}\right|+C$;

(15) $x\arctan(1+\sqrt{x})-\sqrt{x}+\ln(x+2\sqrt{x}+2)+C$;

(16) $2\sin x\,e^{\sin x}-2e^{\sin x}+C$; (17) $x\tan\dfrac{x}{2}+C$;

(18) $2x\sqrt{e^x-2}-4\sqrt{e^x-2}+4\sqrt{2}\arctan\sqrt{\dfrac{e^x}{2}-1}+C$;

(19) $\dfrac{x}{2}[\sin(\ln x)-\cos(\ln x)]+C$;

(20) $\dfrac{1}{2}\arctan[(1+x^2)\ln(1+x^2)-x^2-3]-\dfrac{x}{2}\ln(1+x^2)+C$;

(21) $\dfrac{2}{9}\ln|x^9+8|-\ln|x|+C$; (22) $\dfrac{x+x^3}{8(1-x^2)^2}-\dfrac{1}{16}\ln\left|\dfrac{1+x}{1-x}\right|+C$;

(23) $\dfrac{1}{\sqrt{2}}\arctan\left(\dfrac{x-\dfrac{1}{x}}{\sqrt{2}}\right)+C$; (24) $\dfrac{1}{4}\left(\ln\left|\dfrac{1+x}{1-x}\right|+\dfrac{1}{1-x}-\dfrac{1}{1+x}\right)+C$;

(25) $\ln|x-1|-\dfrac{1}{2}\ln(x^2+1)-\arctan x+\dfrac{1}{x^2+1}+C$;

(26) $\dfrac{1}{\sqrt{2}}\arctan\left(\dfrac{\tan x}{\sqrt{2}}\right)+C$;

(27) $\dfrac{1}{4\cos x}+\dfrac{1}{4}\ln\left|\tan\dfrac{x}{2}\right|+\dfrac{1}{4}\tan x+C$; (28) $\tan x-\dfrac{1}{\cos x}+C$;

(29) $2x+\ln|2\sin x+\cos x|+C$;

(30) $\dfrac{1}{6}\ln\left|\dfrac{1+\sin x}{1-\sin x}\right|+\dfrac{1}{3\sqrt{2}}\arctan\left(\dfrac{\sin x}{\sqrt{2}}\right)+C$;

(31) $\dfrac{1}{2}\ln\left|\dfrac{1-\cos x}{1+\cos x}\right|+\dfrac{1}{2+\cos x}+C$;

(32) $\dfrac{1}{6}\ln\dfrac{\tan^2 x-\tan x+1}{(1+\tan x)^2}+\dfrac{1}{\sqrt{3}}\arctan\left(\dfrac{2\arctan x-1}{\sqrt{3}}\right)+C$;

(33) $\dfrac{2}{\sqrt{3}}\arctan\left(\dfrac{2\tan^2 x-1}{\sqrt{3}}+C\right)$;

(34) $\dfrac{x}{2}+\ln\sec\dfrac{x}{2}-\ln\left|1+\tan\dfrac{x}{2}\right|+C$;

(35) $\ln\left|\dfrac{\sqrt{x-1}+\sqrt{x+1}}{\sqrt{x-1}-\sqrt{x+1}}\right|-2\arctan\sqrt{\dfrac{x+1}{x-1}}+C$;

(36) $\sqrt{x}+\dfrac{x}{2}-\dfrac{1}{2}\sqrt{x^2+x}-\dfrac{1}{4}\ln\left(x+\dfrac{1}{2}+\sqrt{x^2+x}\right)+C$.

2. (1) $f(x)=\dfrac{1}{4}(x-1)^4+2x-2$; (2) $f(\ln x)=\begin{cases}\ln x+C, & 0<x\leqslant 1,\\ x+C, & 1<x<+\infty;\end{cases}$

(3) $f(x)=\dfrac{1}{\sqrt{2x}(1+x)}$; (4) $\int\dfrac{dx}{y^2}=\dfrac{3y}{x}-2\ln\left|\dfrac{y}{x}\right|+C$;

(6) $f(x)=2\left(\dfrac{x^2}{2}-x\right)+3=x^2-2x+3$; (7) $\dfrac{1}{2}\left[\dfrac{f(x)}{f'(x)}\right]^2+C$.

Answers

Exercises 6-1

4. $\dfrac{63}{2}$.

Exercises 6-2

1. (1) 0;　(2) 1;　(3) $\int_1^3 f(x)\,dx$.

2. (1) $\int_0^1 x\,dx \geqslant \int_0^1 x^2\,dx$;　(2) $\int_0^{\frac{\pi}{2}} x\,dx \geqslant \int_0^{\frac{\pi}{2}} \sin x\,dx$;

　(3) $\int_1^e \ln x\,dx \geqslant \int_1^e \ln^2 x\,dx$;　(4) $\int_1^2 x\,dx \leqslant \int_1^2 x^4\,dx$.

3. (1) $6 \leqslant \int_1^4 (x^2+1)\,dx \leqslant 51$;　(2) $\pi \leqslant \int_{\frac{\pi}{4}}^{\frac{5\pi}{4}} (1+\sin^2 x)\,dx \leqslant 2\pi$;

　(3) $0 \leqslant \int_0^{-2} xe^x\,dx \leqslant \dfrac{2}{e}$;　(4) $1 \leqslant \int_0^1 e^{x^3}\,dx \leqslant e$;

　(5) $\dfrac{\pi}{21} \leqslant \int_{\frac{\pi}{4}}^{\frac{\pi}{3}} \dfrac{dx}{1+\sin^2 x} \leqslant \dfrac{\pi}{18}$;　(6) $\dfrac{2}{5} \leqslant \int_1^2 \dfrac{x}{1+x^2}\,dx \leqslant \dfrac{1}{2}$.

4. $A = -\int_0^1 x(x-1)(2-x)\,dx + \int_1^2 x(x-1)(2-x)\,dx$.

Exercises 6-3

1. (1) $\dfrac{\sqrt{2}}{2}$;　(2) $2x\sqrt{1+x^4}$;　(3) 1;　(4) $\dfrac{\pi^2}{4}$;　(5) $\dfrac{1}{3}$.

2. (1) $\dfrac{7}{2}$;　(2) $45\dfrac{1}{6}$;　(3) $\dfrac{\pi}{6}$;　(4) $\dfrac{\pi}{3}$;　(5) $\dfrac{8}{3}$;

　(6) $\dfrac{495}{\ln 10}$;　(7) $\pi - \dfrac{4}{3}$;　(8) $\dfrac{1}{2}(\ln\sqrt{3})^2$;　(9) $2 - \sqrt{3} + \dfrac{\pi}{6}$;　(10) $2\sqrt{2}$.

3. (1) $\dfrac{11}{6}$;　(2) 4.

4. $\Phi(x) = \begin{cases} \dfrac{1}{2}x^2 + x, & 0 \leqslant x < 1, \\ \dfrac{1}{6}x^3 + \dfrac{4}{3}, & 1 \leqslant x \leqslant 2, \end{cases}$　$\Phi(x)$ is continuous at $x=1$, $\Phi(x)$ is continuous on $[0,2]$.

5. $\Phi(x) = \begin{cases} 0, & x < 0, \\ \dfrac{1}{2}\sin^2 x, & 0 \leqslant x \leqslant \pi, \\ 1, & x > \pi. \end{cases}$

Exercises 6-4

1. (1) $10\dfrac{2}{3}$;　(2) $\sqrt{3} - \dfrac{\pi}{3}$;　(3) $\dfrac{1}{5}(e^\pi - 1)$;　(4) $\dfrac{\pi}{2}$;

　(5) $1 - \ln(1+e) + \ln 2$;　(6) $2(\sqrt{3}-1)$;　(7) $2 + 2\ln\dfrac{2}{3}$;

　(8) $\dfrac{\sqrt{2}}{2}$;　(9) $\dfrac{\pi}{2} - \dfrac{4}{3}$;　(10) $\dfrac{3}{2}$;　(11) $\dfrac{5}{3} - 2\ln 2$;

　(12) π;　(13) $20\sqrt{2}$.

2. (1) 0;　(2) $\dfrac{3}{8}\pi x^4$;　(3) $\dfrac{\pi^3}{324}$;　(4) 0.

4. (1) $\dfrac{\pi}{4} - \dfrac{1}{2}\ln 2$;　(2) 0;　(3) $\dfrac{1}{3}(e^\pi - 1)$;　(4) 0;

(5) $2\left(1-\dfrac{1}{e}\right)$; (6) $1-\dfrac{\sqrt{3}}{6}\pi$; (7) $1-2e^{-1}$; (8) $4(2\ln 2-1)$;

(9) $\dfrac{e}{2}(\sin 1-\cos 1)+\dfrac{1}{2}$; (10) $\left(1+\dfrac{1}{e}\right)\ln\left(1+\dfrac{1}{e}\right)-\dfrac{1}{e}$.

5. 3. 6. $n^2\pi$.

Exercises 6-5

1. (1) converges, $\dfrac{1}{5}$; (2) diverges; (3) converges, $\dfrac{1}{a}$; (4) diverges;

(5) converges, 1; (6) diverges; (7) converges, $\dfrac{\pi}{2}$; (8) converges, 1;

(9) converges, $\ln 2$; (10) converges, $\dfrac{2\sqrt{3}}{27}\pi$; (11) converges, $\dfrac{1}{\ln 2}\left(1+\dfrac{1}{2\ln 2}\right)$.

2. $\sqrt{3}+\sqrt{5}$. 3. The maximum value of $f(x)$ is $1+e^{-2}$, the minimum value is 0.
4. (1) diverges; (2) converges.

Quiz

1. (1) B; (2) A; (3) D; (4) D.
2. (1) $2xe^{x^2}$; (2) 0; (3) 0; (4) $\dfrac{4}{7}$;

(5) $\dfrac{1}{3}$.

3. (1) -2; (2) $4-2\ln 3$; (3) 1; (4) $2\sqrt{3}-2$;
(5) 2.

4. (1) $\dfrac{5}{6}$; (3) $F(x)=\begin{cases}-x-1, & -1\leqslant x<0,\\ -1, & x=0,\\ x-1, & 0<x\leqslant 1,\end{cases}$ $F'(x)=\begin{cases}-1, & -1\leqslant x<0,\\ 1, & 0<x\leqslant 1.\end{cases}$

Exercises

1. (1) $\dfrac{2}{3}$; (2) $\dfrac{1}{2}\ln\dfrac{4}{3}$; (3) 1; (4) $af(a)$.
2. (1) $\dfrac{3}{2}e^{\frac{5}{2}}$; (2) $\dfrac{\pi}{2}$; (3) $4\sqrt{2}-1$; (4) π.
4. (1) $x+2$; (2) $\dfrac{1}{2}+\ln 2$; (3) $f'(1)$ doesn't exist; (4) $2e^2$; (5) $\dfrac{1}{x}$; (6) $\dfrac{3}{4}$.

Exercises 7-2

1. (1) $\dfrac{4}{3}$; (2) $\dfrac{1}{3}$; (3) $\dfrac{3}{2}-\ln 2$; (4) 1;

(5) $\dfrac{1}{6}$; (6) $10\dfrac{2}{3}$; (7) $\dfrac{7}{6}$; (8) $2\sqrt{2}-2$.

2. $\left(6\pi-\dfrac{4}{3}\right):\left(2\pi+\dfrac{4}{3}\right)$. 3. $\dfrac{9}{4}$. 4. $\dfrac{8}{3}$.

5. (1) πa^2; (2) $\dfrac{\pi a^2}{4}$.

6. (1) $\dfrac{3}{10}\pi$; (2) $\dfrac{\pi^2}{2}, 2\pi^2$; (3) $\dfrac{128}{7}\pi, \dfrac{64}{5}\pi$.

7. (1) $e+\dfrac{1}{e}-2$; (2) $\dfrac{\pi}{2}(e^2+e^{-2}-2)$. 8. $\dfrac{\pi R^2 h}{2}$.

9. (1) $1+\dfrac{1}{2}\ln\dfrac{3}{2}$;   (2) $\dfrac{8}{27}(10\sqrt{10}-1)$;

   (3) $\dfrac{1}{27}(80\sqrt{10}-13\sqrt{13})$;   (4) $8a$;

   (5) $\dfrac{\sqrt{1+a^2}}{a}(e^{a\varphi_0}-1)$.

10. $6a$.

Exercises 7-3

1. The increment of total cost is 100 ten-thousand Yuan, and the production is 6 hundred units that minimizes the average cost.

2. $R=100Qe^{-\frac{Q}{10}}$.   3. $666\dfrac{1}{3}$.   4. (1) 9 950;   (2) 19 600.

Quiz

1. $\dfrac{7}{6}$.   2. πab.   3. $\dfrac{5}{4}\pi$.

4. $1, (e-2)\pi$.   5. $\dfrac{4}{3}\pi a^3$.   6. $6a$.

7. (1) $C(x)=\dfrac{1}{3}x^3-\dfrac{5}{2}x^2+40x+10$;   (2) $R(x)=50x-x^2$;

   (3) $L(x)=10x+\dfrac{3}{2}x^2-\dfrac{1}{3}x^3-10$;   (4) $-\dfrac{23}{6}$.

8. The production is 500 units that maximizes the profit, if producing 50 more units on the basis of this production, the profit will decrease 25 Yuan.

Exercises

1. (1) $\dfrac{1}{12}\left(1-\dfrac{\sqrt{6}}{3}\right)$;   (2) $\dfrac{7}{6}$;   (3) $\dfrac{1}{2}+\ln 2$;   (4) $\dfrac{9}{2}$.

2. $\dfrac{\pi-1}{4}a^2$.   3. $\dfrac{1}{5}\pi$.   4. $\dfrac{3}{10}\pi$.   5. $a=-\dfrac{5}{3}, b=2, c=0$.

6. $\sqrt{6}+\ln(\sqrt{2}+\sqrt{3})$.   7. 126 424.06(Yuan).